M000276565

CONTROL OF SPACECRAFT AND AIRCRAFT

Control of Spacecraft and Aircraft

ARTHUR E. BRYSON, JR.

PRINCETON UNIVERSITY PRESS

PRINCETON, NEW JERSEY

Copyright © 1994 by Princeton University Press

Published by Princeton University Press, 41 William Street,
Princeton, New Jersey 08540
In the United Kingdom: Princeton University Press, Chichester, West Sussex

Library of Congress Cataloging-in-Publication Data

Bryson, Arthur E. (Arthur Earl)
    Control of spacecraft and aircraft / Arthur E. Bryson, Jr.
      p. cm.
    Includes bibliographical references and index.
    ISBN 0-691-08782-2 (alk. paper)
    1. Space vehicles—Attitude control systems.
  2. Airplanes—Control systems. I. Title.
TL3260.B78   1993
629.45—dc20   93-35565

This book has been composed in Linotype Times Roman

Princeton University Press books are printed on acid-free
paper, and meet the guidelines for permanence and durability
of the Committee on Production Guidelines for Book Longevity of
the Council on Library Resources

Printed in the United States of America

10 9 8 7 6 5 4 3 2 1

To Helen

Contents

List of Figures

List of Tables

Preface and Acknowledgments

This book is about the best possible control of spacecraft and aircraft, that is, the limits of control. The minimum output error responses of controlled vehicles to specified initial conditions, specified output commands, and specified disturbances are determined with specified limits on control authority. To do this, the linear-quadratic-regulator (LQR) method of feedback control synthesis with full state feedback is used, emphasizing two-input, two-output systems.

LQR designs produce coordinated controls that yield graceful flight paths comparable to those produced by expert pilots. They represent the best performance the control engineer can expect, given accurate sensors and a low-noise environment. Human pilots, with their optical and force sensors, combined with their mental signal processing, act in many ways like optimal controllers. Furthermore, on many current (and future) spacecraft and aircraft, all the states are (or will be) measured, and the sensors often contain their own filters; thus it is not unreasonable to consider using state feedback directly (keeping in mind the possible phase shifts associated with the sensor-filter dynamics).

An interesting aspect of determining the limits of performance for a given aircraft is that one "learns" how to fly that aircraft in the sense of seeing how the controls should be moved to make graceful maneuvers. However, as I have discovered, it is one thing to know how the controls *should* be moved and quite another thing to actually move them that way in flight.

Most feedback control systems use a limited set of sensors—not all states are sensed. *Compensators* must then be designed for feedback control. The objective of these designs is to come as close as possible to the responses of the LQR designs. Compensator design is still largely done in practice with classical methods of successive loop closure using many individual lead, lag, and notch compensators combined with frequency-domain and root-locus analysis. Compensators can also be designed using state estimators (also called observers), however, feeding back the estimated states as if they were measured. Linear-quadratic-gaussian (LQG) methods and worst-case-design methods (such as H-infinity) are in this category.

One of the main goals of compensator design is to provide robustness to unmodeled dynamics and parameter variations within the model. Much research has focused on these topics in the last twenty years, and satisfactory algorithms are emerging. This is a fascinating and essential part of control design, but I have not treated it here because it would have doubled the size of the book and clouded its main message about the limits of performance. I believe compensator design should be done *after* determining the limits of performance with LQR design.

This book was developed, in connection with a graduate-level course at Stanford University, as an overview and summary of flight control. I assume the reader has knowledge comparable to that covered in intermediate-level courses in rigid-body mechanics, structural dynamics, linear dynamic systems, and automatic control (including digital and optimal control).

In order to design a good control system, one must have an accurate mathematical model of the behavior of the vehicle. Hence there is an emphasis in the book on modeling. The aerospace control engineer must know something about the dynamics of rigid and deformable bodies, aerodynamics, sensors, actuators, and propulsion.

Guidance, performance, and navigation are other important parts of planning and following flight paths; only some parts of them are covered here. However, "guidance" is distinguished from "control" mainly by its longer time scales, and it usually deals with motions of the center of mass of the vehicle, whereas control deals with attitude motions of the vehicle. These two areas often overlap, as in aircraft control and in missile guidance where aerodynamic forces and moments couple the translational and rotational motions. "Performance" has traditionally been concerned with open-loop design of flight paths for minimizing such things as time, fuel used, or direct operating cost. My previous book *Applied Optimal Control* (AOC) with Y. C. Ho deals with the calculus of variations, one of the methods for synthesizing optimal flight paths. "Navigation" deals with estimating position and velocity from various kinds of sensors; we deal with estimation in AOC, and I treat time-invariant logic for estimation here in connection with feedback of the estimated states of a vehicle.

I assume that the reader has access to a digital computer and specialized control software such as MATLAB[®1], Control-C, or Matrix-X. A disk is available (Ref. MAT) with MATLAB M-files for all of the figures in the book that are plots, and for many of the problems. This is a helpful tool for study, since the parameters of the control design are then easily changed and the user can quickly see the resulting changes in the responses.

Chapters 1–9 are about spacecraft, and chapters 10–15 are about aircraft. Appendix A is a summary of various representations of linear dynamic systems currently used by control engineers and the methods for transforming

[1]MATLAB is a registered trademark of The MathWorks, Inc.

from one representation to another. Appendixes B–D summarize methods for synthesizing analog and digital feedback control logic, with an emphasis on linear-quadratic methods. Appendix E summarizes methods for linear simulation of time response for both deterministic and random inputs. Appendix F summarizes methods for modeling flexible systems.

I have tried to give credit to the people who contributed the new ideas and concepts that have arisen in the past few decades in this field. Where I have gone astray, I apologize and I would be pleased to receive suggestions for corrections. I should like to thank my colleagues at Stanford—especially John V. Breakwell, Robert H. Cannon, Daniel B. DeBra, Gene Franklin, Thomas Kailath, and J. David Powell—first of all for making Stanford such a fascinating and pleasant place to work, and secondly for rounding out my education in this rather complicated field. I would also like to thank the many graduate students who have read through drafts of the book as it evolved. They have corrected many errors and suggested many improvements. As the references indicate, many graduate students have contributed new ideas to the field, and working with them has taught me a great deal.

This book would have been quite different had it not been for Dean William Kays who provided personal computers for the senior engineering faculty at Stanford in 1979. I thank him very much for his generosity and foresight. I would also like to thank Integrated Systems Inc. and The MathWorks for their invaluable software Matrix-X and MATLAB, which makes light work of things that were very heavy work only a decade ago.

Finally, I would like to thank Maribel Calderon, Michelle Whitman, and Carolyn Edwards for their skillful work in putting the text on the computer (using Stanford Professor Donald Knuth's marvelous typesetter ,TEX), and for doing it with patience and good humor. I subsequently converted their work to LATEX, which allowed me to make changes more easily on my IBM PC AT. Kathy Mills skillfully "MacDrew" the figures that are not plots; the figures and plots were printed on a laser printer using a special command developed at Stanford that converts MATLAB.m files to PostScript files.

Arthur E. Bryson, Jr.
Stanford, California

CONTROL OF SPACECRAFT AND AIRCRAFT

1

Natural Motions of Rigid Spacecraft

1.1 Translational Motions in Space

The natural motions of the center of mass of a body in space are described by Newton's equations:

$$m \overset{I}{\vec{v}} = \vec{F}, \tag{1.1}$$

$$\overset{I}{\vec{r}} = \vec{v}, \tag{1.2}$$

where

m = mass of the spacecraft,

$(\vec{v}, \vec{r})$ = (velocity, position) of the center of mass with respect to inertial space,

$\overset{I}{(\ )}$ = time rate of change with respect to inertial space,

$\vec{F}$ = sum of external forces.

In the absence of external forces, the velocity stays constant, and the position changes linearly with time.

For motions taking place in times very much less than an orbit period (either around the Sun or the Earth), neglecting gravity yields a useful approximation for the translational motions.

1.2 Translational Motions in Circular Orbit

In circular orbit, centrifugal force balances gravitational force. For *small deviations from circular orbit*, the equations of motion of the center of mass are conveniently written in locally-horizontal-vertical (LHV) coordinates that rotate with the orbital angular velocity, n (see Fig. 1.1):

$$\delta \overset{I}{\vec{v}} \equiv \delta \overset{L}{\vec{v}} + \vec{\omega}_L \times \delta\vec{v} = \delta\vec{F}/m, \tag{1.3}$$

$$\delta \overset{I}{\vec{r}} = \delta \overset{L}{\vec{r}} + \vec{\omega}_L \times \delta\vec{r} = \delta\vec{v}, \tag{1.4}$$

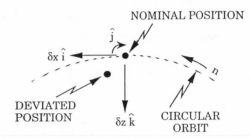

Figure 1.1. Locally-horizontal-vertical (LHV) coordinates for position deviation from circular orbit.

where

$\delta \vec{v}$ = velocity deviation,

$\delta \vec{r}$ = position deviation,

$\delta \vec{F}$ = force deviation,

$\overset{L}{(\ )}$ = denotes time derivative of components with respect to LHV axes,

$\vec{\omega}_L$ = angular velocity of LHV axes with respect to inertial axes.

Following NASA standard notation, δx is in-track position deviation (positive in direction of orbital velocity), δy is cross-track position deviation, and δz is vertical deviation (positive down). Thus,

$$\delta \vec{v} = \hat{\imath} \delta u + \hat{\jmath} \delta v + \hat{k} \delta w, \tag{1.5}$$

$$\delta \vec{r} = \hat{\imath} \delta x + \hat{\jmath} \delta y + \hat{k} \delta z, \tag{1.6}$$

$$\vec{\omega}_L = -n \hat{\jmath}, \tag{1.7}$$

where $n \overset{\Delta}{=} \sqrt{g/R}$ = orbital angular velocity and $(\hat{\imath},\ \hat{\jmath},\ \hat{k})$ are unit vectors along the $(x,\ y,\ z)$ axes. The only force is the inverse-square gravitational force

$$\vec{F} = mg \left(\frac{R}{R - \delta z} \right)^2 \hat{k}, \tag{1.8}$$

where g = gravitational force per unit mass at radial distance, R, from the attracting center. Thus the deviation in force is given by

$$\delta \vec{F} = 2mg \frac{R^2 \delta z}{(R - \delta z)^3} \hat{k} + mg \left(\frac{R}{R - \delta z} \right)^2 \left(\frac{\partial \hat{k}}{\partial x} \delta x + \frac{\partial \hat{k}}{\partial y} \delta y \right), \tag{1.9}$$

and

$$\frac{\partial \hat{k}}{\partial x} = -\frac{1}{R}\hat{i}, \tag{1.10}$$

$$\frac{\partial \hat{k}}{\partial y} = -\frac{1}{R}\hat{j}. \tag{1.11}$$

Substituting (5)–(7) and (9)–(11) into (1) and (2), we obtain the equations of motion for small deviations from circular orbit. They decouple into a set governing *cross-track motions*

$$\begin{bmatrix} \delta\dot{v} \\ \delta\dot{y} \end{bmatrix} = \begin{bmatrix} 0 & -n^2 \\ 1 & 0 \end{bmatrix} + \begin{bmatrix} T_y/m \\ 0 \end{bmatrix} \tag{1.12}$$

and a set governing *in-track/radial motions*

$$\begin{bmatrix} \delta\dot{u} \\ \delta\dot{w} \\ \delta\dot{x} \\ \delta\dot{z} \end{bmatrix} = \begin{bmatrix} 0 & n & -n^2 & 0 \\ -n & 0 & 0 & 2n^2 \\ 1 & 0 & 0 & n \\ 0 & 1 & -n & 0 \end{bmatrix} \begin{bmatrix} \delta u \\ \delta w \\ \delta x \\ \delta z \end{bmatrix} + \begin{bmatrix} T_x/m \\ T_z/m \\ 0 \\ 0 \end{bmatrix} \tag{1.13}$$

where (T_x, T_y, T_z) = thrust components.

The characteristic equation of the system (12) is

$$s^2 + n^2 = 0, \tag{1.14}$$

so there is one purely oscillatory mode at frequency n. The natural motion is of the form

$$\begin{bmatrix} \delta v \\ \delta y \end{bmatrix} = c \begin{bmatrix} n \\ 0 \end{bmatrix} \cos(nt + \beta) - c \begin{bmatrix} 0 \\ -1 \end{bmatrix} \sin(nt + \beta), \tag{1.15}$$

where c and β are arbitrary constants, and the complex eigenvector corresponding to $s = nj$ is $(n, -j)^T$. The real part of this eigenvector is the coefficient of

$\cos(nt + \beta)$, while the imaginary part is the coefficient of $[-\sin(nt + \beta)]$. The motion may be interpreted as a slight change in orbit plane, so that the spacecraft crosses the reference orbital plane twice per revolution, and thus appears to oscillate right-left with orbital frequency n.

The characteristic equation of the system (13) is

$$s^2(s^2 + n^2) = 0, \tag{1.16}$$

so there is one purely oscillatory mode at frequency n, and two stationary modes. The natural motions are of the form

$$\begin{bmatrix} \delta u \\ \delta w \\ \delta x \\ \delta z \end{bmatrix} = c_1 \begin{bmatrix} 0 \\ n \\ 2 \\ 0 \end{bmatrix} \cos(nt + \beta)$$

$$- c_1 \begin{bmatrix} n \\ 0 \\ 0 \\ 1 \end{bmatrix} \sin(nt + \beta) + (c_2 + c_3 nt) \begin{bmatrix} 0 \\ n \\ 1 \\ 0 \end{bmatrix} + c_3 \begin{bmatrix} n/3 \\ 0 \\ 0 \\ 2/3 \end{bmatrix}, \tag{1.17}$$

where c_1, β, c_2, and c_3 are arbitrary constants. The first two column vectors in (17) are the real and imaginary parts of the eigenvector corresponding to $s = nj$; the third and fourth column vectors are the principal and secondary eigenvectors corresponding to $s = 0$ (see Appendix A). The motion of the perturbed spacecraft in each of these three modes, as observed from an unperturbed spacecraft in the same circular orbit, is shown in Fig. 1.2.

The first mode ($c_1 \neq 0$) corresponds to a slightly elliptic orbit so that the spacecraft goes above the circular orbit for half a period and slows down, then goes below and speeds up for the other half. The second mode ($c_2 \neq 0$) corresponds to the spacecraft being in the same circular orbit but slightly ahead of (or behind) the reference point. The third mode ($c_3 \neq 0$) corresponds to the spacecraft being in a lower (or higher) circular orbit that has a faster (or slower) orbital velocity.

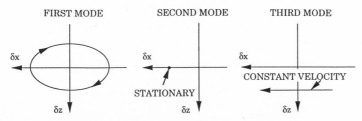

Figure 1.2. Modes of translational motion in circular orbit.

1.3 Rotational Motions in Space

The rotational (attitude) motions of a rigid spacecraft in space are described by Euler's equations:

$$\overset{I}{\vec{H}} \equiv \overset{B}{\vec{H}} + \vec{\omega} \times \vec{H} = \vec{Q}, \qquad (1.18)$$

where

$\vec{H} = \overset{\rightrightarrows}{I} \cdot \vec{\omega}$ = moment of momentum of spacecraft about the c.m.,

$\overset{\rightrightarrows}{I}$ = moment of inertia dyadic with respect to the c.m. of the S/C,

$\vec{\omega}$ = angular velocity of S/C with respect to inertial space,

$\overset{B}{(\ )}$ = time rate of change with respect to body-fixed axes,

$\vec{Q}$ = resultant external torque.

Since the moment of inertia dyadic, $\overset{\rightrightarrows}{I}$, is constant in body-fixed axes, Equation (18) is usually used in body-axis components. If the body-axis components of the angular velocity are $(p, q, r) \Rightarrow \vec{\omega} = p\hat{i} + q\hat{j} + r\hat{k}$ and we use principal axes so that $\overset{\rightrightarrows}{I} = I_x\hat{i}\hat{i} + I_y\hat{j}\hat{j} + I_z\hat{k}\hat{k}$, where $(\hat{i}, \hat{j}, \hat{k})$ are unit vectors along the (x, y, z) principal body axes, then (1) becomes

$$I_x\dot{p} - (I_y - I_z)qr = Q_x, \qquad (1.19)$$

$$I_y\dot{q} - (I_z - I_x)rp = Q_y, \qquad (1.20)$$

$$I_z\dot{r} - (I_x - I_y)pq = Q_z, \qquad (1.21)$$

where (Q_x, Q_y, Q_z) are the body-axis components of the torque.

1.3.1 *Kinematic Equations*

Angular position may be described by three *Euler angles* (or by four Euler parameters (quaternions), or nine direction cosines (cf. Kane)). The NASA

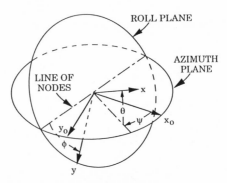

Figure 1.3. Aerospace euler angles (NASA standard).

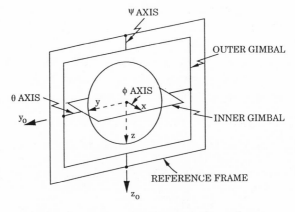

Figure 1.4. Two-gimbal interpretation of aerospace euler angles.

standard Euler angles are shown in Fig. 1.3; the first rotation, ψ, is about the body z-axis; the second rotation, θ, is about the new position of the body y-axis; the third rotation, ϕ, is about the new position of the body x-axis. Fig. 1.4 shows a two-gimbal interpretation of these Euler angles. If the body-axis components of the angular velocity are $(p,\ q,\ r)$, then the Euler angle rates are

$$\begin{bmatrix} \dot{\phi} \\ \dot{\theta} \\ \dot{\psi} \end{bmatrix} = \frac{1}{\cos\theta} \begin{bmatrix} \cos\theta & \sin\theta\sin\phi & \sin\theta\cos\phi \\ 0 & \cos\theta\cos\phi & -\cos\theta\sin\phi \\ 0 & \sin\phi & \cos\phi \end{bmatrix} \begin{bmatrix} p \\ q \\ r \end{bmatrix}. \qquad (1.22)$$

1.3.2 *Small Attitude Changes of a Nonspinning Spacecraft*

For small attitude changes of a nonspinning spacecraft with respect to inertial space, that is, magnitudes of ϕ, θ, ψ small compared to unity, the nonlinear equations of motion (19)–(22) are well approximated by the following *linearized equations of motion*:

$$I_x \dot{p} \cong Q_x, \tag{1.23}$$

$$I_y \dot{q} \cong Q_y, \tag{1.24}$$

$$I_z \dot{r} \cong Q_z, \tag{1.25}$$

$$\dot{\phi} \cong p, \tag{1.26}$$

$$\dot{\theta} \cong q, \tag{1.27}$$

$$\dot{\psi} \cong r. \tag{1.28}$$

Thus angular motion about each of the three principal axes is uncoupled from motion about the other two principal axes. Eliminating (p, q, r) from (23)–(28), we obtain three independent second-order systems:

$$I_x \ddot{\phi} \cong Q_x, \tag{1.29}$$

$$I_y \ddot{\theta} \cong Q_y, \tag{1.30}$$

$$I_z \ddot{\psi} \cong Q_z. \tag{1.31}$$

1.4 Rotational Motions in Circular Orbit

1.4.1 *Gravity Torque*

The very small torque acting on a rigid body in an inverse-square gravitational field may be written as

$$\vec{Q}_g = 3\frac{g}{R}\hat{r} \times (\vec{I} \cdot \hat{r}), \tag{1.32}$$

where

$\quad g \;=\;$ gravitational force per unit mass at radial distance R,

$\quad \hat{r} \;=\;$ unit vector in the radial direction,

$\quad \vec{I} \;=\;$ moment-of-inertia dyadic of the rigid body.

This torque is zero whenever a principal axis is parallel to $\hat{r}$, since $\vec{I} \cdot \hat{r} = \lambda\hat{r}$ and $\hat{r} \times (\lambda\hat{r}) = 0$.

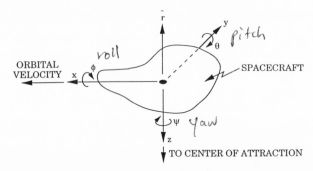

Figure 1.5. Nominal attitude and small-deviation angles for spacecraft in circular orbit.

If the body principal axes (x, y, z) are aligned as shown in Fig. 1.5, then

$$\vec{I} = I_x \hat{\imath}\hat{\imath} + I_y \hat{\jmath}\hat{\jmath} + I_z \hat{k}\hat{k}, \tag{1.33}$$

where (I_x, I_y, I_z) are the principal moments of inertia and $(\hat{\imath}, \hat{\jmath}, \hat{k})$ are unit vectors pointing along the body (x, y, z) axes.

For small angular departures from this nominal position

$$\hat{r} \cong \theta\hat{\imath} - \phi\hat{\jmath} - \hat{k},$$

$$\Rightarrow \vec{I} \cdot \hat{r} \cong I_x\theta\hat{\imath} - I_y\phi\hat{\jmath} - I_z\hat{k},$$

Negligible

$$\Rightarrow \hat{r} \times (\vec{I} \cdot \hat{r}) \cong (I_z - I_y)\phi\hat{\imath} + (I_z - I_x)\theta\hat{\jmath} + (I_x - I_y) \overbrace{\phi\theta\hat{k}}. \tag{1.34}$$

1.4.2 *Relationship between Euler Angle Rates and Angular Velocity*

Let (p, q, r) be body-axis components of the angular velocity of the body with respect to inertial axes, and let (ϕ, θ, ψ) be Euler angles of the body axes with respect to locally horizontal axes, (LHA). Then

$$
\begin{bmatrix} p \\ q \\ r \end{bmatrix} = \begin{bmatrix} \dot{\phi} \\ 0 \\ 0 \end{bmatrix} + T_\phi \begin{bmatrix} 0 \\ \dot{\theta} \\ 0 \end{bmatrix} + T_\phi T_\theta \begin{bmatrix} 0 \\ 0 \\ \dot{\psi} \end{bmatrix} + T_\phi T_\theta T_\psi \begin{bmatrix} 0 \\ -n \\ 0 \end{bmatrix}, \tag{1.35}
$$

where n = angular velocity of LHA with respect to inertial axes and

$$
T_\phi = \begin{bmatrix} 1 & 0 & 0 \\ 0 & c\phi & s\phi \\ 0 & -s\phi & c\phi \end{bmatrix}, \; T_\theta = \begin{bmatrix} c\theta & 0 & -s\theta \\ 0 & 1 & 0 \\ s\theta & 0 & c\theta \end{bmatrix}, \; T_\psi = \begin{bmatrix} c\psi & s\psi & 0 \\ -s\psi & c\psi & 0 \\ 0 & 0 & 1 \end{bmatrix}.
$$

(1.36)

A useful mnemonic for these rotations (thanks to Professor Holt Ashley) is $-s(\;)$ is above the row with a 1 in it, cyclically.

From (35) and (36) it follows that

$$
\begin{bmatrix} p \\ q \\ r \end{bmatrix} = \begin{bmatrix} 1 & 0 & -s\theta \\ 0 & c\phi & c\theta s\phi \\ 0 & -s\phi & c\theta c\phi \end{bmatrix} \begin{bmatrix} \dot{\phi} \\ \dot{\theta} \\ \dot{\psi} \end{bmatrix} - \begin{bmatrix} s\psi c\theta \\ s\psi s\theta s\phi + c\psi c\phi \\ s\psi s\theta c\phi - c\psi s\phi \end{bmatrix} n.
$$

(1.37)

If (ψ, θ, ϕ) are small in magnitude compared to 1 radian, then (37) may be approximated by

$$
\begin{aligned}
p &\cong \dot{\phi} - n\psi - \dot{\psi}\theta, \\
q &\cong \dot{\theta} - n + \dot{\psi}\phi, \\
r &\cong \dot{\psi} + n\phi - \dot{\theta}\phi.
\end{aligned}
$$

(1.38)

The average value, over an orbit period, of the magnitude of the nonlinear terms $(\dot{\psi}\theta, \dot{\psi}\phi, \dot{\theta}\phi)$ in (38) is small compared to the average magnitude of the linear terms if ϕ and θ are oscillatory about zero or if the magnitudes of $\dot{\psi}$ and $\dot{\theta}$ are small compared to the orbit rate n. In such cases, (38) may be approximated as

$$
\begin{aligned}
\dot{\phi} &\cong p + n\psi, \\
\dot{\theta} &\cong q + n, \\
\dot{\psi} &\cong r - n\phi.
\end{aligned}
$$

(1.39)

1.4.3 *Moment of Momentum*

The angular velocities of a rigid spacecraft may be determined from Euler's equations for moment of momentum:

$$
\overset{I}{\dot{\vec{H}}} \equiv \overset{B}{\dot{\vec{H}}} + \vec{\omega}^{B-I} \times \vec{H} = \vec{Q},
$$

(1.40)

where

$$\vec{H} = \vec{I} \cdot \vec{\omega} = \text{moment of momentum,}$$

$$\vec{\omega} = \vec{\omega}^{B-I} = \text{angular velocity body with respect to inertial space,}$$

$$\vec{Q} = \text{external torques on body.}$$

1.4.4 *Equations of Motion, Earth-Pointing Satellite*

Assuming that the body axes are principal axes, the moment of momentum of the spacecraft is

$$\vec{H} = I_x p \hat{\imath} + I_y q \hat{\jmath} + I_z r \hat{k}. \tag{1.41}$$

From (32), (34), (39), and (40), the equations of motion for small perturbations from locally horizontal axes decouple into two sets, one for pitch and one for roll/yaw. The pitch set is:

$$I_y \dot{q} \cong -3n^2(I_x - I_z)\theta + Q_y, \tag{1.42}$$

$$\dot{\theta} - n \cong q. \tag{1.43}$$

The roll/yaw set is

$$I_x \dot{p} + n(I_y - I_z)r \cong -3n^2(I_y - I_z)\phi + Q_x, \tag{1.44}$$

$$I_z \dot{r} - n(I_y - I_x)p \cong Q_z, \tag{1.45}$$

$$\dot{\phi} - n\psi \cong p, \tag{1.46}$$

$$\dot{\psi} + n\phi \cong r, \tag{1.47}$$

where $n \triangleq \sqrt{g/R}$ = orbital angular velocity and (Q_x, Q_y, Q_z) are external torques. The gyroscopic coupling terms, $n(I_y - I_z)r$ in (44) and $n(I_y - I_x)p$ in (45), arise from the rotation of the locally horizontal axes at orbit rate n.

1.4.5 *Pitch Librations*

The characteristic equation of the pitch system (Equations (42) and (43)) in circular orbit is

$$s^2 + 3n^2 \frac{I_x - I_z}{I_y} = 0. \tag{1.48}$$

If $I_x > I_z$, the system has undamped oscillations (librations) at a frequency

$$\omega_p = \sqrt{\frac{3(I_x - I_z)}{I_y}} \, n, \tag{1.49}$$

which is called the pitch libration frequency. If $I_x < I_z$, the system is unstable.

1.4.6 *Roll/Yaw Librations*

Eliminating p and r from Equations (44)–(47), the Laplace transform of the roll/yaw equations may be written as

$$\begin{bmatrix} s^2 + 4an^2 & (a-1)ns \\ -(b-1)ns & s^2 + bn^2 \end{bmatrix} \begin{bmatrix} \phi(s) \\ \psi(s) \end{bmatrix} = \begin{bmatrix} Q_x(s)/I_x \\ Q_z(s)/I_z \end{bmatrix}. \tag{1.50}$$

where

$$a \overset{\Delta}{=} (I_y - I_z)/I_x, \tag{1.51}$$

$$b \overset{\Delta}{=} (I_y - I_x)/I_z. \tag{1.52}$$

The characteristic equation of (50) is

$$\left(\frac{s}{n}\right)^4 + (3a + ab + 1)\left(\frac{s}{n}\right)^2 + 4ab = 0. \tag{1.53}$$

The magnitudes of a and b are less than or equal to unity since

$$a = \frac{\int\int\int(x^2 + z^2)dm - \int\int\int(x^2 + y^2)dm}{\int\int\int(y^2 + z^2)dm},$$

$$= \frac{\int\int\int(z^2 - y^2)dm}{\int\int\int(z^2 + y^2)dm},$$

and similarly

$$b = \frac{\int\int\int(x^2 - y^2)dm}{\int\int\int(x^2 + y^2)dm}.$$

For b = constant, (53) may be written in Evan's root locus form as

$$-a(3 + b) = \frac{s^2(s^2 + 1)}{s^2 + 4b/(3 + b)}, \tag{1.54}$$

where s is in units of n. Fig. 1.6 shows the root loci vs. a for fixed values of b, and Fig. 1.7 summarizes the results. The spacecraft roll/yaw motions are oscillatory for $a > 0$, $b > 0$, and for a small region when $a < 0$, $b < 0$; elsewhere the motions are unstable.

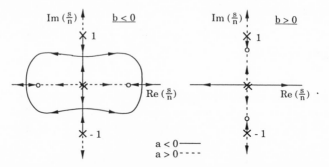

Figure 1.6. Root loci vs. moment-of-inertia ratio $a = (I_y - I_z)/I_x$ for roll/yaw motions.

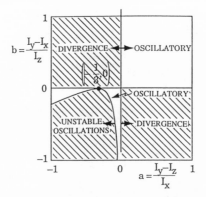

Figure 1.7. Unstable regions for roll/yaw motions of spacecraft in circular orbit.

1.4.7 Roll/Yaw Librations of Axially Symmetric Spacecraft

Symmetry Axis Nominally Vertical

Here $I_x = I_y$, which implies $b = 0$, and $a = 1 - I_z/I_x$. For $0 < a < 1$, the spacecraft is prolate whereas for $-1 < a < 0$ the spacecraft is oblate. The characteristic equation for this case ((53) with $b = 0$)) is

$$s^2[s^2 + (3a + 1)n^2] = 0. \tag{1.55}$$

The double pole at $s = 0$ represents the yaw mode, which is uncoupled to roll motions. The roll mode is oscillatory if $a > -1/3$ or $I_z/I_x < 4/3$ (prolate spacecraft); it is unstable if $I_z/I_x > 4/3$ (oblate spacecraft).

Symmetry Axis Nominally In-Track

Here $I_y = I_z$, which implies $a = 0$, and $b = 1 - I_x/I_z$. The characteristic equation for this case is

$$s^2(s^2 + n^2) = 0. \tag{1.56}$$

One of the $s = 0$ poles represents a roll-angle-indifference mode, that is, $\phi =$ constant $\neq 0$, $\psi = 0$. The other $s = 0$ pole represents a mode with constant roll rate, $\dot{\phi}$, and a constant yaw angle ψ, where

$$\psi = \frac{b-1}{b}\frac{\dot{\phi}}{n}, \tag{1.57}$$

that is, a stationary attitude of the symmetry axis with respect to the locally horizontal axes. The other mode is oscillatory with frequency $= n$, that is, a stationary attitude of the symmetry axis with respect to inertial axes.

Axis of Symmetry Nominally Cross-Track

Here $I_x = I_z$, which implies $a = b = I_y/I_x - 1$. The characteristic equation for this case is

$$s^4 + (3a + a^2 + 1)n^2 s^2 + 4a^2 n^4 = 0. \tag{1.58}$$

The motions are oscillatory if $a > -.146$, or $I_y/I_x > .854$ (spacecraft oblate).

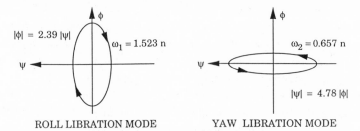

ROLL LIBRATION MODE   YAW LIBRATION MODE

Figure 1.8. Path of tip of symmetry axis for librating oblate spacecraft with symmetry axis nominally cross-track.

Problem 1.4.1 – *Roll/Yaw Libration Modes for an Oblate Axially Symmetric Spacecraft*

Consider the roll/yaw libration motions of an oblate axially symmetric spacecraft with its symmetry axis nominally cross-track, where $I_x = I_z = 2/3I_y \Rightarrow a = b = 1/2$.

Show that the natural motions as viewed by an observer who is stationary with respect to the locally horizontal-vertical axes in circular orbit consist of a superposition of two purely oscillatory modes with frequencies $\omega_1 = 1.523n$, $\omega_2 = .657n$, where the tip of the symmetry axis moves in elliptical paths as shown in Fig. 1.8. Note tip rotates in opposite directions for the two modes.

1.5 Disturbances

In the solar system, away from any planets, a spacecraft is in orbit about the sun. The main disturbance is solar radiation pressure, which is 4.4×10^{-6} Newtons per square meter at the earth's distance from the sun. By careful design, the solar radiation torque can be reduced to 10^{-5} or even 10^{-6} Newton-meters at the earth's distance from the sun.

Another significant disturbance is gas leakage from reaction control systems. Again, by careful design, this torque can be held to values on the order of 10^{-5} Newton-meters.

Moving mechanical parts and crew movements cause internal torques that disturb the attitude of the spacecraft and may cause significant flexural vibrations.

Fig. 1.9 (Ref. DE) shows typical disturbance torques on a spacecraft as a function of altitude above the earth's surface. Gravity torque is caused by the earth's gravity gradient (see Section 4.2). Magnetic torque is caused by the earth's magnetic field acting on the residual magnetic dipole moment of the spacecraft (see Section 4.4). Aerodynamic torque becomes significant below an altitude of 1000 kilometers.

All five of the disturbance torques in Fig. 1.9 can also be used as control torques. A controlled gas leak is a thruster; a solar sail uses radiation pressure; gravity, magnetic, and aerodynamic torques can be (and have been) used for desaturating reaction or momentum wheels (see Chapters 4 and 6).

During thrust maneuvers, the thrust misalignment torque is orders of magnitude larger than the torques mentioned here. Consequently, a special attitude control system is required during thrust maneuvers.

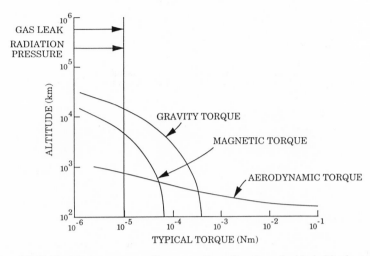

Figure 1.9. Typical torques on a small spacecraft as a function of orbital altitude.

2

Spacecraft Sensors and Attitude Determination

2.1 Introduction

In this chapter, we discuss a few of the many sensors that have been used for attitude determination, and some of the methods used to estimate attitude angles that are not directly sensed. A very thorough treatment is given in the book edited by Wertz (Ref. WE).

Most spacecraft in circular or elliptic orbits are stabilized with one body axis pointing downward so that cameras or antennas point toward the earth. Horizon sensors can then be used to measure the pitch and roll angles of the S/C. Adding roll-rate and yaw-rate gyros allows the yaw angle to be estimated using the knowledge of kinematic roll/yaw coupling (a process called "orbital gyro-compassing").

Many spacecraft have solar cells for generating electrical power from the sun's radiation, so some means must be provided for pointing solar panels toward the sun while the antennas/cameras point toward the earth; this requires a slow rotation of the panels with respect to the S/C, which changes the moments of inertia of the overall S/C.

Some spacecraft, particularly those whose missions involve scientific exploration, are stabilized so that the attitude is fixed with respect to inertial space. Thus, in circular orbit, the attitude with respect to the earth is constantly changing, like the cars on a ferris wheel.

Spinning S/C usually have the spin axis perpendicular to the orbit plane, except in transfer orbits. In order to have cameras or antennas pointing toward the earth and still have the advantage of spin stabilization, part of the S/C is de-spun (thus the term "dual-spin" S/C). Alternatively, a nonspinning S/C may contain a "bias momentum wheel" that provides spin stabilization.

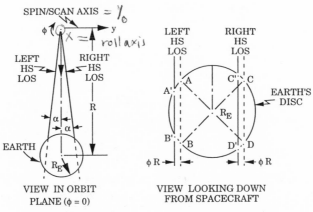

Figure 2.1. Geometry of spacecraft with two scanning horizon sensors.

2.2 Infra-Red, Optical, and Radar Sensors

2.2.1 *Horizon Sensors*

The earth's horizon is distinguished most sharply in infrared (IR) wavelengths, and IR sensors are less susceptible to sun interference than optical sensors.

A scanning mechanism is required because of the large angle subtended by the earth in circular orbit. For spinning S/C, the sensor is simply mounted on the S/C body. For nonspinning S/C, the sensor may either be mounted on a bias momentum wheel (which is also used for S/C attitude stabilization), or scanned with oscillating mirrors.

For spin-scan sensors, the times at which the sensor axis enters and leaves the earth's disc can be used to determine both pitch and roll angles. At the average of these two times, the sensor axis has zero pitch angle, so the angle between the z-axis of the de-spun part of the S/C and the sensor axis at that time is the S/C pitch angle θ. The difference between the two times is the time to cross the disc; knowing the altitude of the S/C, and the spin/scan speed, the magnitude (but not the sign) of the roll angle can be calculated.

More accurate sensing of the roll angle, ϕ, along with its sign, can be obtained with two spinning sensors whose axes diverge slightly (see Fig. 2.1). At zero roll angle the lines of sight (LOS) of the two sensors pass over the earth's disc with equal chords ($AB = CD$). When ϕ is not zero, one chord ($C'D'$ in Fig. 2.1) is longer than the other ($A'B'$). From measurements of the entry and exit times, ϕ can be determined quite accurately.

Problem 2.2.1 – *Analysis of Scanning Horizon Sensor*

Fig. 2.1 was simplified somewhat for clarity. The horizon sensor line of sight (LOS) actually moves on the surface of a cone, which intersects the (almost) spherical (earth + atmosphere). The entry and exit points (A', C' and B', D') occur when the LOS is just tangent to the sphere.

Express the pitch angle θ and the roll angle ϕ in terms of the entry and exit times t_A, t_C and t_B, t_D, the spin/scan angular velocity, the radius of the orbit R, and the sphere radius R_E.

2.2.2 V-Beam Optical Slit Sensors ⟨ Sun earth n moon ⟩ sensors

A V-beam slit sensor consists of two photocells mounted in slits that form a V as shown in Fig. 2.2. The photocells are illuminated ("pulsed") by the sun (or the earth or the moon) once each rotation of a spinning spacecraft (or rotor section of a dual-spin spacecraft). The time difference between successive pulses of photocell S_1 (the "spin-angle sensor") is the rotation period, which determines the spin rate very precisely. The time difference between pulses of photocell S_1 and photocell S_2 (the "ϕ-angle sensor") is a function of the spin rate and the angle that the spin axis makes with the sun line, ϕ. Hence it can be used, with the spin rate, to determine ϕ.

The spin-angle sensor S_1 is pulsed when the sun's rays are perpendicular to the normal to the plane DS_1E. The ϕ-angle sensor S_2 is pulsed when the sun's rays are perpendicular to the normal to plane CS_2E.

Fig. 2.2 shows the case when the spin axis is perpendicular to the sun's rays ($\phi = 90$ deg), while Fig. 2.3 shows the case when the spin axis makes an angle

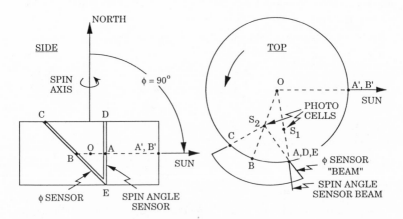

Figure 2.2. V-beam slit sensor on a spinning spacecraft—spin axis normal to the sun's rays.

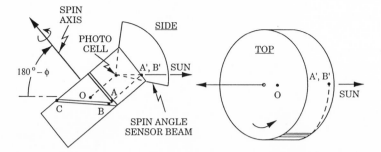

Figure 2.3. V-beam slit sensor on a spinning spacecraft—spin axis not normal to the sun's rays.

> 90 deg with the sun line. The photocells are pulsed when the lines OA and OB are parallel to the sun line in postion OA' and OB'. The larger ϕ is, the shorter the time between the pulses.

Analysis of the V-Beam Slit Sensor

With reference to Fig. 2.4, let β = the azimuthal angle between the two photocells, $\vec{S}$ = sun line vector, $\vec{N_1}$ = normal to the spin angle sensor slit, $\vec{N_2}$ = normal to the ϕ angle sensor slit, $\alpha = \pi/2-$ angle between the spin axis and $\vec{N_2}$. The sun pulses sensor 1 when $\lambda = 0$ ($\vec{S} \cdot \vec{N_1} = 0$); it pulses sensor 2 when ($\vec{S} \cdot \vec{N_2} = 0$), that is, when

$$0 = \left(\hat{\imath} \sin \phi + \hat{k} \cos \phi\right) \cdot \left[\cos \alpha \cos \left(\frac{\pi}{2} + \lambda - \beta\right) \hat{\imath} + ()\hat{\jmath} + \sin \alpha \hat{k}\right],$$

or

$$0 = - \sin \phi \cos \alpha \sin(\lambda - \beta) + \cos \phi \sin \alpha,$$

or

$$\lambda = \beta + \sin^{-1} \frac{\tan \alpha}{\tan \phi}. \tag{2.1}$$

The difference in time between pulses on S_1 and S_2 is therefore

$$\Delta t = \frac{1}{\omega} \left[\beta + \sin^{-1} \frac{\tan \alpha}{\tan \phi}\right], \tag{2.2}$$

where ω = spin rate. Knowing ω and Δt, we can determine ϕ:

$$\phi = \tan^{-1} \frac{\tan \alpha}{\sin(\omega \Delta t - \beta)}. \tag{2.3}$$

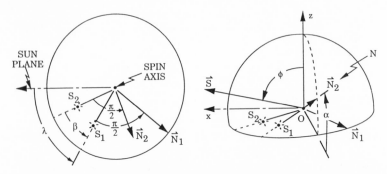

Figure 2.4. Nomenclature for analyzing V-beam slit sensor on a spinning spacecraft.

As an example, suppose that $\alpha = \beta = 35$ deg, $\omega = 16.96$ rad/sec (2.7 rev/sec). Then

Δt	ϕ
10^{-3} sec	deg
0	129.3
11.48	120.0
36.01	90.0
73.05	50.0

The accuracy of slit sensors in determining ϕ varies from .001 to 1.0 deg.

ϕ Locates the Spin Axis on a Cone

The angle ϕ determined by the V-beam slit sensor locates the spin axis on a cone that makes an angle ϕ with the direction to the sun (see Fig. 2.5). Another sensor is needed to locate the spin axis on the cone. For example, the first geosynchronous communication satellite, COMSAT II (see Fig. 5.1), had a "pancake antenna beam," which passed over an earth receiving station once every 12 hours (while the spin axis was in the orbit plane, before being precessed to its final position normal to the orbit plane); at the time of peak radiation intensity, the spin axis is perpendicular to the line of sight (LOS) $\vec{OE}$ in Fig. 2.5; hence it lies in the plane OAA' normal to the LOS. Even very coarse dead reckoning resolves the ambiguity between the two possible spin-axis locations OA and OA'.

Problem 2.2.2 – *Satellite with Pancake Antenna Pattern*

Fig. 2.6 shows a satellite in an equatorial geosynchronous orbit whose antenna axis (coincident with its spin axis) is $\vec{TA}$. The beam has a "pancake" shape (see Fig. 5.1).

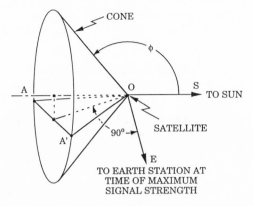

Figure 2.5. Determination of the spin axis orientation from a V-beam slit sensor and a pancake antenna beam.

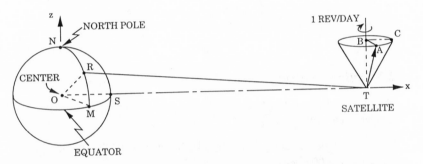

Figure 2.6. Geometrical relations of antenna axis to earth receiving station.

Thus angle $ATR = \alpha$ determines signal strength at an earth receiving station R. Maximum signal strength occurs when $\alpha = 90$ deg.

Let $\vec{TB}$ be parallel to the earth's axis $\vec{ON}$. To an observer at R the satellite spin axis $\vec{TA}$ appears to rotate around $\vec{TB}$ once per day in a clockwise direction (looking from above), because it is fixed in inertial space and the line $\vec{OT}$ rotates counterclockwise once per day with respect to inertial space. Let $\vec{BC}$ be parallel to $\vec{OT}$, and angle $CBA(= \theta_T)$; then $\dot{\theta}_T = 2\pi$ rad/day.

Let $\phi_R =$ angle $ROM =$ latitude of R, $\theta_R =$ angle $SOM =$ difference in longitude of R and S where S is the subsatellite point, $\phi_T =$ angle BTA, $r_T = \vec{OT}$, $r_E = \vec{OS}$.

(a) *Show that*

$$\cos\alpha = a\cos\phi_T + \sin\phi_T(b\sin\theta_T - c\cos\theta_T),$$

where

$$a = \frac{r_E}{\ell}\sin\phi_R,$$

$$b = \frac{r_E}{\ell} \cos \phi_R \sin \theta_R,$$

$$c = \frac{r_T}{\ell} - \frac{r_E}{\ell} \cos \phi_R \cos \theta_R,$$

$$\ell = \left| \vec{TR} \right| = \left(r_T^2 + r_E^2 - 2r_E r_T \cos \phi_R \cos \theta_R \right)^{1/2},$$

$$\frac{r_T}{r_E} = 6.628.$$

(b) Suppose the receiving station R is at Los Angeles ($\phi_R = 34$ deg N, longitude = 118 deg W) and the subsatellite point is on the equator at longitude = 40 deg W. *Show that*

$$\cos \alpha = .08563 \cos \phi_T + \sin \phi_T (.1241 \sin \theta_T - .9885 \cos \theta_T),$$

$$= .08563 \cos \phi_T + .9963 \sin \phi_T \sin(\theta_T - 82.84°).$$

Suppose observations of signal strength at Los Angeles indicate maximum strength ($\alpha = 0$) at midnight and noon and no signal at all from 2 AM to 10 AM and 2 PM to 10 PM. The half-width of the antenna beam is known to be 6 deg. Estimate ϕ_T and θ_T at midnight in Los Angeles.
Answer: $\phi_T = 11.9$ deg, $\theta_T = 58.8$ or 286.9 deg.

2.2.3 Optical Telescope Sensors

Attitude Determination Using Two Lines of Sight to Celestial Objects

Optical telescope measurements provide the most accurate determination of attitude (.1 to .0005 deg).

In space, as opposed to in orbit, attitude is usually described relative to the ecliptic plane (the plane of the earth's orbit about the sun), and the vernal equinox (the line from the sun to the earth on March 21). This may be done by giving the Euler angles of a set of body-fixed axes with respect to this "celestial sphere" (see Fig. 2.7), or a set of direction cosines.

If an optical telescope is pointed at a known star, then the declination (elevation angle above the ecliptic plane) and the right ascension (azimuth angle relative to the vernal equinox) of the telescope axis is known from a star table. Let us call this the x' axis (see Fig. 2.7). If the telescope is then rotated in the body to point at another known star (preferably about 90 deg away from the first star), then a z' axis may be defined perpendicular to the plane determined by the two star lines. Finally a y' axis (in the star line plane) is perpendicular to the x' and z' axes.

Attitude may also be determined using the sun line and a star line. This requires that the spacecraft position be known so that the declination and right

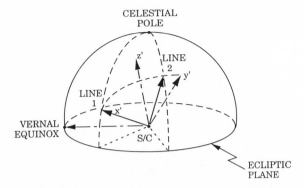

Figure 2.7. Attitude determination using two lines to celestial objects.

ascension of the sun line can be determined. If the spacecraft is in the ecliptic plane, the declination of the sun line is, of course, zero. Many spacecraft use the star Canopus, since it has a high declination and is readily identified.

Planet lines may also be used, but they require a knowledge of the planet position and the spacecraft postion at the time of the sighting so that the declination and right ascension of the planet line can be determined.

Position Determination Using Angles between Celestial Objects

In interplanetary space, position is given in polar spherical coordinates, radial distance from the sun, declination relative to the ecliptic plane, and right ascension from the vernal equinox. Position may be sensed by optical sightings on the sun, the planets, and the stars.

Measurements of the angle between the sun and a planet (or between two planets) locate the spacecraft on a toroidal surface of position, generated by rotating a circle on the planet-sun (or planet-planet) line, as shown in Fig. 2.8. If two other surfaces of position are found, the spacecraft position is determined by the intersection of the three surfaces (there may be more than one possible location).

Measurement of the angle between the sun (or planet) and a star whose direction is known relative to the ecliptic plane and the vernal equinox locates the spacecraft on a conical surface of position generated by rotating the spacecraft-sun (or spacecraft-planet) line about the line of sight (LOS) to the star from the sun (or planet), as shown in Fig. 2.9.

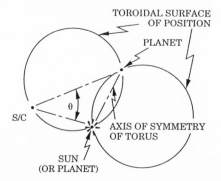

Figure 2.8. Toroidal surface of position for a given angle measurement between the sun and a planet.

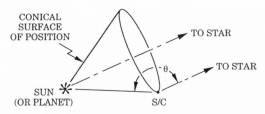

Figure 2.9. Conical surface of position for a given angle measurement between the sun (or planet) and a star.

2.2.4 *Doppler Velocity Sensors*

The velocity of a spacecraft along the line of sight from the earth can be determined very accurately (±1 cm/sec) by measuring frequency of received electromagnetic radiation from the earth (or vice versa). By the use of Kalman filtering, which involves integrating the measured velocity, radial position can be estimated to an accuracy of a few hundred kilometers a year after the spacecraft has left the earth.

2.3 Orbital Gyrocompassing

Horizon sensors measure roll angle, ϕ, and pitch angle, θ, but they obviously cannot measure yaw angle, ψ. However, if roll-rate and yaw-rate gyros are added (measuring p and r), then it is possible to estimate the yaw angle, using the kinematic roll/yaw coupling described in Equations (1.46) and (1.47):

$$\dot{\phi} = p + n\psi, \tag{2.4}$$

$$\dot{\psi} = r - n\phi. \tag{2.5}$$

If the orbit rate, n, is known, then ψ is observable from (4):

$$\psi = (\dot{\phi} - p)/n. \tag{2.6}$$

Now, the horizon sensor measurement of ϕ is noisy, so we cannot really differentiate the ϕ signal to find $\dot{\phi}$. However, we can construct a kinematic estimator (a steady-state Kalman-Bucy filter), which uses both (4) and (5), along with the measurements of ϕ, p, and r (call them ϕ_m, p_m, r_m), as follows:

$$\dot{\hat{\phi}} = p_m + n\hat{\psi} + K_\phi(\phi_m - \hat{\phi}), \tag{2.7}$$

$$\dot{\hat{\psi}} = r_m - n\hat{\phi} + K_\psi(\phi_m - \hat{\phi}), \tag{2.8}$$

where $(\hat{\phi}, \hat{\psi})$ are the estimates of (ϕ, ψ). This estimator does not require us to differentiate ϕ_m.

The estimate errors, $\tilde{\phi} = \hat{\phi} - \phi$ and $\tilde{\psi} = \hat{\psi} - \psi$, can be predicted by subtracting (4) and (5) from (7) and (8). If we assume that (ϕ_m, p_m, r_m) are reasonably accurate, then

$$\dot{\tilde{\phi}} = n\tilde{\psi} - K_\phi\tilde{\phi}, \tag{2.9}$$

$$\dot{\tilde{\psi}} = -n\tilde{\phi} - K_\psi\tilde{\phi}. \tag{2.10}$$

The characteristic equation of (9)–(10) is

$$s(s + K_\phi) + n(n + K_\psi) = 0. \tag{2.11}$$

If we wish the error-decay eigenvalues to be $(-n, -n)$, the desired characteristic equation is:

$$(s + n)^2 \equiv s^2 + 2ns + n^2 = 0. \tag{2.12}$$

Comparing coefficients of s in (11) and (12) gives the required gains:

$$K_\phi = 2n, \tag{2.13}$$

$$K_\psi = 0. \tag{2.14}$$

2.4 Gyros

The angular velocity of a spacecraft with respect to inertial space can be determined using three *rate gyros*. If the gyros are attached rigidly to the spacecraft, then the three measured angular velocities are in body-axis components. If they are mounted on a gimbaled platform, the platform can be servo-stabilized to have nearly zero angular velocity with respect to inertial space. If the platform is initially aligned to be parallel to the ecliptic plane, with one axis pointing toward the vernal equinox, the gimbal angles of this platform are then estimates of the Euler angles of the spacecraft body axes with respect to the celestial sphere. Current inertial platforms drift away from their initial positions at a rate of .01 to .001 deg/hr. Periodic sightings on celestial objects can be used to estimate the platform misalignment; in between sightings the platform provides a good estimate of current spacecraft attitude.

2.5 Inertial Measurement Units

2.5.1 *Inertially Fixed Platform*

For this type of inertial measurement unit (IMU) a platform is mounted in gimbals relative to the spacecraft. Three gyros and three specific force sensors are mounted on the platform with their sensitive axes mutually orthogonal (two in the plane of the platform and one perpendicular to the platform). Using the gyro signals, the gimbals are torqued to keep the angular velocity of the platform as close to zero as possible. The gimbal angles are then the Euler angles of the spacecraft with respect to inertial space (see Fig. 1.3).

The specific force sensors measure the external forces per unit mass on the spacecraft, except for gravitational force per unit mass (since this force acts on both the platform and the proof mass of the sensor). Using the estimated position of the spacecraft, the gravitational force per unit mass can be estimated and added to the signals from the specific force sensors. These signals are then integrated once to give the estimated velocity of the spacecraft with respect to inertial space. These signals in turn are integrated to give estimated position. Note this signal processing involves feedback since the gravitational force per unit mass is determined from the estimated position.

This is a "dead-reckoning" attitude and position system, based on knowledge of initial position and velocity plus measurements of angular velocity and specific force. Since the measurements and the servos are not perfect, the system develops errors that grow with time. The accuracy of inertial-quality sensors is on the order of .01 deg/hr for gyros and 10^{-6} earth g's for specific force sensors. Position or velocity measurements are often used to "update" the IMU, either continuously or at discrete time intervals.

Detailed treatment of IMUs is given in Refs. WR, WE, and KAY.

2.5.2 *Locally Horizontal Platform*

In circular orbit, it is usually more convenient to maintain the platform in a locally horizontal attitude. To do this, the gimbals are torqued to keep roll and yaw rates as close to zero as possible and the the pitch rate equal to the negative of the estimated orbit rate $(-n)$. The gimbal angles are then the Euler angles of the spacecraft with respect to the local horizontal and the normal to the orbit plane (see Fig. 1.4).

Alternatively, an horizon sensor is mounted on the platform to measure pitch and roll angles directly, and the gimbals are torqued to keep them as close to zero as possible. The yaw gimbal is torqued to keep the yaw angular velocity as close to zero as possible. A third alternative is to update the first system from time to time with the horizon sensor (instead of continuously).

2.5.3 *Strapdown IMUs*

If the gyros and specific force sensors are mounted rigidly to the spacecraft, instead of on a gimbaled platform, this forms a *strapdown IMU*. A computer is used to estimate the gimbal angles of a "virtual platform" by integrating the rate gyro signals (using the inverse of Equation 1.37). The specific force measurements are resolved into the locally horizontal frame using the estimated Euler angles of the spacecraft.

3

Attitude Control with Thrusters

3.1 Fast versus Slow Attitude Control

If the desired attitude control bandwidth is large compared to orbit rate n, then gravity or magnetic torques are small compared to the required control torques. Hence, they may be treated as disturbance torques, using a free-space model of the spacecraft dynamics. We shall call this *fast attitude control*.

If, on the other hand, the satellite mission is such that a control bandwidth comparable to the orbit rate is acceptable, then gravity or magnetic torques can be used with reaction wheels (or gimbaled momentum wheels; see Chapter 6) to stabilize the satellite *without the use of thrusters*. We shall call this *slow attitude control*. The use of slow attitude control requires careful configuration design to keep disturbance torques acceptably small.

We shall not discuss slow attitude control with passive libration dampers. This type of damping was used in the early days of the space era (Ref. SY) but has been succeeded by improved methods. It is difficult to build lightweight passive dampers that will dissipate energy at the low libration frequencies.

3.2 Fast Attitude Control Using Proportional Thrusters

If the available control torques are large compared to the disturbance torques, and we wish to *stabilize* the spacecraft attitude *with respect to inertial space*, then the attitude motions about the three principal axes are nearly uncoupled, and stabilization about each principal axis may be treated separately (see Fig. 3.1).

A convenient inertial reference system in interplanetary space is the celestial sphere.

In this section, we consider the somewhat unrealistic case where *proportional thrusters* are available to produce torques about each of the three principal axes.

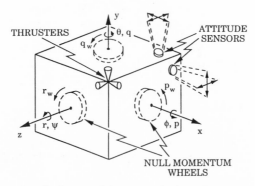

Figure 3.1. Three-axis attitude control thrusters, wheels, and sensors.

Proportional thrusters are *not* easily achieved in practice; we treat them here for pedagogical reasons, since control logic for such thrusters is simpler than control logic for the more practical on-off thrusters, which are treated in the next section.

If *both attitude and attitude rate are sensed* (e.g., using a sun sensor, a star sensor, and three rate gyros), then stabilization about each principal axis may be obtained by feeding back a linear combination of attitude deviation and attitude rate to the torquer. For example, about the body y-axis,

$$Q_y = -D\dot{\theta} - K\theta, \tag{3.1}$$

where

$$I_y\ddot{\theta} = Q_y. \tag{3.2}$$

Clearly, for $D > 0$, $K > 0$, the motion is stabilized.

If *only attitude is sensed* (gyros consume power, reduce reliability, and they cost and weigh more than the devices needed to implement the control logic), then stabilization about each principal axis may be obtained by feeding back attitude deviation with *lead compensation* to the torquer. For example, about the body y-axis,

$$Q_y = -K(\theta - \xi), \tag{3.3}$$

$$\dot{\xi} + b\xi = (b - a)\theta. \tag{3.4}$$

In *transfer function notation*, (2)–(4) become

$$Q_y(s) = -K\frac{s + a}{s + b}\theta(s), \tag{3.5}$$

$$\theta(s) = \frac{1}{I_y s^2}Q_y(s). \tag{3.6}$$

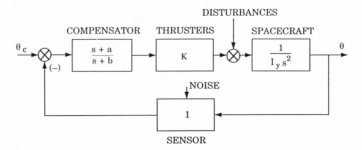

Figure 3.2. Attitude control system with proportional thrusters and attitude sensor.

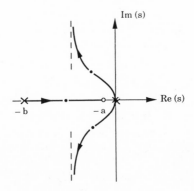

Figure 3.3. Locus of closed-loop roots vs. feedback gain K for system of Fig. 3.2.

Fig. 3.2 shows a block diagram of the system, including the possibility of commanding a nonzero value of θ, called θ_c.

The characteristic equation of the system (5) and (6), in Evans's form, is

$$-\frac{K}{I_y} = \frac{s^2(s+b)}{s+a}. \tag{3.7}$$

Fig. 3.3 shows a root locus versus K for $b > a > 0$. Clearly the attitude is stabilized.

Problem 3.2.1 – *Second-Order Estimator Using Attitude Angle*

Show that attitude-rate q, and attitude θ, may be estimated using a noisy measurement of attitude z_y by the system

$$\dot{\hat{q}} = \frac{Q_y}{I_y} + k_q(z_y - \hat{\theta}),$$

$$\dot{\hat{\theta}} = \hat{q} + k_\theta(z_y - \hat{\theta}),$$

where $k_q > 0, k_\theta > 0$.

Problem 3.2.2 – *Second-Order Compensator Synthesis Using an Estimator*

(a) If we use estimated-state feedback, that is, modify (1) to

$$Q_y = -D\hat{q} - K\hat{\theta},$$

where $\hat{q}$ and $\hat{\theta}$ are obtained with the estimator of Problem 3.2.1, *show that* this yields a second-order compensator, $Q_y(s)/z_y(s)$.

(b) Choose D and K to place the poles of the regulator, (1) and (2) at $s = -\omega_o \pm \omega_o j$, and choose k_q and k_θ in Problem 3.2.1 to place the estimate-error poles at $s = -\omega_o \pm \omega_o j$.

(c) Show that the closed-loop system has double poles at $s = -\omega_o \pm \omega_o j$.

(d) Show that the compensator is

$$\frac{Q_y(s)}{I_y} = -8\omega_o^3 \frac{s + \omega_o/2}{(s + 2\omega_o)^2 + (2\omega_o)^2} z_y(s).$$

(e) Multiply the right-hand side of the equation in (d) by k_o and sketch a root locus versus k_o of the closed-loop system with this compensator. Note for $k_o = 1$, the root locus must have double poles at $s = -\omega_o \pm \omega_o j$.

Problem 3.2.3 – *First-Order Estimator Using Attitude Angle*

Show that attitude rate q can be estimated from a relatively noise-free measurement of attitude z_y, by the system

$$\hat{q} = q^* + \ell z_y,$$

$$\dot{q}^* = -\ell\hat{q} + Q_y/I_y,$$

where $\ell > 0$. Note that

$$\dot{\hat{q}} = Q_y/I_y + \ell(\dot{z}_y - \hat{q}).$$

Problem 3.2.4 – *First-order Compensator Synthesis Using an Estimator*

(a) If we use $Q_y = -D\hat{q} - Kz_y$, where $\hat{q}$ is obtained with the estimator of Problem 3.2.3, show that this yields a classical lead compensator, $Q_y(s)/z_y(s)$.

(b) Choose D and K to place the poles of the regulator (1) and (2) at $s = -\omega_o \pm \omega_o j$, and choose ℓ in Problem 3.2.3 to place the estimate-error pole at $s = -\omega_o$.

(c) Show that the closed-loop system has poles at $s = -\omega_o \pm \omega_o j, -\omega_o$.

(d) Show that the compensator is

$$\frac{Q_y(s)}{I_y} = -4\omega_o^2 \frac{s + \omega_o/2}{s + 3\omega_o} z_y(s).$$

(e) Multiply the right-hand side of the equation in (d) by k_o and sketch a root locus versus k_o of the closed-loop system with the compensation. Note for $k_o = 1$, the root locus must pass through $s = -\omega_o \pm \omega_o j$ and $s = -\omega_o$.

3.3 Fast Attitude Control Using On-Off Thrusters

3.3.1 On-Off Thrusters

Proportional gas jets are difficult to build; they usually have a large amount of hysteresis. In addition, proportional valves need open only a small amount to produce the small torques required for control; as a result, dirt and ice particles tend to stick in the valve openings and they do not close completely. The resulting leakage causes the opposing jets to open, and the gas supply dwindles rapidly! Consequently, control techniques have been developed where the valves are either completely open or completely closed ("on-off" or "bang-bang" control). Large springs may be used to hold them shut, thereby reducing leakage; the closing "bang" jars loosen any ice or dirt particles.

Valves can be operated to stay open as little as a few milliseconds and can be fired over a million times reliably, but the valves must stay open a finite length of time, and therefore there is a discrete angular velocity change with each actuation of the valve. As a result, it is not possible to get zero residual angular velocity as it is in principle with a proportional valve.

To prevent opposing jets from fighting each other, there must be a deadband in a system using on-off control. When the vehicle is in the deadband, no control action is taken. When the error signal—which is made up of vehicle attitude and attitude-rate information—exceeds the deadband, then the gas valves are appropriately modulated.

3.3.2 Bang-Bang Control (No Deadband)

A simple on-off reaction jet control scheme is obtained using a "linear switching function." If the control torque has only two values (Q_o and $-Q_o$), we can bring θ and $\dot{\theta}$ almost to zero by the control logic

$$Q = -Q_o \; sgn(\theta + \tau \, \dot{\theta}), \tag{3.8}$$

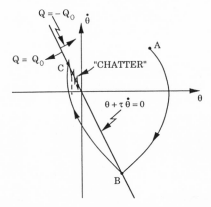

Figure 3.4. Phase plane path of a bang-bang control system with a linear switching function.

where

$$sgn(x) \overset{\Delta}{=} \begin{cases} 1 & x > 0 \\[2ex] -1 & x < 0 \end{cases}$$

and $\theta + \tau \dot{\theta}$ is the *switching function*, which, in this case, is a linear combination of θ and $\dot{\theta}$. A phase plane diagram of the system behavior is shown in Fig. 3.4.

For Q = constant,

$$\dot{\theta} = \frac{Q}{I}(t - t_o) + \dot{\theta}_o, \tag{3.9}$$

$$\theta = \frac{1}{2}\frac{Q}{I}(t - t_o)^2 + \dot{\theta}_o(t - t_o) + \theta_o. \tag{3.10}$$

Eliminating $t - t_o$,

$$\theta - \theta_o = \frac{I}{2Q}(\dot{\theta} - \dot{\theta}_o)^2 + \frac{I}{Q}\dot{\theta}_o(\dot{\theta} - \dot{\theta}_o) \equiv \frac{I}{2Q}(\dot{\theta}^2 - \dot{\theta}_o^2), \tag{3.11}$$

$$\Rightarrow \dot{\theta}^2 = \frac{2Q}{I}\left(\theta - \theta_o + \frac{I}{2Q}\dot{\theta}_o^2\right), \tag{3.12}$$

which is a parabola in the θ, $\dot{\theta}$ phase space. The path in the phase space is made up of parabolas until a point like C is reached where the next parabola does not occur to the right of the switching line but starts off to the left; due to the finite time (Δ) necessary to switch the jets off and on, the path actually overshoots the switching line a small amount, and the torque reversal sends the path back

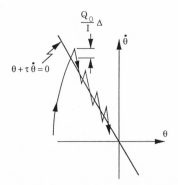

Figure 3.5. Chatter along the switching line in the phase plane.

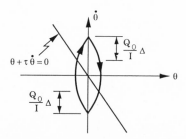

Figure 3.6. Limit cycle around $\theta = \dot{\theta} = 0$, a nonlinear oscillator.

across the switching line, reversal occurs, etc., giving rise to a high-frequency chatter shown in Fig. 3.5.

The path essentially follows $\tau\dot{\theta} + \theta \cong 0 \Rightarrow \theta \cong \theta_c exp(-(t - t_c)/\tau)$, so the system moves toward $\theta = \dot{\theta} = 0$.

The chatter is caused here by the time delay (Δ) in switching the jets on and off. Unfortunately, the point $\theta = \dot{\theta} = 0$ is never reached; instead a *limit cycle* around that point occurs as shown in Fig. 3.6. Obviously this wastes fuel, so some other scheme with a *dead zone* (where $Q = 0$ is used) is desired.

3.3.3 Bang-Off-Bang Control

To make effective use of a dead-zone we also use *hysteresis*; dead zone and hysteresis are combined in a scheme called a Schmitt trigger, shown in Fig. 3.7.

It is simple to use the Schmitt trigger with a *linear switching function* as shown in Fig. 3.8.

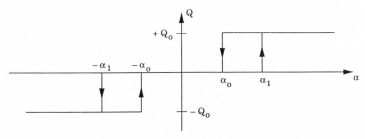

Figure 3.7. Output vs. input with dead zone and hysteresis (Schmitt trigger).

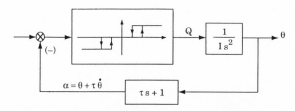

Figure 3.8. Block diagram of Schmitt trigger used with a linear switching function.

This control system cannot bring θ and $\dot{\theta}$ to zero, but at least it can bring them to acceptably small values, ending with a low-frequency limit cycle, which does not use as much fuel as the bang-bang scheme. It is straightforward to show that the limit cycle period and amplitude are given by (see Fig. 3.9):

$$\text{Period} = 4\tau \left(\frac{\alpha_1 + \alpha_o}{\alpha_1 - \alpha_o} + \frac{\alpha_1 - \alpha_o}{2N\tau^2} \right), \quad N \triangleq \frac{Q_o}{I}, \tag{3.13}$$

$$\text{Amplitude} = \frac{1}{2}(\alpha_1 + \alpha_o) + \frac{1}{8}\frac{(\alpha_1 - \alpha_o)^2}{N\tau^2}. \tag{3.14}$$

Problem 3.3.1 – *Limit Cycle Period and Amplitude*

(a) Given the pitch dynamics
$$I\ddot{\theta} = Q$$
and the Schmitt trigger of Fig. 3.7, verify the expressions for limit cycle period and an amplitude given in equations (13) and (14).

(b) For $\alpha_o = 1$ degree, $\alpha_1 = 3$ degrees, $N \equiv Q_o/I = 1/3$ degree/sec^2, $\tau = 5$ sec, determine the period and amplitudes of θ and $\dot{\theta}$ in the limit cycle, and plot the limit cycle in the $(\dot{\theta}, \theta)$ plane.

(c) Using the data in (b), calculate the response for $\theta(0) = 30°$, $\dot{\theta}(0) = 0$ and plot the response in the $(\dot{\theta}, \theta)$ plane.

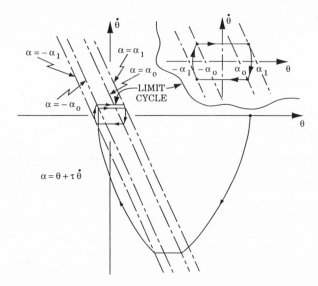

Figure 3.9. Phase plane path; attitude control using deadband and hysteresis (Schmitt trigger).

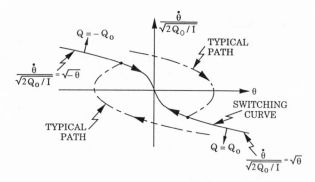

Figure 3.10. Minimum time bang-bang control of attitude (quadratic switching function).

Problem 3.3.2 – *Minimum Time Bang-Bang Control of Attitude*

Minimum-time control of attitude to $\theta = \dot{\theta} = 0$ using two on-off torquers (one for positive torque, one for negative torque) is achieved using the *nonlinear switching* function shown in Fig. 3.10 (see, e.g., Ref. BR-1, pp. 112–13); only one switch is required.

Starting with $\theta(0) = 30°$, $\dot{\theta}(0) = 0$, calculate the time (in units of $1/\sqrt{2Q_o/I}$) to $\theta = \dot{\theta} = 0$ using the linear switching function with $\frac{1}{\tau} = \sqrt{2Q_o/I}$, and with the minimum-time switching function.

4

Attitude Control with
Reaction Wheels

4.1 Fast Control

Pointing accuracy with on-off thrusters is limited to about 0.1 to 1.0 degrees (amplitude of limit cycles). If more accuracy is required (e.g., pointing a telescope), reaction wheels may be used (see Fig. 3.1). The concept here is to put unwanted spacecraft angular momentum into the wheels. This can be done with *proportional* electromagnetic torquers (such as D.C. motors), so very precise control of spacecraft attitude is possible. By careful design of the spacecraft geometry and mass distribution, in connection with a careful prediction of disturbance torques (mainly solar pressure and gas leaks), the average disturbance torques may be brought close to zero. In this way the average wheel torque is close to zero and the average angular velocity of each wheel will be close to zero; hence they are sometimes called null momentum wheels. This cannot be done perfectly, so occasionally, thrusters must be fired to "desaturate" the wheel, that is, to bring its angular velocity back close to zero (see Problem 4.1.1).

For fast attitude control, motion about each principal axis is nearly decoupled from motion about the other two axes. Let us consider control about the pitch (y) axis, and let q_w = angular velocity of the wheel with respect to inertial space. Then the equations of motion are

$$I_y \dot{q} = Ni + Q_f + Q_{dy}, \qquad (4.1)$$

$$\dot{\theta} = q, \qquad (4.2)$$

$$J\dot{q}_w = -Ni - Q_f, \qquad (4.3)$$

$$Ri = e - N(q - q_w), \qquad (4.4)$$

$$Q_f = -c(q - q_w). \qquad (4.5)$$

where

Q_{dy} = external disturbance torque,

Q_f = wheel-bearing friction torque,

I_y = moment of inertia of spacecraft,

J = moment of inertia of wheel plus motor,

i = armature current in D.C. motor,

e = armature voltage (the control input),

R = armature resistance,

N = torque per unit current ≡ back emf per unit angular velocity,

c = viscous friction coefficient.

The total angular momentum of the S/C plus reaction wheel is

$$H = I_y q + J q_w. \tag{4.6}$$

Adding (1) and (3) and using (6) gives

$$\dot{H} = Q_{dy}. \tag{4.7}$$

Thus H *is disturbable by the external torque* Q_{dy} *but is not controllable by* e, *since* e *controls only the internal torque* Ni. However, using e, angular momentum can be transferred back and forth between the spacecraft and the reaction wheel.

Using θ, q, H as states, the normalized equations of motion are

$$\dot{\theta} = q, \tag{4.8}$$

$$\dot{q} = -q + H/(1 + \epsilon) + e + Q_{dy}, \tag{4.9}$$

$$\dot{H} = Q_{dy}, \tag{4.10}$$

where

$$\epsilon q_w = H - q. \tag{4.11}$$

The normalization uses time in units of $1/\sigma, (q, q_w)$ in σ, H in $I_y\sigma, Q_{dy}$ in $I_y\sigma^2, e$ in $RI_y\sigma^2/N$, where

$$\sigma = (c + N^2/R)(1/J + 1/I_y), \tag{4.12}$$

$$\epsilon = J/I_y. \tag{4.13}$$

Fast Control of θ

Consider the case where q and q_w are both zero and the spacecraft has the desired orientation with respect to inertial space, which we shall take as $\theta = 0$. Suppose the spacecraft then receives an impulsive external disturbance torque (e.g., from a tiny meteorite impact). We wish to use e to transfer the spacecraft angular momentum to the wheel *and* bring the spacecraft attitude back to $\theta = 0$ in the process. The angular momentum will be constant after the impact:

$$H = q(0+). \tag{4.14}$$

The equations of motion become just (8) and (9) with H = constant, $Q_{dy} = 0$, and initial conditions

$$\theta(0) = 0, \quad q(0) = H. \tag{4.15}$$

This leads us to consider stabilizing control logic of the form

$$e = -H/(1 + \epsilon) + \delta e \tag{4.16}$$

and a performance index

$$J = \int_0^\infty (A\theta^2 + (\delta e)^2)dt. \tag{4.17}$$

Fig. 4.1 shows a locus of the closed-loop poles vs. the weighting factor A.

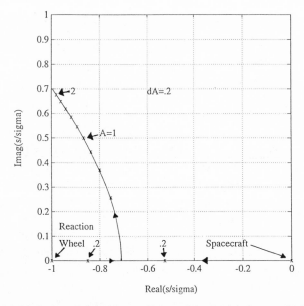

Figure 4.1. Locus of closed-loop poles vs. A for fast pitch control using a reaction wheel.

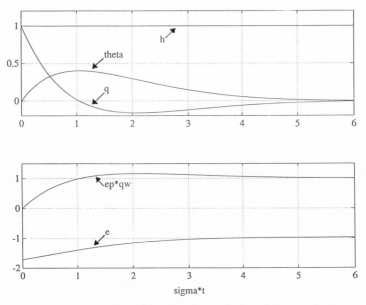

Figure 4.2. Response to an impulsive disturbance for fast pitch control using a reaction wheel.

To achieve fast control, that is, a control bandwidth comparable to the motor bandwidth, Fig. 4.1 indicates a reasonable choice is $A = 1$. LQR synthesis gives the state feedback control law:

$$\delta e = -1.000\theta - .732q. \qquad (4.18)$$

Note that q is measureable with a rate gyro, $q_w - q$ is measureable with a tachometer, and θ is measureable with an horizon sensor; H can then be found from (6). Thus state feedback is possible, but it requires three sensors.

Fig. 4.2 shows the simulated response of the closed-loop system to an external disturbance torque impulse of magnitude $I_y\sigma$. H is impulsively changed to $I_y\sigma$ and q to σ; control action then transfers the angular momentum from the spacecraft to the reaction wheel, overshooting slightly so that some of the excess momentum in the wheel can be transferred back to bring θ to zero. The equilibrium control voltage (e) balances the motor back emf and produces just enough current to overcome the friction torque.

Problem 4.1.1 – *Desaturation Using Thrusters*

External thrusters can be used to desaturate the reaction wheel, that is, bring its angular velocity to zero. The normalized equations of motion are

$$\dot{\theta} = q,$$
$$\dot{q} = -q + H/(1+\epsilon) + e - Q_t,$$
$$\dot{H} = -Q_t,$$

where Q_t = torque due to thrusters.

(a) Assuming Q_t = constant and $\theta(0) = 0$, $q(0) = 0$, $H(0) = H$, find a control law for e that will keep θ small during the time $t_f = H/Q_t$ needed to bring H to zero.

(b) Plot the time history of the states and the control during desaturation.

Problem 4.1.2 – *Response to a Constant Disturbance Torque*

(a) Find and plot the response of the closed-loop system of Fig. 4.2 to a constant disturbance torque of magnitude $I_y \sigma^2$ with zero initial conditions. Note this produces a ramp in H and that, asymptotically, $\theta \to$ nonzero constant, while $|e|$ and $|q_w|$ increase linearly with time. Hence an external control torque would eventually have to be used to keep the wheel from saturating.

(b) Design a control system using integral-error feedback on θ so that $\theta \to 0$ with a constant disturbance torque. Plot the response to a constant disturbance torque of magnitude $I_y \sigma^2$ with zero initial conditions.

4.2 Slow Pitch Control Using Gravity Torque Desaturation

In the previous section, we discussed fast attitude control using reaction wheels as storage elements for unwanted angular momentum. If the wheel angular velocity becomes too large, thrusters are fired to desaturate the wheels. In low earth orbit, there are external torques available for desaturation, namely gravity and magnetic torques. Magnetic torques are produced by electric currents on the spacecraft interacting with the earth's magnetic field (also by residual magnetic dipoles, which are then disturbance torques.) These are both very small torques, and desaturation time constants are on the order of an orbit period. To use gravity torque in this manner, the spacecraft pitch angle must be perturbed slightly from the desired equilibrium attitude. Magnetic torques have the advantage that the pitch angle can be held at zero; however, the current coils must be oriented properly with respect to the earth's magnetic field, which adds some complication. We shall discuss only gravity desaturation here.

For fast control, small attitude perturbations about each axis are uncoupled from each other. For slow control in circular orbit, the pitch motion is decoupled, but the roll and yaw motions are coupled; control of roll/yaw is discussed in the next section.

If we let the y-axis be the pitch axis (see Fig. 3.1), small perturbations in pitch angle from the locally horizontal axes are described by (1.42), where we break Q_y into

$$Q_y = Ni + Q_f + Q_{dy}, \tag{4.19}$$

with Ni = internal torque from a D.C. motor on the pitch momentum wheel, Q_f = wheel-bearing friction torque, and Q_{dy} = external disturbance torque. The equations of motion of the wheel are the same as in (3) and (4). Thus the pitch equations are

$$I_y \dot{q} = -3n^2(I_x - I_z)\theta + Ni + Q_f + Q_{dy}, \tag{4.20}$$

$$\dot{\theta} = q + n, \tag{4.21}$$

$$J\dot{q}_w = -Ni - Q_f, \tag{4.22}$$

$$Ri = e + N(q_w - q), \tag{4.23}$$

$$Q_f = -c(q - q_w). \tag{4.24}$$

The rate of change of the total angular momentum, H_y, of the spacecraft plus momentum wheel is obtained by adding (20) and (22); the internal torques Ni and Q_f cancel out, leaving

$$\dot{H} = -3n^2(I_x - I_z)\theta + Q_{dy}, \tag{4.25}$$

where

$$H = I_y q + J q_w. \tag{4.26}$$

If we use H as the third state variable instead of q_w, eliminate i and Q_f using (23) and (24), and consider perturbations about the equilibrium condition $\theta = 0$, $q = q_w = -n$, $H = -(I_y + J)n$, then the system equations become

$$\dot{\theta} = \delta q, \tag{4.27}$$

$$\delta \dot{q} = -\delta q + \delta H/(1 + \epsilon) - \omega_p^2 \theta + e + Q_{dy}, \tag{4.28}$$

$$\delta \dot{H} = -\omega_p^2 \theta + Q_{dy}, \tag{4.29}$$

where time is in units of $1/\sigma$ and $(\delta q, H/I_y, \omega_p)$ are in units of σ, e is in units of $RI_y \sigma^2/N$, and

$$\delta q = q - (-n),$$

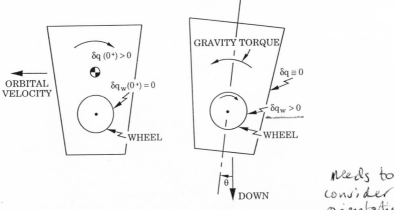

Figure 4.3. Slow pitch control using a reaction wheel and gravity desaturation. On left—just after impulsive disturbance torque. On right—using gravity torque to desaturate the reaction wheel.

needs to consider orientation of S/C wrt gravity?

$$\delta H = [H - (I_y + J)(-n)],$$
$$\delta q_w = q_w - (-n),$$
$$\equiv (\delta H/I_y - \delta q)/\epsilon,$$
$$\sigma = (c + N^2/R)(1/I_y + 1/J),$$
$$\epsilon = J/I_y,$$
$$\omega_p = n\sqrt{3(I_x - I_z)/I_y}.$$

Unlike the fast control case, H is controllable here because of the gravitational torque. For example, consider an *impulsive disturbance torque*, which starts the spacecraft rotating with $\delta q > 0$ (see Fig. 4.3). The excess angular momentum can be transferred to the momentum wheel and the S/C rotated back to a small positive θ; the resulting negative gravity torque on the S/C is approximately balanced by a motor torque; the equal but opposite torque on the momentum wheel starts to slow it down. The deviation in wheel speed and the spacecraft pitch angle slowly tend to zero together.

There are two very different time scales in this system: the motor time constant $1/\sigma$ and the *libration period* $2\pi/\omega_p$. The shorter one $(1/\sigma)$ corresponds to the time to transfer angular momentum to the reaction wheel, the longer one $(2\pi/\omega_p)$, to the desaturation of the wheel and the attenuation of the pitch angle to zero.

Suppose we choose a motor with a time constant $1/\sigma$, that is, 1/100 of the libration period $2\pi/\omega_p$.

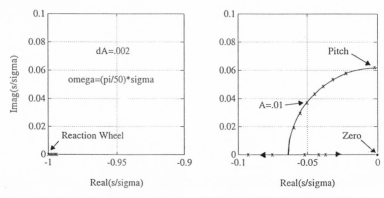

Figure 4.4. Locus of closed-loop poles vs. A for slow pitch control using a reaction wheel and gravity desaturation.

If we use a performance index

$$J = \int_0^\infty (A\theta^2 + e^2)dt, \tag{4.30}$$

a locus of closed-loop poles vs. A is shown in Fig. 4.4.

To desaturate the wheel without using too large values of θ we choose $A = .01$, which (from Fig. 4.4) provides about .707 damping ratio for the pitch libration mode. The corresponding state feedback law from LQR synthesis is

$$e = -.014\theta - .099\delta q. \tag{4.31}$$

Note there is zero gain on H, which is quite different from the fast control law where the gain was on the order of unity.

Fig. 4.5 shows the response of the closed-loop system to a pitch torque disturbance impulse of magnitude $I_y\sigma$. The angular momentum is transferred to the reaction wheel by $\omega_p t/2\pi \approx .2$ where θ reaches its peak value; then follows the desaturation period, which is effectively completed in one libration period $\omega_p t/2\pi \approx 1$. The maximum pitch angle deviation is 17 times larger than in the fast control case, and the attenuation time is 17 times longer, which is the penalty we pay for using the free gravity torque instead of costly fuel for thruster desaturatation.

Problem 4.2.1 – *Controller for an Unstable Spacecraft*

If $I_x < I_z$ then $-\omega_p^2$ is replaced by $\mu_p^2 = 3n^2(I_z - I_x)/I_y$ in (28)-(29), and the uncontrolled S/C is in an *unstable* equilibrium when $\theta = 0$.

(a) Design control logic to stabilize the S/C using a motor with $\sigma = 50\mu_p/\pi$.

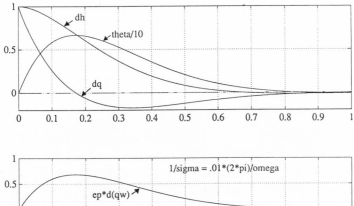

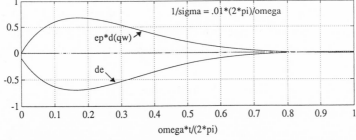

Figure 4.5. Response to an impulsive disturbance for slow pitch control using a reaction wheel and gravity desaturation.

(b) Plot the closed-loop response to an impulsive disturbance torque of magnitude $I_y \mu_p$.

4.3 Slow Roll/Yaw Control Using Gravity Desaturation

Roll and yaw motions of an earth-pointing spacecraft are coupled in orbit. While it is possible to stabilize both roll and yaw with only a roll reaction wheel, the yaw response to yaw disturbances may be unsatisfactory (see Problem 4.3.2). Hence we consider the case with reaction wheels on both the roll axis and the yaw axis; the roll gravitational torque can be used to desaturate both roll and yaw reaction wheels.

From (22–24) and (1.44–47) the equations of motion for such a system are

$$I_x \dot{p} = -n(I_y - I_z)r - 3n^2(I_y - I_z)\phi + N_x i_x + Q_{fx} + Q_{dx}, \quad (4.32)$$

$$I_z \dot{r} = n(I_y - I_x)p + N_z i_z + Q_{fz} + Q_{dz}, \quad (4.33)$$

$$\dot{\phi} = n\psi + p, \quad (4.34)$$

$$\dot{\psi} = -n\phi + r, \quad (4.35)$$

$$J_x \dot{p}_w = -N_x i_x - Q_{fx}, \quad (4.36)$$

$$J_z \dot{r}_w = -N_z i_z - Q_{fz}, \tag{4.37}$$

$$R_x i_x = e_x + N_x(p_w - p), \tag{4.38}$$

$$R_z i_z = e_z + N_z(r_w - r), \tag{4.39}$$

$$Q_{fx} = -c_x(p - p_w), \tag{4.40}$$

$$Q_{fz} = -c_z(r - r_w), \tag{4.41}$$

where (e_x, e_z) are the armature voltages of the (roll,yaw) reaction wheel motors, and the control variables here.

Let us use $H_x = I_x p + J_x p_w =$ total roll angular momentum and $H_z = I_z r + J_z r_w =$ total yaw angular momentum, as state variables in place of p_w and r_w (the angular velocities of the roll and yaw reaction wheels), so that

$$p_w = (H_x - I_x p)/J_x, \tag{4.42}$$

$$r_w = (H_z - I_z r)/J_z. \tag{4.43}$$

Adding the $\dot{p}$ and $\dot{p}_w$ equations and the $\dot{r}$ and $\dot{r}_w$ equations eliminates the internal torques $N_x i_x, N_z i_z, Q_{fx}, Q_{fz}$, giving

$$\dot{H}_x = -n(I_y - I_z)r - 3n^2(I_y - I_z)\phi + Q_{dx}, \tag{4.44}$$

$$\dot{H}_z = n(I_y - I_x)p + Q_{dz}. \tag{4.45}$$

Substituting (38)–(43) into (32)–(33) gives

$$\dot{p} = -ar - 3a\phi - \sigma_x p + \sigma_x(1 - \epsilon_x)H_x/I_x + b_x e_x + Q_{dx}/I_x, \tag{4.46}$$

$$\dot{r} = bp - \sigma_z r + \sigma_z(1 - \epsilon_z)H_z/I_z + b_z e_z + Q_{dz}/I_z, \tag{4.47}$$

where time is in units of $1/n$, (p, r) in n, (H_x, H_z) in $(I_x n, I_x n)$, (e_x, e_z) in $(I_x n^2 R_x/N_x, I_z n^2 R_z/N_z)$, (Q_{dx}, Q_{dz}) in $(I_x n^2, I_z n^2)$, and

$$\sigma_x = (c_x + N_x^2/R_x)(1/J_x + 1/I_x)/n^2,$$

$$\sigma_z = (c_z + N_z^2/R_z)(1/J_z + 1/I_z)/n^2,$$

$$\epsilon_x = J_x/I_x,$$

$$\epsilon_z = J_z/I_z.$$

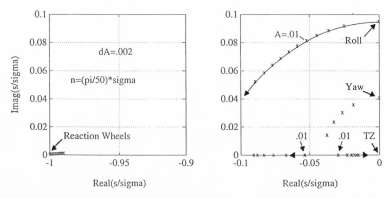

Figure 4.6. Locus of closed-loop poles vs. A for slow roll/yaw control using roll and yaw reaction wheels and gravity desaturation.

The motor time constants $1/\sigma_x$ and $1/\sigma_z$ will, in general, be much smaller than the orbit period $2\pi/n$, so that we again have two very different time scales to consider. The shorter one corresponds to the time to transfer angular momentum to the reaction wheels; the longer one to the desaturation of the wheels and the attenuation of the roll and yaw angles to zero.

Example

Consider the *oblate, axially symmetric S/C* of Problem 1.4.1 where $a = b = 1/2$, with its symmetry axis cross-track. The uncontrolled S/C has two undamped oscillatory modes with frequencies $1.523n$ and $.657n$. Since they involve primarily roll and yaw motions respectively, we shall call them the roll and yaw modes.

We shall use the performance index

$$ J = \int_0^\infty [A(\phi^2 + \psi^2) + ex^2 + ez^2])dt. \tag{4.48} $$

For $n = \pi\sigma/50$, the locus of closed-loop poles vs. A is shown in Fig. 4.6. The reaction wheel poles are changed only slightly for the range of A used in the plot, while the roll and yaw mode poles are changed significantly. Note this is *not* an Evans type of root locus because A enters quadratically into the symmetric root characteristic equation.

We chose $A = .01$, for which the corresponding state feedback control law is

$$
\begin{bmatrix} e_x \\ e_z \end{bmatrix} = \begin{bmatrix} .010 & .095 & 0 & .006 & 0 & .001 \\ -.006 & 0 & -.001 & .010 & .095 & 0 \end{bmatrix} \begin{bmatrix} \phi \\ p \\ H_x \\ \psi \\ r \\ H_z \end{bmatrix}.
$$

Fig. 4.7 shows the response of the closed-loop system to a roll disturbance torque of magnitude $I_x n$. The roll angular momentum is transferred to the roll reaction wheel by $nt/2\pi = .15$ where ϕ reaches its peak value; then follows the desaturation period, which is essentially completed in one orbit ($nt = 2\pi$). The yaw coupling is small but non-negligible.

Fig. 4.8 shows the response of the closed-loop system to a yaw disturbance

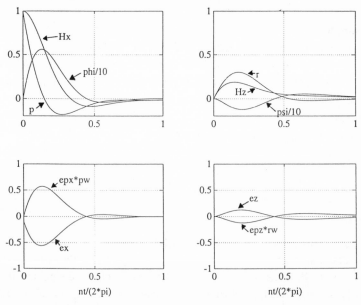

Figure 4.7. Response to an impulsive roll disturbance for slow roll/yaw control using roll and yaw reaction wheels and gravity desaturation.

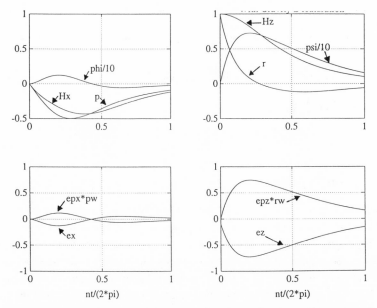

Figure 4.8. Response to an impulsive yaw disturbance for slow roll/yaw control using roll and yaw reaction wheels and gravity desaturation.

Fig. 4.8 shows the response of the closed-loop system to a yaw disturbance torque of magnitude $I_z n$. The yaw angular momentum is transferred to the yaw reaction wheel by $nt/2\pi = .16$ where ψ reaches its peak value; then follows the desaturation period, which is completed in about 1.5 orbits. The roll coupling is large, since the gravity torque acts only in roll, not in yaw.

Problem 4.3.1 – *Controller for an Unstable Spacecraft*

Consider roll/yaw stabilization of a prolate S/C with symmetry axis crosstrack, where $a = b = -1/2$, which has an *unstable equilibrium* at $\phi = \psi = 0$.

(a) Design state feedback (LQR) logic to stabilize the S/C using roll and yaw reaction wheels with motors having $\sigma = 50n/\pi$. Assume $\epsilon = .02$.

(b) Plot the closed-loop response to impulsive roll and yaw disturbance torques with magnitudes $I_x n$ and $I_z n$ respectively.

Problem 4.3.2 – *Roll/Yaw Controller Using Only a Roll Reaction Wheel*

Consider roll/yaw stabilization of the oblate S/C of the example using only a roll reaction wheel. Note that the yaw mode is attenuated very slowly compared to the roll mode.

5

Attitude Stabilization
with Spin

5.1 Spin Stabilization

Imparting spin to a body is a simple, passive method of stabilizing its attitude. A rigid body with its angular velocity parallel to its major axis (the principal axis passing through the center of mass, having maximum moment of inertia) will maintain this axis in a fixed direction with respect to inertial space in the absence of external torques. The earth is, approximately, such a spinning body.

The first communication satellite placed in a synchronous orbit (1963) used spin stabilization. The satellite was spun up about its major axis by tangential jets at the same time that it was injected into an elliptic transfer orbit from its low earth orbit around the equator (see Fig. 5.1). This spin maintained the direction of the axis of symmetry during the transfer orbit. When this orbit became tangent

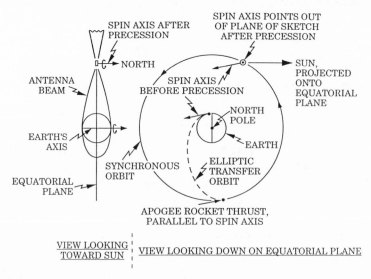

Figure 5.1. Use of spin stabilization to orient a communication satellite.

to the synchronous orbit, the apogee rocket, whose axis was along the axis of symmetry, was fired to inject the satellite into synchronous orbit (an orbit whose period is 24 hours). Some time later the direction of the spin axis was changed by 90 degrees so that it was perpendicular to the equatorial plane. The axially symmetric antenna beam then covered the earth at all times. This concept was originated by D. D. Williams and H. Rosen of the Hughes Aircraft Company (see Aviation Week, 10/15/59, pp. 26–27, "U.S. Plans Gyro-Stabilized Solar Satellite."

5.1.1 *Nutation*

If the angular velocity of a rigid body is *not* parallel to its major or minor axis, the body is said to be nutating. The angular momentum vector, $\vec{H}$, is fixed with respect to inertial space if there are no external torques, but the angular velocity vector, $\vec{\omega}$, rotates around the angular momentum vector.

For *axially symmetric bodies*, $\vec{H}, \vec{\omega}$, and the axis of symmetry are co-planar (see. Fig. 5.2). The symmetry axis (hence also the $\vec{\omega}$ vector) rotates about the fixed $\vec{H}$ vector at a constant rate called the *nutation frequency*. If the axis of geometric symmetry is not a principal axis, the body is said to be *dynamically unbalanced*; even when nutation is absent, such a body will appear to *wobble*.

Let the y-axis be the axis of symmetry (the nominal spin axis). We do this because the spin axis of most satellites is perpendicular to the orbit plane and the NASA standard coordinates take this as the y-axis (pitch axis). Let $I_S = $ moment of inertia about the spin axis and $I_T = $ moment of inertia about the two transverse axes, x and z. We shall use *nonspinning body axes as* coordinate axes, that is, axes that roll ($p \neq 0$) and yaw ($r \neq 0$) with the body but do not spin (pitch) with it; the components of the angular velocity of these coordinate axes with respect to inertial space resolved onto themselves is thus

$$\vec{\omega}_c = [\, p \;\; 0 \;\; r \,]^T. \tag{5.1}$$

The components of angular momentum, resolved onto these axes, are

$$\vec{H} = [\, I_T p \;\; I_S \omega_s \;\; I_T r \,]^T, \tag{5.2}$$

where $\omega_s = $ spin angular velocity. Consequently, Euler's law,

$$\overset{I}{\vec{H}} \equiv \overset{c}{\vec{H}} + \vec{\omega}_c \times \vec{H} = 0, \tag{5.3}$$

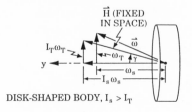

DISK-SHAPED BODY, $I_s > I_T$

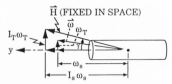

ROD-SHAPED BODY, $I_T > I_s$

Figure 5.2. Disk-shaped and rod-shaped spin-stabilized spacecraft.

in terms of these components is given by

$$I_T \dot{p} - (I_S \omega_s) r = 0, \tag{5.4}$$

$$I_S \dot{\omega}_s = 0, \tag{5.5}$$

$$I_T \dot{r} + (I_S \omega_s) p = 0. \tag{5.6}$$

(5) implies that

$$\omega_s = \text{constant}, \tag{5.7}$$

so that (4) and (6) may be written as

$$\dot{p} - \omega_n r = 0,$$
$$\dot{r} + \omega_n p = 0, \tag{5.8}$$

where

$$\omega_n \triangleq \frac{I_S}{I_T} \omega_s. \tag{5.9}$$

is the *nutation frequency* as observed in these nonspinning coordinates. Hence it is also the nutation frequency as seen by an observer at rest *in an inertial reference frame*. The general solution to (8) is

$$p = \omega_T \sin(\omega_n t + \beta),$$
$$r = \omega_T \cos(\omega_n t + \beta), \tag{5.10}$$

where

ω_T = transverse component of angular velocity (see Fig. 5.2),

β = phase angle.

For a *solid circular cylinder* of radius r and length ℓ,

$$\frac{I_S}{I_T} = \frac{r^2/2}{r^2/4 + \ell^2/12} = \frac{2}{1 + \ell^2/3r^2}. \tag{5.11}$$

Thus, for a *disk-shaped body* ($\ell < \sqrt{3}r$), spin is about the *major axis*, and it follows from (9) that

$$\omega_s < \omega_n < 2\omega_s, \tag{5.12}$$

that is, the nutation frequency is higher than the spin rate. For a *rod-shaped body* ($\ell > \sqrt{3}r$), spin is about the minor axis, and

$$0 < \omega_n < \omega_s, \tag{5.13}$$

that is, the nutation frequency is lower than the spin rate. Consequently, to an observer in spinning body axes, the nutation of a rod-shaped body appears to be in the opposite direction to the spin.

For small roll and yaw Euler angles (ϕ, ψ), of the nonspinning body axes,

$$\dot\phi \cong p,$$
$$\dot\psi \cong r. \tag{5.14}$$

Substituting (10) into (14) and integrating gives

$$\phi \cong -\frac{\omega_T}{\omega_n} \cos(\omega_n t + \beta),$$
$$\psi \cong \frac{\omega_T}{\omega_n} \sin(\omega_n t + \beta), \tag{5.15}$$

so the *nutation angle*, γ, is approximately

$$\gamma \cong \frac{\omega_T}{\omega_n} \equiv \frac{I_T \omega_T}{I_S \omega_s},$$

which checks Fig. 5.2 since the nutation angle is the angle between $\vec{H}$ and the y-axis.

Problem 5.1.1 − *Nutation as Viewed on the Spinning Body*

(a) Using *spinning body axes* as coordinate axes for a body with axial symmetry about its y-axis, show that Euler's equation may be written as

$$I_T \dot{p} - (I_S - I_T)\omega_s r = 0,$$
$$I_S \dot{\omega}_s = 0,$$
$$I_T \dot{r} + (I_S - I_T)\omega_s p = 0,$$

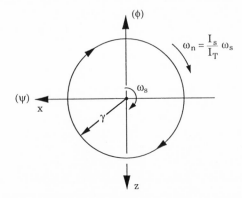

Figure 5.3. Path of tip of symmetry axis in nutation mode as viewed from inertial space.

where the angular velocity of the body is (p, ω_s, r) and the angular momentum is $(I_T p, I_S \omega_s, I_T r)$.

(b) Deduce from (a) that ω_s = constant and the nutation frequency as viewed by an observer on the rotating body is

$$\left(\frac{I_S}{I_T} - 1\right) \omega_s \equiv \omega_n - \omega_s,$$

where ω_n = nutation frequency as viewed by an observer in inertial space.

Problem 5.1.2 – *Roll/Yaw Nutation Mode*

Show that the roll/yaw nutation as viewed by an observer who is stationary with respect to inertial space consists of a circular motion of the tip of the symmetry axis at the nutation frequency as shown in Fig. 5.3.

5.2 Nutation Damping

If a part on a spinning spacecraft is arranged such that relative motion occurs between the part and the spacecraft when nutation is present (angular velocity not parallel to angular momentum), and no relative motion occurs when nutation is absent, the nutation will be damped out if the relative motion dissipates energy, provided that the spin axis is the principal axis with *maximum moment of inertia*. A damped wheel, pendulum, or spring mass can be used, or a viscous fluid in an appropriately shaped container.

As an example, consider a damped wheel (say a conducting wheel in a magnetic field, an eddy-current damper), where the wheel axis is perpendicular to

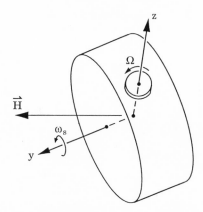

Figure 5.4. Spinning spacecraft with nutation damping wheel.

the spin axis as shown in Fig. 5.4. In body-fixed (spinning) coordinates, the angular velocity of the spacecraft is (p, ω_s, r) and the angular momentum is

$$\vec{H} = [\ I_T p \quad I_S \omega_s \quad I_T r + I_W \Omega \ ],$$

where (I_T, I_S, I_T) are the principal moments of inertia of the spacecraft including the damper wheel at $\Omega = 0$, I_W = moment of inertia of damper wheel about its free axis (the spacecraft z-axis), and Ω = angular velocity of damper wheel relative to the spacecraft.

Spacecraft angular momentum is changed only by external disturbance torques so $\overset{B}{\dot{\vec{H}}} + \vec{\omega}_B \times \vec{H} = \vec{Q}_d$ gives

$$I_T \dot{p} - (I_S - I_T)\omega_s r + I_W \omega_s \Omega \ = \ Q_{dx}, \qquad (5.16)$$

$$I_S \dot{\omega}_s - I_W \Omega p \ = \ Q_{dy}, \qquad (5.17)$$

$$I_T \dot{r} + (I_S - I_T)\omega_s p + I_W \dot{\Omega} \ = \ Q_{dz}. \qquad (5.18)$$

The components of damper wheel angular momentum in body-fixed axes are

$$[I_{wT} p, I_{wT} \omega_s, I_W(r + \Omega)],$$

and the wheel has a damping torque $-D\Omega$. Thus, about the z-axis,

$$I_W(\dot{r} + \dot{\Omega}) = -D\Omega. \qquad (5.19)$$

The term $I_W \Omega p$ in (17) is usually negligible so that $\omega_s \approx$ constant if $Q_{dy} \approx 0$. Thus, the system equations may be written as

$$
\begin{bmatrix} 1 & 0 & 0 \\ 0 & 1 & \epsilon \\ 0 & 1 & 1 \end{bmatrix} \begin{bmatrix} \dot{p} \\ \dot{r} \\ \dot{\Omega} \end{bmatrix} = \begin{bmatrix} 0 & \lambda - 1 & -\epsilon \\ -(\lambda - 1) & 0 & 0 \\ 0 & 0 & -D \end{bmatrix} \begin{bmatrix} p \\ r \\ \Omega \end{bmatrix} + \begin{bmatrix} Q_{dx} \\ Q_{dz} \\ 0 \end{bmatrix}, \quad (5.20)
$$

where time is in units of $1/\omega_s$, (p, r, Ω) are in units of ω_s, D is in units of $I_W \omega_s$, and

$$
\epsilon \overset{\Delta}{=} I_W / I_T,
$$

$$
\lambda \overset{\Delta}{=} I_S / I_T.
$$

Note $(\lambda - 1)\omega_s =$ the nutation frequency as viewed in the spinning body axes.

The characteristic equation of (20) may be written in Evans's form as

$$
-\frac{D}{1 - \epsilon} = \frac{s(s^2 + z^2)}{s^2 + (\lambda - 1)^2}, \quad (5.21)
$$

where

$$
z^2 \overset{\Delta}{=} \frac{(\lambda - 1)(\lambda - 1 + \epsilon)}{1 - \epsilon}.
$$

For disc-shaped spacecraft, $I_S > I_T$, which implies $(\lambda - 1)^2 < z^2$, that is, the zero of the root locus is between the rigid-body pole and the nutation pole; a root locus versus D (see Fig. 5.5) shows that the spacecraft is stabilized, that is, the nutation is damped.

Fig. 5.6 shows the response of the system to an impulsive roll disturbance torque for the case $\lambda = 1.5, \epsilon = .06, D = .5$

For rod-shaped spacecraft, $I_S < I_T$, which implies $(\lambda - 1)^2 < z^2$, that is, the zero of the root locus is above the nutation pole; a root locus versus D (see Fig. 5.7) shows that the spacecraft is destabilized.

Fig. 5.8 shows the response of a rodlike spacecraft to an impulsive roll disturbance torque for the case $\lambda = .5, \epsilon = .06, D = .5$.

Since no real body is actually rigid, energy dissipation due to structural deformations will occur (structural damping). Many spacecraft carry liquid rocket fuel, and as this fuel is used up, fuel sloshing will occur, which also dissipates energy. It is evident from the preceding discussion that single-spin spacecraft can be spin stabilized passively only about a major axis. From the

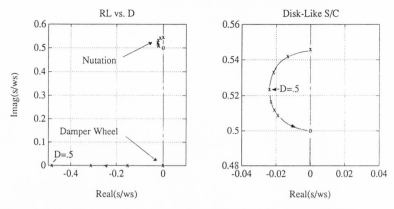

Figure 5.5. Root locus vs. damping constant D for disk-like spacecraft with nutation damping wheel.

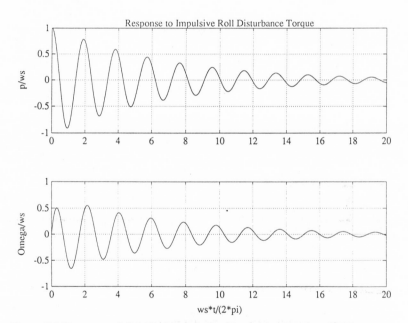

Figure 5.6. Response of disk-like spacecraft with nutation damping wheel to an impulsive roll disturbance torque.

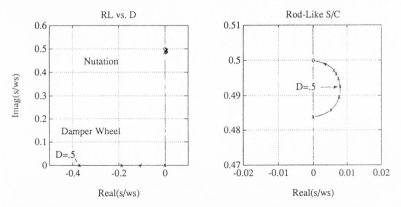

Figure 5.7. Root locus vs. damping constant D for rod-like spacecraft with nutation damping wheel.

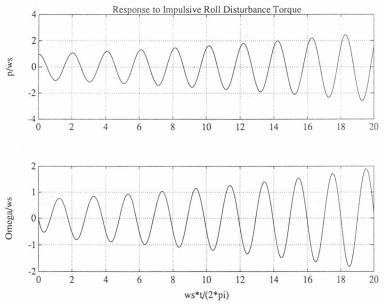

Figure 5.8. Response of rod-like spacecraft with nutation damping wheel to an impulsive roll disturbance torque.

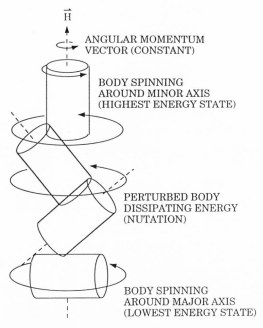

Figure 5.9. Instability of a spacecraft when spun about its minor axis.

energy viewpoint, a body spinning about its minor axis has maximum rotational energy; energy dissipation causes the total energy to decrease until the body has minimum rotational energy, which corresponds to spin about the body's major axis. This is illustrated in Fig. 5.9.

The spacecraft Explorer I, launched in 1958, was spun about a minor axis and had "whip" antennas that introduced substantial structural damping. After only one orbit it began to nutate and, after several orbits, it had tumbled out of control. The explanation of this phenomenon was given by Bracewell and Garriott (Ref. BRG).

Problem 5.2.1 – *Active Nutation Damping; Feedback of p Only*

Suppose the damper wheel is torqued by a D.C. electric motor so that (19) becomes

$$I_W(\dot{r} + \dot{\Omega}) = Ni,$$
$$iR = e - N\Omega,$$

where

$$i = \text{armature current},$$
$$e = \text{armature voltage},$$

$$Ni \; = \; \text{torque applied to rotor,}$$
$$N\Omega \; = \; \text{back emf,}$$
$$R \; = \; \text{armature resistance.}$$

(a) Show that modified (19) may be written as

$$\dot{r} + \dot{\Omega} + D\Omega = e,$$

where $D = N^2/RI_W\omega_s$, and e is in units of $RI_W\omega_s^2/N$.

(b) Show by means of a root locus sketch that proportional feedback of p to e will stabilize the spacecraft for $I_S < I_T$ or augment the stability for $I_S > I_T$.

Problem 5.2.2 – *Active Nutation Damping; Full State Feedback*

Using the motor of Problem 5.2.1 to torque the damper wheel, sketch the locus of closed-loop poles versus A (the symmetric root locus) for the state-feedback system that minimizes

$$J = \int_0^\infty \left[A(p^2 + r^2) + e^2 \right] dt,$$

where normalized variables are used. Take $I_S/I_T = 0.5$, $I_W/I_T = .06$, $D = .5$.

5.3 Dual Spin

The constraint of having to fit the spacecraft into a booster makes a rodlike shape distinctly preferable to a disclike shape. Thus it is desirable to be able to spin stabilize about a minor axis. As shown in Section 5.2, this can be done using active nutation damping (see Problem 5.2.1). However, engineers discovered (Refs. IO, JOH, SA) that if a damper was placed on a de-spun part of the spacecraft, it could be spin stabilized passively about a minor axis. Since a de-spun platform is desirable on a communication satellite for placing spot-beam antennas that point toward particular areas on the earth's surface, this method of spin stabilization was attractive. The communication satellite TACSAT I (Ref. JOH), launched in 1969, was the first flight demonstration of passive damping of a dual-spin spacecraft (or "gyrostat").

Fig. 5.10 is a sketch of a dual-spin spacecraft with a mass-spring-dashpot damper mounted on the de-spun platform. (Torsion pendulums were used on TACSAT I and INTELSAT IV.) Note that a nonspinning spacecraft with a fixed momentum wheel is also a "dual-spin" spacecraft. Let the axis of symmetry be the y-axis and let (x_p, z_p) be axes perpendicular to the y-axis, fixed to the de-spun platform (these are nonspinning body axes exactly like those used to discuss nutation in Section 5.1).

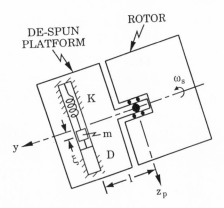

Figure 5.10. Dual-spin spacecraft.

The angular velocity of the platform, resolved onto the platform axes, is $\vec{\omega}_p = [\, p \;\; 0 \;\; r \,]^T$ where the zero component about the y-axis is maintained by active feedback to a motor on the y-axis. The angular velocity of the rotor, resolved onto platform axes, is $\vec{\omega}_R = [\, p \;\; \omega_s \;\; r \,]^T$ where ω_s = spin angular velocity of the rotor. The total angular momentum of the spacecraft is

$$\vec{H} = [\, I_T p + m\ell\dot{\xi} \quad I_S\omega_s \quad I_T r \,]^T, \tag{5.22}$$

where

$$I_T = \text{transverse moment of inertia (about } x_p \text{ and } z_p \text{ axes)}$$
$$\text{of platform + rotor + damper (with } \xi = 0),$$
$$I_S = \text{moment of inertia of rotor about } y\text{-axis.}$$

Euler's equations $\overset{I}{\vec{H}} = \overset{P}{\vec{H}} + \vec{\omega}_p \times \vec{H} = \vec{Q}_d$ give

$$I_T\dot{p} + m\ell\ddot{\xi} - (I_S\omega_s)r = Q_{d_x}, \tag{5.23}$$
$$I_S\dot{\omega}_s + m\ell\dot{\xi}r = Q_{d_y}, \tag{5.24}$$
$$I_T\dot{r} + (I_S\omega_s)p = Q_{d_z}. \tag{5.25}$$

The terms $m\ell\dot{\xi}r$ and Q_{dy} are usually negligible in (24), so $\omega_s \cong$ constant.

The damper mass is constrained to move parallel to the z_p-axis, and its acceleration is given approximately by

$$\ddot{\xi} + \ell\dot{p},$$

so

$$m(\ddot{\xi} + \ell\dot{p}) + D\dot{\xi} + K\xi = 0. \tag{5.26}$$

If we let $\alpha \triangleq I_S\omega_s/I_T$ = nutation frequency, $\epsilon \triangleq m\ell^2/I_T$, and measure time in units of $1/\alpha$, (p, r) in units of α D in units of $m\alpha$, K in units of $m\alpha^2$, Q_d in units of $I_T\alpha^2$, and ξ in units of ℓ, then the system equations may be written as

$$\begin{bmatrix} 1 & 0 & \epsilon & 0 \\ 0 & 1 & 0 & 0 \\ 1 & 0 & 1 & 0 \\ 0 & 0 & 0 & 1 \end{bmatrix} \begin{bmatrix} \dot{p} \\ \dot{r} \\ \dot{v} \\ \dot{\xi} \end{bmatrix} = \begin{bmatrix} 0 & 1 & 0 & 0 \\ -1 & 0 & 0 & 0 \\ 0 & 0 & -D & -K \\ 0 & 0 & 1 & 0 \end{bmatrix} \begin{bmatrix} p \\ r \\ v \\ \xi \end{bmatrix} + \begin{bmatrix} Q_{dx} \\ Q_{dz} \\ 0 \\ 0 \end{bmatrix}. \tag{5.27}$$

The characteristic equation of (27) with $D = 0$ is

$$(s^2 + K)(s^2 + 1) - \epsilon s^4 = 0. \tag{5.28}$$

For $K = 1$ (a "tuned" damper, which was the design choice for INTELSAT IV), the roots of (28) are

$$s = \pm\frac{1}{\sqrt{1 + \sqrt{\epsilon}}}j, \ \pm\frac{1}{\sqrt{1 - \sqrt{\epsilon}}}j. \tag{5.29}$$

The characteristic equation for $D \neq 0$, $K = 1$, in Evans's form, is

$$-D \cong \frac{\left[s^2 + (1/(1 + \sqrt{\epsilon}))\right]\left[s^2 + (1/(1 - \sqrt{\epsilon}))\right]}{s(s^2 + 1)}. \tag{5.30}$$

A root locus versus. D is shown in Fig. 5.11 for $K = 1$, $\epsilon = .06$.

The nutation is damped for $D > 0$, independent of the value of I_S/I_T, so that this form of passive damping is effective even when the spin axis is a minor axis of the spacecraft.

In Fig. 5.11, the nutation damping is maximum for $D \approx .45$. Fig. 5.12 shows the response of the damped system to an impulsive roll disturbance torque for this value of D.

5.3.1 De-Spin Active Nutation Damping

An improved method for nutation damping was discovered shortly after the first dual-spin spacecraft were built (Ref. SA). If the de-spun platform is dynamically

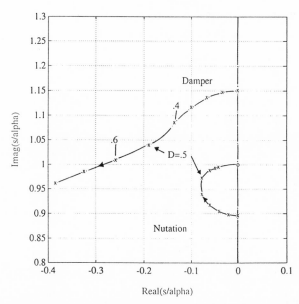

Figure 5.11. Root locus vs. damping constant D for dual-spin spacecraft with tuned damper on de-spun platform.

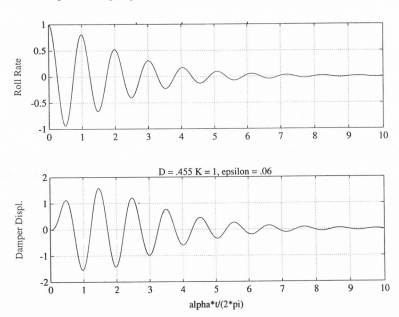

Figure 5.12. Response of a dual-spin spacecraft with a tuned damper on the de-spun platform to an impulsive roll disturbance torque.

unbalanced, the de-spin motor on the rotor axis (that keeps the platform pointing toward the earth) will have a component of torque that can oppose the nutation. If this component is phased properly with respect to the nutation by active feedback, it can damp the nutation quickly with only small oscillations of the platform about its nominal earth-pointing attitude. Fig. 5.13 shows a dual-spin spacecraft with a platform that is deliberately unbalanced, that is, I_{yz}, the product of inertia about the platform center of mass, is not zero.

The angular velocities of the platform and the rotor, in platform-axis components, are

$$\vec{\omega}_P = [\, p \quad q \quad r \,]^T, \tag{5.31}$$

$$\vec{\omega}_R \cong [\, p \quad \omega_s \quad r \,]^T, \tag{5.32}$$

and the angular momenta are

$$\vec{H}_P = \begin{bmatrix} I_x & 0 & 0 \\ 0 & I_y & I_{yz} \\ 0 & I_{yz} & I_z \end{bmatrix} \begin{bmatrix} p \\ q \\ r \end{bmatrix}, \tag{5.33}$$

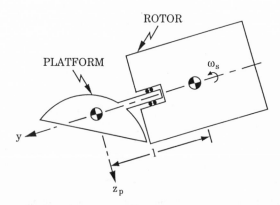

Figure 5.13. Dual-spin spacecraft with unbalanced platform.

$$\vec{H}_R = \begin{bmatrix} I_T & 0 & 0 \\ 0 & I_S & 0 \\ 0 & 0 & I_T \end{bmatrix} \begin{bmatrix} p \\ \omega_s \\ r \end{bmatrix}, \tag{5.34}$$

where I_{yz} is the yz product of inertia of the platform.

Euler's equations for the spacecraft are

$$I_1\dot{p} - I_S\omega_s r = Q_{dx}, \tag{5.35}$$
$$I_S\dot{\omega}_s + I_y\dot{q} + I_{yz}\dot{r} = Q_{dy}, \tag{5.36}$$
$$I_2\dot{r} + I_{yz}\dot{q} + I_S\omega_s p = Q_{dz}, \tag{5.37}$$

where

$$I_1 = I_x + I_T,$$
$$I_2 = I_z + I_T,$$

and all moments and products of inertia are with respect to the overall center of mass. We have assumed platform symmetry with respect to x so that $I_{xy} = I_{xz} = 0$.

Motion of the rotor about the spin axis (y-axis) is described by

$$I_S\dot{\omega}_s = Q, \tag{5.38}$$

where Q is a control torque supplied by the de-spin motor. Hence

$$Q = \frac{N}{R}[e - N(\omega_s - q)] - c(\omega_s - q), \tag{5.39}$$

where c is a viscous friction constant, R = armature resistance, e = armature voltage (the control variable), and N = torque per unit current = back emf per unit angular velocity.

Assuming Q_{dy} negligible, (36) may be integrated to give

$$I_S\omega_s + I_y q + I_{yz}r = h, \tag{5.40}$$

where h = nominal value of $I_S\omega_s$ (where $q = r = 0$). Equations (35) and (37) can then be linearized as

$$I_1\dot{p} - h \cdot r = 0, \tag{5.41}$$
$$I_2\dot{r} + I_{yz}\dot{q} + h \cdot p = 0. \tag{5.42}$$

Eliminating Q and ω_s among (38)–(40) gives

$$I_y\dot{q} + I_{yz}\dot{r} + (N^2/R + c)\left[q - (h - I_y q - I_{yz}r)/I_S\right] = -Ne/R. \quad (5.43)$$

The system (41)–(43) is *completely controllable if I_{yz} is nonzero*. From (43), the steady-state value of e is given by

$$Ne_{ss}/R = (N^2/R + c)h/I_S. \quad (5.44)$$

Let us use p in units of h/I_1, r in h/I_2, q in h/I_{yz}, and time in units of $1/\omega_o$, where

$$\omega_o = h/\sqrt{(I_1 I_2)}.$$

The normalized equations of motion are then

$$
\begin{bmatrix} 1 & 0 & 0 \\ 0 & 1 & 1 \\ 0 & \epsilon_y & 1 \end{bmatrix}
\begin{bmatrix} \dot{p} \\ \dot{r} \\ \dot{q} \end{bmatrix}
=
\begin{bmatrix} 0 & \lambda & 0 \\ -1/\lambda & 0 & 0 \\ 0 & -\epsilon_y\sigma & -\sigma(1+\epsilon_s) \end{bmatrix}
\begin{bmatrix} p \\ r \\ q \end{bmatrix}
+
\begin{bmatrix} 0 \\ 0 \\ -1 \end{bmatrix}
\delta e,
$$

$$(5.45)$$

where

$$
\begin{aligned}
\sigma &= (N^2/R + c)/(I_S\omega_o), \\
\epsilon_y &= I_{yz}^2/I_y I_2, \\
\epsilon_s &= I_S/I_y, \\
\delta e &= e - e_{ss}.
\end{aligned}
$$

The bandwidth of the de-spin motor should be chosen significantly larger than the nutation frequency so that σ is $\gg \omega_o$, $\epsilon_y \ll 1$ and $\epsilon_s < 1$. Fig. 5.14 shows a locus of closed-loop poles vs. A for the case where $\sigma = 5$, $\epsilon_y = .1$, and $\epsilon_s = .25$, and

$$J = \int_0^\infty [A(p^2 + r^2) + (\delta e)^2]dt. \quad (5.46)$$

There is a large frequency separation between the fast modified spin (pitch) mode and the slower modified nutation (roll/yaw) mode. For large A, the modified nutation mode eigenvalues tend toward a zero at $s = 0$ and a "compromise" zero at $s = -1$. We call this a compromise zero since the transfer function to p has zeros at $s = 0$ and $s = \infty$ while the transfer function to r has two zeros

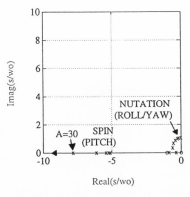 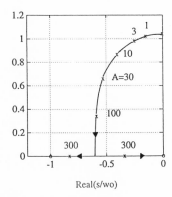

Figure 5.14. Locus of closed-loop poles vs. A for de-spin active nutation damping with $\sigma=5, \epsilon_y = .1, \epsilon_s = .25, \lambda = 1$.

at $s = 0$. A single control cannot control two outputs perfectly, so the LQR algorithm compromises by placing zeros in the symmetric root locus somewhere between the zeros of the two output transfer functions (see Appendix C).

Choosing $A = 30$, the normalized feedback gains are (from Matlab)

$$k = \begin{bmatrix} 7.057 & 3.391 & 0 \end{bmatrix},$$

and the normalized closed loop eigenvalues are

$$s = \begin{bmatrix} -7.826 & -0.527 \pm 0.658j \end{bmatrix}.$$

Fig. 5.15 shows the response to the worst normalized unit initial conditions, for $A = 30$ (see Appendix C for determination of the worst unit initial conditions), which are

$$x0 = [.5926 - .5575 - .5814]'.$$

q and r are very quickly changed by the modified spin (pitch) mode, and then all three angular velocities are attenuated slowly by the modified nutation (roll/yaw) mode.

Problem 5.3.1 – *Damping Time Constant*

The communication satellite INTELSAT IV had $I_T = 500$ slug ft^2 , $I_S = 160$slug ft^2, $\omega_s = 50$rev/min, $\epsilon = 1/60$, $K/m = \alpha^2$, $D/m = 0.7\alpha$.
Show that

$$\alpha = 1.68 \text{ rad sec}^{-1},$$
$$\tau = 50 \text{ sec},$$

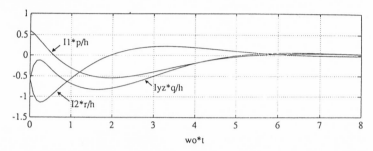

Figure 5.15. Response to worst unit initial condition vector for $A = 30$; de-spin active nutation damping.

where τ = time constant of the damped nutation mode and α = nutation frequency (see Equation 5.27).

Problem 5.3.2 – *Dual-Spin Spacecraft with Damper Wheel*

Instead of a pendulum damper, suppose an eddy-current wheel damper is mounted on the de-spun platform with its axis of rotation parallel to the z_p-axis (cf. Section 5.2).

(a) Show that the linearized equations of roll/yaw motion are

$$\dot{p} - \alpha r = 0,$$
$$\dot{r} + \alpha p + \epsilon \dot{\Omega} = 0,$$
$$\dot{r} + \dot{\Omega} + \sigma \Omega = 0,$$

where

$$\alpha = I_S \omega_s / I_T, \quad \epsilon = I_W / I_T, \quad \sigma = D / I_W.$$

(b) Sketch a locus of system poles versus σ for $\epsilon \ll 1$.

5.4 Offset Axial Thruster

5.4.1 *Need for an External Torque*

The damping of nutation using internal torques, discussed in Sections 5.2 and 5.3, does *not* change the total angular momentum of the spacecraft. Thus the direction of the spin axis drifts when an external torque acts on a spacecraft with nutation dampers. In some cases, small drift angles may be tolerable. However, these changes may build up so that some re-orientation is necessary from time to time. This can only be accomplished by using another external torque. Gravity

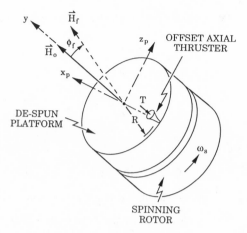

Figure 5.16. Offset axial thruster on a dual-spin spacecraft.

torque, being an external torque, can be used, in combination with dampers, to stabilize an earth-pointing satellite in orbit (see Chapter 4), but the damping time constants are on the order of one orbit period. The torque from an *offset axial thruster* can be used to change the attitude of a spinning vehicle rapidly and precisely (see Fig. 5.16). The price to be paid for this is the weight of the fuel for the thruster.

5.4.2 *Changing Spin Axis Direction with a Nonspinning Offset Axial Thruster*

Fig. 5.16 shows a dual-spin spacecraft with its symmetry axis, its angular velocity vector, and its angular momentum vector, $\vec{H}_o$, all along the same line in space (pure spin, no nutation). Suppose we wish to change the direction of the symmetry axis with respect to inertial space through the angle ϕ_f and again have pure spin in this new orientation, with the same magnitude of spin angular velocity. Let the new angular momentum vector be $\vec{H}_f$ where $H_f = H_o$ (see Fig. 5.16). One way to accomplish this change is to place the offset axial thruster on the de-spun section of the vehicle as shown in Fig. 5.16. Let the symmetry axis be the y-axis, and choose the platform z_p-axis to be in the plane determined by $\vec{H}_o$ and $\vec{H}_f$; then the thruster should be located on the negative x_p-axis near the rim of the platform. If the thrust is T and the axis of the thruster is at a distance R from the x-axis, then a torque of magnitude RT will be produced about the z_p-axis.

From Section 5.3 the roll/yaw equations of motion for ω_s = constant are

$$\dot{p} - \alpha r = 0, \tag{5.47}$$

$$\dot{r} + \alpha p = RT(t)/I_T, \tag{5.48}$$

$$\dot{\phi} = p, \tag{5.49}$$

$$\dot{\psi} = r, \tag{5.50}$$

where

$$\alpha \triangleq \frac{I_S}{I_T}\omega_s = \text{inertially observed nutation frequency,}$$

$(\phi, \psi) = $ Euler angles describing the direction of the symmetry axis as small rotations about the (x_p, z_p) axes.

The maneuver of changing the spin-axis direction may be described with these state variables as the following two-point boundary-value problem. Find $T(t)$ so that

$$p(0) = p(t_f) = 0,$$
$$r(0) = r(t_f) = 0,$$
$$\phi(0) = \phi(t_f) - \phi_f = 0,$$
$$\psi(0) = \psi(t_f) = 0.$$

The simplest maneuver uses *constant thrust*, $T = T_o$. Equations (47)–(50) are easily solved for this case:

$$p(t) = \alpha\phi_o(1 - \cos\alpha t), \tag{5.51}$$

$$r(t) = \alpha\phi_o \sin\alpha t, \tag{5.52}$$

$$\phi(t) = \phi_o(\alpha t - \sin\alpha t), \tag{5.53}$$

$$\psi(t) = \phi_o(1 - \cos\alpha t), \tag{5.54}$$

where $\phi_o = RT_o/I_T\alpha^2$.

The motion consists of a *precession* at constant angular velocity $= \alpha\phi_o$ of the angular momentum vector about the x_p-axis and a *nutation* (rotation of the symmetry axis about the angular momentum vector) at angular velocity α. The path of the tip of the symmetry axis is a cycloid in (ϕ, ψ) space, that is, the path traced by a point on a circle of radius ϕ_o that is rolling on the bottom side of the ϕ-axis with angular velocity α (see Fig. 5.17). When $\alpha t = 2\pi$ (a full nutation cycle), $p = r = \psi = 0$, and $\phi = 2\pi\phi_o$, which satisfies the desired end conditions if $\phi_o = \phi_f/2\pi$, or $T_o = I_T\alpha^2\phi_f/2\pi R$.

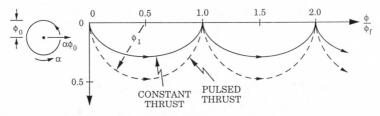

Figure 5.17. Path of tip of symmetry axis with offset axial thrust.

Another simple maneuver is to *pulse the thruster* at twice the nutation frequency. By pulsing, we mean high thrust for a time very short compared to the nutation period. If we let $P = \int_o^{o+} T dt$ = thrust impulse from one pulse of the thruster, then the solution to Equations (47)–(50) is

$$p(t) = \alpha\phi_1 \sin \alpha t, \tag{5.55}$$

$$r(t) = \alpha\phi_1 \cos \alpha t, \tag{5.56}$$

$$\phi(t) = \phi_1(1 - \cos \alpha t), \tag{5.57}$$

$$\psi(t) = \phi_1 \sin \alpha t, \tag{5.58}$$

where $\phi_1 = RP/I_T\alpha$.

This motion is *pure nutation*; however, when $\alpha t = \pi$ (half a nutation cycle), $p = \psi = 0, r = -\alpha\phi_1, \phi = 2\phi_1$, so another pulse at this instant, equal in magnitude to the first pulse, will produce $\Delta r = \alpha\phi_1 \rightarrow r = 0$, that is, the desired end conditions are satisfied if $\phi_1 = \phi_f/2$, or $P = I_T\alpha\phi_f/2R$. The path of the tip of the symmetry axis is a semicircle (see Fig. 5.17).

The thrust impulse for the constant thrust maneuver is

$$T_o\frac{2\pi}{\alpha} \equiv \frac{I_T\alpha}{R}\phi_f, \tag{5.59}$$

and the thrust impulse for the pulsed-thrust maneuver (two pulses) is

$$2P = \frac{I_T\alpha}{R}\phi_f, \tag{5.60}$$

so exactly the same thrust impulse is required for both maneuvers. However, the pulsed-thrust maneuver takes only half a nutation cycle whereas the constant-thrust maneuver takes a whole nutation cycle.

If the constant thrust is not shut off at $\alpha t = 2\pi$, the cycle will be repeated as shown in Fig. 5.17. Similarly, if the thrust is pulsed repeatedly at intervals of $\alpha t = \pi$, its cycle will be repeated as shown in Fig. 5.17; note the spin axis is stationary for half a nutation cycle between semicircles. Large changes in the direction of the spin axis can thus be accomplished by a sequence of small changes of the type shown in Fig. 5.17.

5.4.3 Changing Spin Axis Direction with a Spinning Offset Axial Thruster

For a single-spin spacecraft the offset axial thruster spins with the spacecraft, which makes the maneuver of changing the spin-axis direction appear to be more complicated (but the spacecraft is simpler). However, the pulsed thruster maneuver can still be used (Ref. WH), provided the thruster is pulsed only when it passes through the correct position—through a line perpendicular to $\vec{H}_o$ and $\vec{H}_f$ as shown in Fig. 5.16. To an inertial observer the axial thruster rotates at the spin angular velocity, ω_s, whereas the symmetry axis rotates (nutates) at angular velocity $\alpha = I_S\omega_s/I_T$. Thus the axial thruster will be in the correct firing position if

$$\omega_s t = 2\pi k, \tag{5.61}$$

when

$$\frac{I_S}{I_T}\omega_s t = 2\pi m + \pi, \tag{5.62}$$

where k and m are integers. Eliminating $\omega_s t$ between (61) and (62) gives the condition

$$\frac{I_S}{I_T} = \frac{2m+1}{2k} = \frac{\text{odd integer}}{\text{even integer}}. \tag{5.63}$$

For $I_S/I_T = 3/2$ (a disc-shaped body), the condition (63) is satisfied for $k = m = 1$, that is, for 1 revolution of the axial jet and 1.5 revolutions of the symmetry axis. For $I_S/I_T = 1/2$ (a rod-shaped body), the condition is satisfied for $k = 1, m = 0$, that is, for 1 revolution of the jet and 1/2 a revolution of the body.

This appears to be the technique used (with $I_S/I_T = 1.5$) to change the direction of the spin axis of the first communication satellite by 90 deg after it reached synchronous orbit (see Fig. 5.1); many small pulses were used, checking orientation with a sun sensor.

Problem 5.4.1 – *Precession with Damped Nutation—Nonspinning Thruster*

If a dual-spin spacecraft has nutation dampers and an offset axial thruster on the de-spun section, then constant thrust produces a constant precession and a damped nutation. After the nutation is damped out, only the constant precession remains.

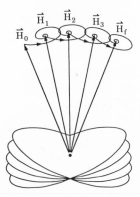

Figure 5.18. Multipulse change of spin-axis direction; vehicle with nutation dampers.

(a) Modify Equation 5.23 to include a thruster torque Q_z as in Equations (5.49)–(5.52) and find a state variable form of this equation.

(b) Plot p vs. r and ϕ vs. ψ for $\epsilon/(1 - \epsilon) = .04$, $D = .2$, $K = 1$, and Q_z = unit step function.

Problem 5.4.2 − *Precession with Damped Nutation - Spinning Thruster*

If a single-spin spacecraft is spun about its major axis and has nutation dampers, an offset axial thruster can be pulsed many times, waiting a few revolutions between each pulse to let the nutation damp out. As before, it must be pulsed when it passes through a line perpendicular to both current $\vec{H}$ and $\vec{H}_f$.

Show that the path of the tip of the symmetry axis using this technique is a sequence of helices as sketched in Fig. 5.18. Use $\epsilon/(1 - \epsilon) = .06$, $\sigma = 1$, $\eta = 2$.

6

Attitude Control with a Gimbaled Momentum Wheel

6.1 Introduction

Spin with nutation damping is a relatively simple way to stabilize the attitude of a spacecraft, but the direction of the spin axis still drifts when an external disturbance torque acts on the spacecraft. A pulsed offset axial thruster can eliminate this drift, but it uses precious fuel.

If the mean value of the external disturbance torques can be made approximately zero by careful configuration design, then a gimbaled momentum wheel (GMW) (also called a control moment gyro) can be used to keep the spacecraft attitude in a desired orientation at all times with great precision and without the use of fuel. The external disturbance torques are counteracted by internal torques on the spacecraft from the GMW outer gimbal. The equal but opposite reaction torques on the GMW gimbal cause the direction of the spin axis of the GMW rotor to drift. Thus, if the mean value of the external disturbance torques is zero, and the GMW spin angular momentum is large enough, the GMW spin-axis direction deviates only through small angles from its nominal direction. If the GMW spin axis drifts too far from its nominal direction, an external torque (thrusters, gravity, magnetic) must be used to "desaturate" the GMWs, that is, bring the spin axis back to its nominal direction (see Ref. CA-1).

6.2 Fast Control

6.2.1 *Control Concept*

A two-gimbal momentum wheel (GMW), can be used to provide three-axis attitude control (see Fig. 6.1). Angular momentum imparted to the spacecraft by disturbances can be transferred to the GMW by torquing the two gimbals

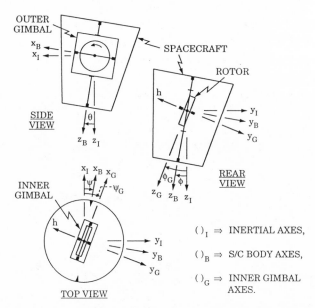

Figure 6.1. Spacecraft with gimbaled momentum wheel (GMW); actual GMW smaller than shown.

how the transfer is done ?

(changing the direction of the spin axis) and the wheel itself (changing the wheel spin rate).

If a constant disturbance torque acts on the spacecraft and the angular momentum is transferred to a GMW, the wheel axis gradually lines up with the direction of this torque and the wheel speed steadily increases. Consequently an external torque provided by thrusters must be used from time to time to desaturate the GMW—bring the wheel speed and spin direction back to nominal conditions.

Fig. 6.1 shows a GMW with its nominal spin direction parallel to the spacecraft y-axis (pitch axis). Spacecraft pitch attitude is controlled by increasing or decreasing the wheel spin rate and requires a pitch attitude sensor. Thrusters must be fired from time to time to keep the wheel speed near its nominal value. This is identical to pitch control using a reaction wheel (Section 4.1), except that the nominal wheel speed is nonzero here.

Spacecraft roll/yaw attitude is controlled by applying torques to the space- *how* craft from the inner and outer gimbals. The equal but opposite torques that are applied to the gimbals by the spacecraft cause the wheel spin axis to precess relative to inertial space. Using feedback control with roll and yaw angle sensors, the roll and yaw angles of the spacecraft can be restored to their desired values.

Passive control (without any sensors), using dampers between the GMW outer gimbal and the spacecraft, transfers the S/C angular momentum to the GMW but leaves small offsets in S/C roll and yaw angles.

If the gimbals are locked to the spacecraft, the GMW becomes a bias momentum wheel (see Section 5.3 on dual spin). A disturbance torque impulse causes the spacecraft to nutate about the new angular momentum vector, so a separate nutation damper is required.

6.2.2 *Equations of Motion of the Spacecraft*

These equations were developed in Section 1.3 and are repeated here for convenience:

$$I_x \dot{p} \cong Q_{cx} + Q_{dx}, \tag{6.1}$$

$$\dot{\phi} \cong p, \tag{6.2}$$

$$I_y \dot{q} \cong Q_{cy} + Q_{dy}, \tag{6.3}$$

$$\dot{\theta} \cong q, \tag{6.4}$$

$$I_z \dot{r} \cong Q_{cz} + Q_{dz}, \tag{6.5}$$

$$\dot{\psi} \cong r, \tag{6.6}$$

which are reasonable approximations for small angles. Note Q_c is the control torque and Q_d is the external disturbance torque. (Q_{cx}, Q_{cz}) are applied to the spacecraft from the (inner, outer) gimbals, Q_{cy} by a motor on the wheel axis.

6.2.3 *Equations of Motion of the GMW*

Let the Euler angles of the inner gimbal with respect to inertial space be $[\phi_G, \theta_G, \psi_G]$, and let the angular velocity of the inner gimbal with respect to inertial space be $[p_G, q_G, r_G]$. The angular momentum of the GMW is then

$$\vec{H} = [J_x p_G, J_y q_G - h, J_z r_G], \tag{6.7}$$

where

$$h = 2J_w \omega_s = \text{spin angular momentum of rotor,}$$

$$J_x = J_w + J_{IGx},$$

$$J_y = J_w + J_{IGy} + J_{OGy},$$

$$J_z = J_w + J_{IGz} + J_{OGz},$$

$$J_w = \text{moment of inertia wheel about a diameter,}$$

and J_{IG} = moment of inertia of inner gimbal, J_{OG} = moment of inertia of outer gimbal.

From Euler's law,

$$\overset{I}{\vec{H}} = \overset{G}{\vec{H}} + \vec{\omega}_G \times \vec{H} = \vec{Q},$$

it follows that

$$J_x \dot{p}_G + hr_G = -Q_{cx}, \qquad (6.8)$$

$$J_y \dot{q} - \dot{h} = -Q_{cy}, \qquad (6.9)$$

$$J_z \dot{r}_G - hp_G = -Q_{cz}, \qquad (6.10)$$

and, for small angles

$$\dot{\phi}_G \cong p_G, \qquad (6.11)$$

$$\dot{\theta}_G \cong q_G, \qquad (6.12)$$

$$\dot{\psi}_G \cong r_G. \qquad (6.13)$$

6.2.4 System Pitch Equations

Pitch motions are uncoupled from roll/yaw motions for small angular deviations:

$$-\dot{h} = -Q_{cy}, \qquad (6.14)$$

$$I_y \ddot{\theta} = Q_{cy} + Q_{dy}. \qquad (6.15)$$

Designing a feedback from (h, θ) to Q_{cy} to keep h close to a nominal (nonzero) value and θ close to zero is almost identical to the design considered in Section 4.1, except that there the nominal value of h was zero.

6.2.5 System Roll/Yaw Equations

The roll and yaw motions of the GMW are coupled by the spin angular momentum, h (see Equations (8), (10)). The roll/yaw motions of the GMW are coupled to the roll/yaw motions of the spacecraft by the control torques (Q_{cx}, Q_{cz}), since a torque on the spacecraft produces an equal but opposite torque on the GMW:

$$I_x \ddot{\phi} \cong Q_{cx} + Q_{dx}, \qquad (6.16)$$

$$I_z \ddot{\psi} \cong Q_{cz} + Q_{dz}, \qquad (6.17)$$

$$J_x \ddot{\phi}_G + h\dot{\psi}_G \cong -Q_{cx}, \qquad (6.18)$$

$$J_z \ddot{\psi}_G - h\dot{\phi}_G \cong -Q_{cz}. \qquad (6.19)$$

Adding (16) to (18) and (17) to (19) eliminates the control torques:

$$J_x \dot{\phi}_G + h \psi_G + I_x \dot{\phi} \triangleq H_x, \tag{6.20}$$

$$\dot{H}_x = Q_{dx}, \tag{6.21}$$

$$J_z \dot{\psi}_G - h \phi_G + I_z \dot{\psi} \triangleq H_z, \tag{6.22}$$

$$\dot{H}_z = Q_{dz}. \tag{6.23}$$

Equations (20)–(23) imply, for a stable control ($\dot{\phi}_G \rightarrow 0$, $\dot{\psi}_G \rightarrow 0$), that impulsive disturbance torques yield

$$h \psi_G \rightarrow H_x, \tag{6.24}$$

$$H_x = \int_o^t Q_{dx} dt, \tag{6.25}$$

$$-h \phi_G \rightarrow H_z, \tag{6.26}$$

$$H_z = \int_o^t Q_{dz} dt, \tag{6.27}$$

that is, the roll and yaw components of the total spacecraft angular momentum are *transferred to the GMW*.

6.2.6 *Passive Roll/Yaw Stabilization*

Passive stabilization of spacecraft roll/yaw attitude can be obtained by connecting the gimbals to the spacecraft with viscous dampers:

$$Q_{cx} = -D(\dot{\phi} - \dot{\phi}_G),$$
$$Q_{cz} = -D(\dot{\psi} - \dot{\psi}_G). \tag{6.28}$$

Equations (21) and (23) are quadratures for determining (H_x, H_z). (18)–(20) and (22) are four equations for ($\phi_G, \psi_G, \phi, \psi$) given ($H_x, H_z$). The characteristic equation of this damped system shows that for $J \leq I_x, I_z$, there are two complex modes with eigenvalues that differ by several orders of magnitude. For $J_x = J_z = J$ the faster set of eigenvalues is

$$s_3, s_4 \cong -\frac{D}{J} \pm \frac{h}{J} j, \tag{6.29}$$

which corresponds to a *damped nutation mode*. The other mode may be analyzed by treating the nutation mode as quasi-steady, that is, by putting $J_x = J_z = 0$

in (18) and (19). Thus, if we use time in units of I_x/h, (H_x, H_z, D) in units of h, I_z in units of I_x, the equations of motion are

$$
\begin{bmatrix} \dot{\phi}_G \\ \dot{\psi}_G \\ \dot{\phi} \\ \dot{\psi} \end{bmatrix} \cong \begin{bmatrix} \mu/I_z & \mu D & 0 & 0 \\ \mu D/I_z & -\mu & 0 & 0 \\ 0 & -1 & 0 & 0 \\ 1/I_z & 0 & 0 & 0 \end{bmatrix} \begin{bmatrix} \phi_G \\ \psi_G \\ \phi \\ \psi \end{bmatrix} + \begin{bmatrix} \mu D & -\mu/I_z \\ \mu & \mu D/I_z \\ 1 & 0 \\ 0 & -1/I_z \end{bmatrix} \begin{bmatrix} H_x \\ H_z \end{bmatrix},
$$

(6.30)

where $\mu = D/(1 + D^2)$.

The characteristic equation of (30) is

$$
s^2[s^2 + \mu(1 + 1/I_z)s + \mu^2(D^2 + 1)/I_z] = 0.
$$

(6.31)

For the special case $I_z = 1$, the eigenvalues are

$$
s = (0, 0, -\mu \pm \mu Dj).
$$

(6.32)

The *asymptotic response* of the damped system to a roll or yaw *disturbance* torque impulse may be obtained from (30):

$$
\begin{bmatrix} \phi \\ \psi \\ \phi_G \\ \psi_G \end{bmatrix} \rightarrow \begin{bmatrix} 1/D & -1/h \\ 1/h & 1/D \\ 0 & -1/h \\ 1/h & 0 \end{bmatrix} \begin{bmatrix} H_x \\ H_z \end{bmatrix},
$$

(6.33)

where (H_x, H_z) are the roll and yaw disturbance torque impulse magnitudes. Thus, the torque impulse is transferred to the GMW, but (ϕ, ψ) have offsets which depend on D and h. If D and H are large compared to anticipated values of H_x and H_z, then the offset angles are small. Such a system might be of interest as a back-up system.

To bring (ϕ, ψ) to zero after an impulsive disturbance torque, we must use active control, which requires (ϕ, ψ) sensors and torque actuators on the outer gimbal of the GMW (see next subsection).

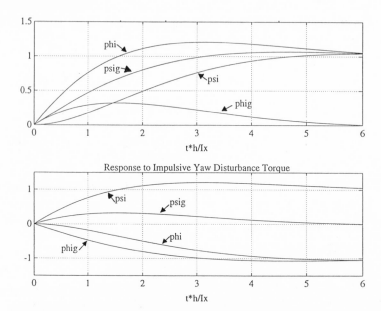

Figure 6.2. Response of S/C with a GMW to impulsive roll and yaw disturbance torques with viscous damper between S/C and outer gimbal.

Fig. 6.2 shows the response to impulsive roll and yaw disturbance torques that produce steps in (H_x, H_z) of magnitude h for the case $D = h, I_z = I_x$ (note the system is linear and we would expect the magnitudes to be very much less than h, so the responses would be scaled down accordingly).

6.2.7 Active Roll/Yaw Control

We shall try combining the passive gimbal dampers with active feedback of (ϕ, ψ) as follows:

$$Q_{cx} = -D(\dot{\phi} - \dot{\phi}_G) - K\phi, \tag{6.34}$$

$$Q_{cz} = -D(\dot{\psi} - \dot{\psi}_G) - K\psi, \tag{6.35}$$

that is, we supply restoring "springs" in roll and yaw to the desired attitude. This would require (ϕ, ψ) sensors and torque actuators on the outer gimbal.

Again, the damping torques are large compared to the D'Alembert torques on the outer gimbal, so we put $J_x = J_z = 0$. Using the same normalization of the variables as in (30), the equations of motion are

$$
\begin{bmatrix} \dot{\phi}_G \\ \dot{\psi}_G \\ \dot{\phi} \\ \dot{\psi} \end{bmatrix} \cong \begin{bmatrix} \mu/I_z & \mu D & \mu K & -K \\ \mu D/I_z & -\mu & K & \mu K \\ 0 & -1 & 0 & 0 \\ 1/I_z & 0 & 0 & 0 \end{bmatrix} \begin{bmatrix} \phi_G \\ \psi_G \\ \phi \\ \psi \end{bmatrix} + \begin{bmatrix} \mu D & -\mu/I_z \\ \mu & \mu D/I_z \\ 1 & 0 \\ 0 & -1/I_z \end{bmatrix} \begin{bmatrix} H_x \\ H_z \end{bmatrix},
$$

$$(6.36)$$

where $\mu = D/(1 + D^2)$ and K is in units of h^2/I_x.

Fig. 6.3 is a locus of the closed-loop roots versus K for the case $D = h, I_z = I_x$. The two poles at the origin are stabilized while the damping of the other two poles is reduced; a good value of normalized K is .25 where the real parts of the two sets of complex poles are equal at $-.25$.

Fig. 6.4 shows the response of the S/C with active control of the GMW to impulsive roll and yaw disturbance torques. Comparing these responses to those using only passive control in Fig. 6.2, we see that here (ϕ, ψ) are returned to

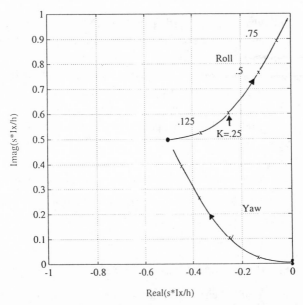

Figure 6.3. Locus of closed-loop poles vs. K for active control of S/C with a GMW.

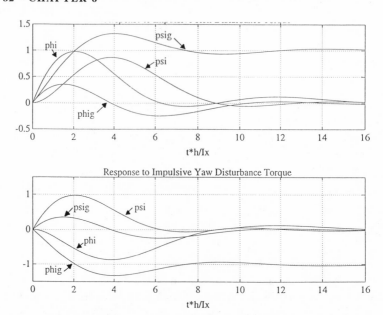

Figure 6.4. Response of S/C with active control of a GMW to impulsive roll and yaw disturbance torques.

zero; this is done by over-transferring angular momentum to the GMW from the spacecraft, then using the excess to bring the S/C attitude back to zero (just as in the case of reaction wheels).

Of course, we do not need to use passive dampers with active feedback; the damping could be introduced by using lead compensation. However, this system has the advantage that, if the active system fails, the S/C will stay close to the desired attitude with the dampers alone, that is, there is a fairly good back-up system.

Problem 6.2.1 – *Roll/Yaw Transient Response of Spacecraft with a Damped GMW*

Determine and plot the roll/yaw transient response of a spacecraft with a damped GMW to a roll disturbance torque impulse with $D = h$ and $I_z = I_x/10$.

6.3 Slow Control Using Gravity Desaturation

Consider a spacecraft in circular orbit with a two-gimbal momentum wheel (GMW) oriented with its spin axis nominally perpendicular to the orbit plane. Referring to Fig. 6.1, (x_I, y_I, z_I) are replaced here by (x_L, y_L, z_L), which are locally horizontal-vertical axes with z_L down, x_L in-track and pointing forward, and y_L cross-track.

The linearized equations of motion of the spacecraft without the GMW are given in Section 1.4.

The equations of motion of the GMW are those of the fast control case with the addition of a roll gravitational torque and Coriolis terms due to rotation of the reference axes at angular velocity n:

$$J_x \dot{p}_G + (h - J_z n) r_G \; = \; -3n^2 (J_y - J_z)\phi_G - Q_{cx}, \tag{6.37}$$

$$J_y \dot{q}_G - \dot{h} \; = \; -Q_{cy}, \tag{6.38}$$

$$J_z \dot{r}_G - (h - J_x n) p_G \; = \; -Q_{cz}, \tag{6.39}$$

$$\dot{\phi}_G - n\psi_G \; \cong \; p_G, \tag{6.40}$$

$$\dot{\theta}_G + n \; \cong \; q_G, \tag{6.41}$$

$$\dot{\psi}_G + n\phi_G \; \cong \; r_G, \tag{6.42}$$

where $(\;)_G$ indicates quantities associated with the GMW, and h = spin angular momentum of the GMW.

The linearized equations for the spacecraft with the GMW decouple into a third order set for the pitch motions and an eighth-order set for the roll/yaw motions.

The *pitch equations* are similar to those treated previously when discussing pitch control using a reaction wheel in circular orbit (see Section 4.2).

The Js are very much less than I_x whereas h is of order $I_x n$. Thus the *roll/yaw* GMW equations are well approximated by

$$\dot{\phi}_G - n\psi_G \; \cong \; Q_{cz}/h, \tag{6.43}$$

$$\dot{\psi}_G + n\phi_G \; \cong \; -Q_{cx}/h, \tag{6.44}$$

The roll/yaw equations for the spacecraft are (1.44)–(1.47) from Section 1.4:

$$\dot{p} \; = \; -nar - 3an^2\phi + Q_{cx}/I_x, \tag{6.45}$$

$$\dot{r} \; = \; nbp + Q_{cz}/I_z, \tag{6.46}$$

$$\dot{\phi} \; = \; p + n\psi, \tag{6.47}$$

$$\dot{\psi} \; = \; r - n\phi, \tag{6.48}$$

where $a = (I_y - I_z)/I_x$, $b = (I_y - I_x)/I_z$.

The spacecraft and the GMW are coupled only through the torques Q_{cx} and Q_{cz}.

6.3.1 *Active Roll/Yaw Control*

For active roll/yaw control, one needs an horizon sensor to measure ϕ, rate gyros to measure (p, r), and sensors on the outer gimbal to measure $\phi - \phi_G$ and $\psi - \psi_G$. Using the ϕ, p, r measurements, ψ can be estimated using "orbital gyrocompassing" as described in Section 2.3.

Using the model of (43)–(48) with time in units of $1/n$, (p, r) in n, Q_{cx}, and Q_{cz} in units of $I_x n^2$, we synthesize a state feedback controller that minimizes

$$J = \int_0^\infty \left\{ A(\phi^2 + \psi^2) + A_G(\phi_G^2 + \psi_G^2) + Q_{cx}^2 + Q_{cz}^2 \right\} dt. \qquad (6.49)$$

As *an example*, consider the Agena spacecraft (Ref. SC) with $I_x = I_y = 10 I_z$; we shall take $h = I_x n$ and $A_G = A/10$. Both A and A_G *must* be nonzero since we are controlling two independent systems with the same two controls. A symmetric root locus versus A is shown in Fig. 6.5.

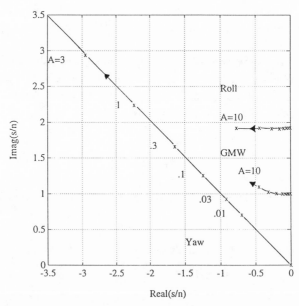

Figure 6.5. Root locus vs. A for active slow roll/yaw control of agena spacecraft with a GMW.

Choosing $A = 3$, the feedback gains are

$$
\begin{bmatrix} Q_{cx} \\ Q_{cz} \end{bmatrix} = - \begin{bmatrix} .749 & -.004 & 1.060 & .646 & 0.58 & -.769 \\ -.059 & .592 & -.383 & 1.670 & -.015 & -.076 \end{bmatrix} \begin{bmatrix} p \\ r \\ \phi \\ \psi \\ \phi_G \\ \psi_G \end{bmatrix}.
$$

(6.50)

Figs. 6.6 and 6.7 show the closed-loop responses to impulsive roll and yaw disturbance torques, that is, for $p(0) = n$ and $r(0) = n$, with all other initial conditions zero; the responses have essentially attenuated in two orbits ($nt = 4\pi$).

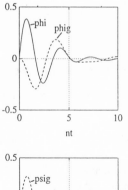

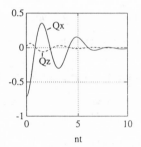

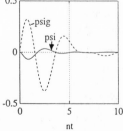

Figure 6.6. Roll/yaw response of agena spacecraft with a GMW to an impulsive roll disturbance torque using LQR with $A = 3$.

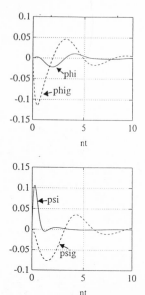

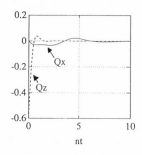

Figure 6.7. Roll/yaw response of agena spacecraft with a GMW to an impulsive yaw disturbance torque using LQR with $A = 3$.

Problem 6.3.1 – *Active Control Using a GMW*

Verify Figs. 6.5 through 6.7.

Problem 6.3.2 – *Stabilization of a S/C in Circular Orbit with a Damped GMW*

Consider the case $I_x = I_z = 2I_y/3$, an *oblate spacecraft with symmetry about the pitch axis.* For this case, $a = b = 1/2$, $I_z = 1$.

(a) Show that the *open-loop roots* are oscillatory:

$$s_{1,2} = \pm1.523j \text{(spacecraft roll mode)},$$
$$s_{3,4} = \pm0.657j \text{(spacecraft yaw mode)},$$
$$s_{5,6} = \pm1.000j \text{(GMW spin axis stationary with}$$
$$\text{respect to inertial space)}$$

(b) With $Q_{cx} = -D(\dot{\phi} - \dot{\phi}_G), Q_{cz} = -D(\dot{\psi} - \dot{\psi}_G)$, plot a root locus versus $D/I_x n$ with $h/I_x n = 1$. Note D appears quadratically in the characteristic equation so the root locus is *not* of the Evans's type.

(c) Calculate and plot the time responses of the spacecraft to roll and yaw disturbance torque impulses where $D/I_x n = 1.75$.

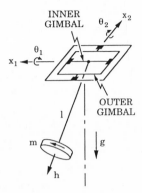

Figure 6.8. Gyropendulum in double gimbal mounting,

Problem 6.3.3 – *The Gyropendulum*

A spinning rotor of mass m and spin angular momentum h is mounted at the end of a pendulum of length ℓ on a double-gimbal suspension as shown in Fig. 6.8.

(a) Neglecting the mass of the support rod, and assuming viscous damping on the gimbal bearings, show that the equations of motion for small oscillations are

$$\ddot{\theta}_1 + D\dot{\theta}_1 + h\dot{\theta}_2 + \theta_1 \cong 0,$$
$$\ddot{\theta}_2 + D\dot{\theta}_2 - h\dot{\theta}_1 + \theta_2 \cong 0,$$

where time is in units of $\sqrt{\ell/g}$, and (D, h) are in units of $m\ell\sqrt{g\ell}$.

(b) Plot a root locus versus h for the case $D = 1$.

Problem 6.3.4 – *The Inverted Gyropendulum (or "Top")*

A spinning rotor of mass m and spin angular momentum h is mounted at the top of an inverted pendulum of length ℓ on a double gimbal suspension (Fig. 6.8 upside down.)

(a) Neglecting the mass of the support rod and neglecting friction in the gimbal bearings, show that the equations of motion for small perturbations from the vertical equilibrium position are

$$\ddot{\theta}_1 + h\dot{\theta}_2 - \theta_1 \cong 0,$$
$$\ddot{\theta}_2 - h\dot{\theta}_1 - \theta_2 \cong 0,$$

where time is in units of $\sqrt{\ell/g}$, and h is in units of $m\ell\sqrt{g\ell}$.

(b) Show, using a root locus sketch, that the inverted gyropendulum is neutrally stable for $h^2 > 4$. (The higher frequency mode is called *nutation*, the lower frequency mode is called *precession*.)

(c) If gimbal bearing friction is approximated by viscous damping, show that the equations of motion become

$$\ddot{\theta}_1 + D\dot{\theta}_1 + h\dot{\theta}_2 - \theta_1 \cong 0,$$
$$\ddot{\theta}_2 + D\dot{\theta}_2 - h\dot{\theta}_1 - \theta_2 \cong 0,$$

where D is in units of $m\ell\sqrt{g\ell}$.

(d) *Show that* the characteristic equation of the damped motion is

$$(s^2 + Ds - 1)^2 + h^2 s^2 = 0,$$

or

$$[s^2 + (D + hj)s - 1][s^2 + (D - hj)s - 1] = 0.$$

Sketch a root locus versus h for $D = .1$. Is the system stabilized by spin?

Problem 6.3.5 – *The Passive Gyroscopic Ship Roll Stabilizer*

This was a historical antecedent to the spacecraft GMW (Ref. TI). Yaw stabilization is not a problem with ships, but roll stabilization is, so a *single-gimbal* gyro was used with a vertical spin axis (see Fig. 6.9). It was not very popular since the torques transmitted to the ship are very large, so that extensive modifications had to be made to the ship's structure. Later in the 1930's, small gyros were used as sensors and hydrofoil "ailerons" were used to provide the rolling moments. Currently hydrofoil ships use this latter method for roll stabilization.

Let ϕ = roll angle of ship, ϵ = gimbal angle of GMW, h = spin angular momentum of GMW, I_x = roll moment of inertia of ship, J = moment of inertia of GMW about its gimbal axis. Assume a ship restoring torque in roll and a pendulous restoring torque on the GMW about its gimbal axis.

(a) Show that the equations of motion may be written as

$$
\begin{bmatrix} s^2 + \omega_o^2 & hs/I_x \\ -hs/Js & s^2 + \omega_g^2 \end{bmatrix}
\begin{bmatrix} \phi(s) \\ \epsilon(s) \end{bmatrix}
=
\begin{bmatrix} Q_{dx}(s)/I_x \\ Q_c(s)/J \end{bmatrix},
$$

where

$$
\begin{aligned}
Q_{dx} &= \text{disturbance roll torque on ship,} \\
Q_c &= \text{control torque on gimbal axis of GMW,} \\
\omega_o &= \text{rolling natural frequency of ship for } h = 0, \\
\omega_g &= \text{pendulum frequency of GMW about its gimbal axis for } h = 0.
\end{aligned}
$$

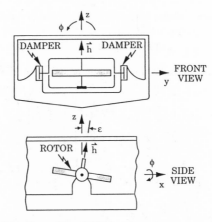

Figure 6.9. Ship roll stabilization with a GMW.

(b) For $\omega_o = .419, \omega_g = 1.83, h/\sqrt{I_x J} = 1.90$ rad/sec, $I_x/J = 5130$ (Foppl's example in Ref. TI), show that the eigensystem is

$$s \quad \pm.289j \quad \pm2.66j \quad \sec^{-1}$$

$$\phi \quad \mp.0831j \quad \pm.0102j \qquad .$$

$$\epsilon \qquad 1 \qquad 1$$

(c) For an impulsive disturbance roll torque, compare amplitudes of ϕ with $h = 0$ and h as in part (b).

(d) For viscous damping on the gimbal axis, $Q_c = -D\dot{\epsilon}$, make a root locus sketch versus D for the case of part (b) to show that the system is stabilized for $D > 0$.

(e) For maximum damping of the lower frequency mode in (d), show that $D/J \cong 16.0 \sec^{-1}$ and the corresponding eigensystem is

$$s \quad -.081 \pm .354j \quad -15.6 \quad -.285 \quad \sec^{-1}$$

$$\phi \quad .102 \mp .063j \quad .0017 \quad .0294 \qquad .$$

$$\epsilon \qquad 1 \qquad 1 \qquad 1$$

Note the higher frequency (nutation) mode is overdamped.

Problem 6.3.6 – *Roll Stabilization of a Monorail Car*

An interesting application of a single-gimbal GMW is to the monorail car (A. Schnerl and P. Schilowsky, 1909 (see Ref. TI)). The rail provides yaw stability so only roll stabilization is needed. However, in contrast to the ship stabilizer of problem 6.3.5, there is no restoring torque on the car (see Fig. 6.10), and the gimbal is an inverted pendulum with no restoring torque about the gimbal axis.

(a) Show that the equations of motion (neglecting friction) are

$$
\begin{bmatrix}
s^2 - \mu_o^2 & hs/I_x \\
-hs/J & s^2 - \mu_g^2
\end{bmatrix}
\begin{bmatrix}
\phi(s) \\
\epsilon(s)
\end{bmatrix}
=
\begin{bmatrix}
Q_{dx}(s)/I_x \\
Q_c(s)/J
\end{bmatrix},
$$

where $\pm\mu_o$ are roll eigenvalues of car for $h = 0$, $\pm\mu_g$ are inverted pendulum eigenvalues of GMW for $h = 0$, and other nomenclature is same as in Problem 6.3.5.

(b) Make a root locus sketch versus $h^2/I_x J$ to show that spin makes the system oscillatory for $|h|/\sqrt{I_x J} > \mu_o + \mu_g$, with one frequency $> \sqrt{\mu_o \mu_g}$, the other less than that.

(c) For $\mu_o = .35\ \sec^{-1}$, $\mu_g = .15\ \sec^{-1}$, $h/\sqrt{I_x J} = 0.65$, and $I_x/J = 5000$, show that the eigensystem is

$$
\begin{array}{ccc}
s & \pm .102j & \pm .515j \quad \sec^{-1} \\[2mm]
\phi & \pm .00703j & \pm .01218j \\[2mm]
\epsilon & 1 & 1
\end{array}
$$

(d) With viscous damping on the gimbal axis, the equations of motion become

$$
\begin{bmatrix}
s^2 - \mu_o^2 & hs/I_x \\
-hs/J & s^2 + Ds/J - \mu_g^2
\end{bmatrix}
\begin{bmatrix}
\phi(s) \\
\epsilon(s)
\end{bmatrix}
=
\begin{bmatrix}
Q_{dx}(s)/I_x \\
Q_c(s)/J
\end{bmatrix}.
$$

For $\mu_o, \mu_g, h^2/I_x J$ as in (c), sketch a root locus versus D/J to show that gimbal damping *destabilizes* the oscillatory system. Alternatively, sketch a root locus versus $h^2/I_x J$ for $D/J = .1\mu_g$ to show that the system cannot be stabilized by spin when there is damping on the gimbal axis.

(e) For $D/J = .070\ \sec^{-1}$ in (d) show that the eigensystem is

$$
\begin{array}{ccc}
s & .018 \pm .099j & -.0528 \pm .518j \quad \sec^{-1} \\[2mm]
\phi & .0070\angle \pm 81.4° & .0122\angle \pm 87.8° \\[2mm]
\epsilon & 1 & 1
\end{array}
$$

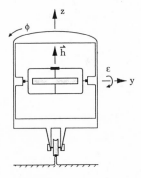

Figure 6.10. Monorail-car roll stabilization with a GMW.

Problem 6.3.7 – *Active Monorail Car Stabilization*

(a) Using the example of Próblem 6.3.6(c), sketch the LQR symmetric root locus versus A/B using the performance index

$$J = \int_o^\infty \left[A \left(\phi^2 + \frac{\epsilon^2}{100} \right) + B \left(\frac{Q_c}{J\mu_o^2} \right)^2 \right] dt.$$

(b) For $A/B = 250$ in (a) show that the regulator eigenvalues are

$$s = -.212 \pm .594j, \; -.215 \pm .117j \; \sec^{-1}$$

and the feedback gains are:

$$\frac{Q_c}{J\mu_o^2} = \begin{bmatrix} 267.4 & 166.1 & -1.408 & 2.239 \end{bmatrix} \begin{bmatrix} \phi \\ \phi/\mu_o \\ \epsilon \\ \epsilon/\mu_o \end{bmatrix}.$$

(c) For the example in (a), sketch the estimate-error SRL versus Q/R assuming a measurement $z = \phi$ plus a white noise process with spectral density R, and Q_{dx} another independent white noise process with spectral density Q.

(d) For $Q/R = 100$ in normalized units (Q_{dx} in units of $I_x\mu_o^2$, time in units of $1/\mu_o$) *show that* the estimator eigenvalues are

$$s = -.212 \pm .594j, \; -.215 \pm .117j \; \sec^{-1}.$$

(e) Combining (b) and (d), show that the LQG compensator is

$$\frac{Q_c(s)}{z(s)} = \frac{J\mu_o^3 \cdot 1321.9(s+.1307)[(s-.0205)^2 + (.4038)^2]}{(s+.1306)(s+1.1045)[(s+.6207)^2 + (1.1308)^2]}.$$

Note the pole/zero cancellation, which reduces the order of the compensator from 4 to 3.

(f) Sketch a root locus versus overall gain, K_o, for the LQG compensator such that $K_o = 1$ corresponds to (d). Note this is a *notch compensator*.

7

Attitude Control during
Thrust Maneuvers

7.1 Using Reaction Jets

During thrust maneuvers to change the velocity of the center of mass of a spacecraft, attitude stabilization is essential to keep the thrust vector in the desired direction.

Even with great care the thrust line of action never passes through the center of mass of the spacecraft (see Fig. 7.1). The thrust misalignment torque, $T\epsilon$, is usually much larger than the disturbance torques due to gravity, solar light pressure, magnetic fields, aerodynamics, etc. Hence, a special attitude control system is required.

Stabilization torques can be supplied by reaction jets, by gimbaling the nozzle of the main engine (see Section 7.2), or by off-modulation of a multinozzle main engine (see Section 7.3).

Alternatively, the spacecraft may be spun about the thrust axis, so that the average thrust misalignment torque in any direction perpendicular to the thrust axis is zero (see Section 7.3).

Treating the spacecraft as rigid with no spin, the equations of planar motion are

$$I_y\dot{\omega} + (-\dot{m})(r_e^2 - k^2)\omega = \ell f + Tr\epsilon, \tag{7.1}$$

$$m(\dot{u} + \omega w) \cong T, \tag{7.2}$$

$$m(\dot{w} - \omega u) \cong T\epsilon + f, \tag{7.3}$$

$$\dot{\theta} = \omega, \tag{7.4}$$

$$\begin{bmatrix} \dot{x}_I \\ \dot{z}_I \end{bmatrix} = \begin{bmatrix} c\theta & s\theta \\ -s\theta & c\theta \end{bmatrix} \begin{bmatrix} u \\ w \end{bmatrix}, \tag{7.5}$$

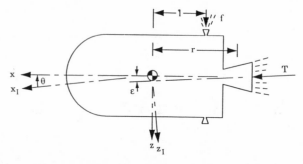

Figure 7.1. Spacecraft using reaction jets for attitude stabilization during a thrust maneuver.

where ϵ = thrust misalignment angle, I_y = pitch moment of inertia $\equiv mk^2$, and the rest of the nomenclature is defined in Fig. 7.1. The term proportional to ω in (1) is a "jet-damping" torque.

The thrust maneuver begins with all state variables = 0, and the attitude control logic should keep $\dot{z}_I$ close to zero. Assuming m and T to be nearly constant and ω, w small during the thrust maneuver implies that $\dot{u}$ is nearly constant.

Since ϵ is not known, we must treat $T\epsilon$ as a random constant disturbance and introduce integral-error feedback to the reaction jet force f, that is, feedback z_I (or an estimate of z_I). Furthermore, the jet-damping term is usually quite small compared to the desired closed-loop damping, so that it can be neglected in synthesizing the control law. The transfer function from f to z_I is then

$$\frac{z_I(s)}{f(s)} \cong \frac{1}{m} \frac{s^2 - T\ell/I_y}{s^4}. \tag{7.6}$$

If we had proportional reaction jets, the synthesis of time-invariant control logic would be relatively straightforward (cf. problem 7.1.1). However, we usually have on-off reaction jets, and the synthesis of time-invariant control logic for this case is considerably more complicated; one method of synthesis involves the use of "describing functions" (see, for example, Ref. GE).

The problem is even more complicated, in that we should like to have w, θ, and $\omega = 0$, when the thruster is turned off. A set of time-varying gains that will do this for proportional reaction jets can be determined using time-varying linear-quadratic control synthesis (see, for example, BR-1). Given $T\epsilon$, one could also calculate an open-loop on-off control history that would give w, θ, and $\omega = 0$ at thrust cut-off.

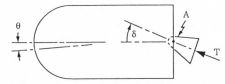

Figure 7.2. Spacecraft with a gimbaled main engine.

Problem 7.1.1

(a) Show that the Laplace transform of the equations of motion can be written in the dimensionless form:

$$
\begin{bmatrix} s^2 & 1 \\ 0 & s^2 \end{bmatrix}
\begin{bmatrix} z_I(s) \\ \theta(s) \end{bmatrix}
=
\begin{bmatrix} 1 \\ 1 \end{bmatrix} f
+
\begin{bmatrix} 1 \\ r/l \end{bmatrix} \epsilon ,
$$

where time is in units of α, $\alpha^2 = Tl/mk^2$, $I = mk^2$, (θ, ϵ) in l^2/k^2.

(b) Verify the transfer funtion (6).

(c) For the quadratic performance index

$$
J = \int_0^\infty (Az_I^2 + f^2)dt,
$$

sketch a locus of the closed-loop poles versus A.

(d) For $A = .01$, determine the state feedback gains and plot the closed-loop response to a constant ϵ with zero initial conditions.

7.2 Using a Gimbaled Engine

Reaction jets sized to handle disturbances with the main engine off do not have enough torque to overcome the thrust misalignment torque. Hence a special additional reaction jet system is required. To avoid this, some spacecraft have used a gimbaled main engine. Many launch vehicles (boosters) have also used gimbaled engines for attitude control. Fig. 7.2 shows such a system for con-trolling pitch motions (the engine must have two gimbals; deflecting the other gimbal controls yaw motions).

To synthesize control logic for this two rigid-body system, we need the equa-tions of motion with a control torque on the gimbal axis. We develop these equa-tions here using D'Alembert's method. Fig. 7.3 shows free-body diagrams of

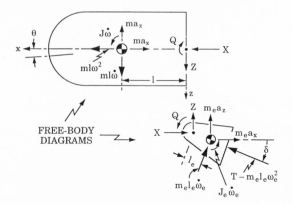

Figure 7.3. Free-body diagrams of spacecraft and main engine.

the spacecraft and the nozzle with the internal control torque Q and the internal forces X and Z that act on the gimbal axis.

Also shown are the D'Alembert forces and torques and the thrust force T. Taking moments about the gimbal axis (A in the diagrams), the equations of motion for the spacecraft are

$$0 = \sum M_A = (J + m\ell^2)\dot{\omega} - ma_z\ell - Q, \tag{7.7}$$

$$0 = \sum F_z = -ma_z + m\ell\dot{\omega} + Z, \tag{7.8}$$

$$0 = \sum F_x = -ma_x + m\ell\omega^2 + X. \tag{7.9}$$

The equations of motion for the gimbaled engine are

$$0 = \sum M_A = (J_e + m_e\ell_e^2)\dot{\omega}_e + m_e a_z\ell_e \cos\delta + m_e a_x\ell_e \sin\delta + Q, \tag{7.10}$$

$$0 = \sum F_z = -m_e a_z - m_e\ell_e\dot{\omega}_e \cos\delta - [T - m_e\ell_e\omega_e^2]\sin\delta - Z, \tag{7.11}$$

$$0 = \sum F_x = -m_e a_x - m_e\ell_e\dot{\omega}_e \sin\delta + [T - m_e\ell_e\omega_e^2]\cos\delta - X, \tag{7.12}$$

and

$$\dot{\delta} = \omega_e - \omega, \tag{7.13}$$

where

(ω, ω_e) = angular velocity of (spacecraft, engine),

(m, m_e) = mass of (spacecraft, engine),

(J, J_e) = moment of inertia about c.m. of (spacecraft, engine),

(a_x, a_z) = spacecraft body-axes components of acceleration of gimbal axis (point A).

Eliminating X between (9) and (12), and Z between (8) and (11), gives the acceleration components:

$$(m + m_e)a_x = m\ell\omega^2 - m_e\ell_e(\omega_e^2\cos\delta + \dot{\omega}_e\sin\delta) + T\cos\delta, \quad (7.14)$$

$$(m + m_e)a_z = m\ell\dot{\omega} - m_e\ell_e(-\omega_e^2\sin\delta + \dot{\omega}_e\cos\delta) - T\sin\delta. \quad (7.15)$$

Substituting for (a_x, a_z) from (14) and (15) into (7) and (10) and repeating (13) gives the following equations of motion:

$$J^*\dot{\omega} + J_c(\dot{\omega}_e\cos\delta - \omega_e^2\sin\delta) = Q - b\sin\delta, \quad (7.16)$$

$$J_e^*\dot{\omega}_e + J_c(\dot{\omega}\cos\delta + \omega^2\sin\delta) = -Q, \quad (7.17)$$

$$\dot{\delta} + \omega - \omega_e = 0, \quad (7.18)$$

where

$$J^* \triangleq J + m^*\ell^2,$$

$$J_e^* \triangleq J_e + m^*\ell_e^2,$$

$$J_c \triangleq m^*\ell\ell_e,$$

$$m^* \triangleq mm_e/(m + m_e),$$

$$b \triangleq m\ell T/(m + m_e).$$

The equations of motion may be linearized for $|\delta| \ll 1$, $J_c\omega_e^2 \ll b$, and $J_c\omega^2 \ll b$. Taking the Laplace transform of these equations gives

$$
\begin{bmatrix} J^*s & J_cs & b \\ J_cs & J_e^*s & 0 \\ 1 & -1 & s \end{bmatrix}
\begin{bmatrix} \omega(s) \\ \omega_e(s) \\ \delta(s) \end{bmatrix}
\cong
\begin{bmatrix} 1 \\ -1 \\ 0 \end{bmatrix} Q(s). \quad (7.19)
$$

From (19), we may deduce the transfer function from Q to ω:

$$\frac{\omega(s)}{Q(s)} = \frac{p^2}{bs}\frac{s^2 + z^2}{s^2 - p^2}, \quad (7.20)$$

where

$$z^2 \triangleq b/(J_e^* + J_c), \quad (7.21)$$

$$p^2 \triangleq b(J_e^* + J_c)/(J^*J_e^* - J_c^2). \quad (7.22)$$

Since $\omega = \dot\theta$, (20) may be written as

$$\frac{\theta(s)}{Q(s)} = \frac{1}{s^2}\frac{s^2 + z^2}{s^2 - 1}, \tag{7.23}$$

where time is in units of $1/p$, z in units of p, and Q in units of b. Note the zeros of this transfer function on the imaginary axis imply that torquing the engine at this frequency would produce no pitch motion of the spacecraft (the torques about the spacecraft center of mass add up to zero; that is, the control torque and the torque due to engine deflection are just balanced by the torque due to the reaction force Z).

With (ω, ω_e) in units of p, time in units of $1/p$, and Q in units of b, (19) may be put into state-variable form:

$$\begin{bmatrix} \dot\omega \\ \dot\omega_e \\ \dot\delta \end{bmatrix} \cong \begin{bmatrix} 0 & 0 & -\epsilon \\ 0 & 0 & 1-\epsilon \\ -1 & 1 & 0 \end{bmatrix} \begin{bmatrix} \omega \\ \omega_e \\ \delta \end{bmatrix} + \begin{bmatrix} 1 \\ -\lambda \\ 0 \end{bmatrix} Q, \tag{7.24}$$

where

$$\epsilon \triangleq J_e^*/(J_e^* + J_c),$$
$$\lambda \triangleq (J^* + J_c)/(J_e^* - J_c),$$

As a *numerical example*, consider $m_e = m/30$, $\ell_e = \ell/10$, $J = \frac{1}{3}m\ell^2$, $J_e = \frac{1}{3}m_e\ell_e^2$, which gives

$$\epsilon = .11851,$$
$$\lambda = 100.78,$$
$$z = 3.3260.$$

Using linear-quadratic synthesis with performance index

$$J = \int_o^\infty (A\theta^2 + Q^2)dt, \tag{7.25}$$

we see the locus of closed-loop poles versus the weighting factor A shown in Fig. 7.4.

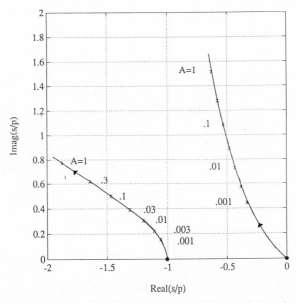

Figure 7.4. Spacecraft with gimbaled engine; locus of LQ regulator poles vs. weighting factor A.

Choosing $A = .01$, the closed-loop poles are at

$$s = -.434 \pm .731j , \quad -1.200 \pm .306j$$

and the control law is

$$Q = [-.1000, -.3061, .0294, .0514] \begin{bmatrix} \theta \\ \omega \\ \omega_e \\ \delta \end{bmatrix}.$$

Fig. 7.5 shows the response of the closed-loop system to an initial pitch rate. To calculate the inertial position (x_I, z_I) of the point A on the spacecraft, given $\omega(t), \omega_e(t), \theta(t)$, and $\delta(t)$, we may use (14) and (15) with

$$\dot{u} + \omega w = a_x, (theta = \omega) \qquad (7.26)$$

$$\dot{w} - \omega u = a_z, \qquad (7.27)$$

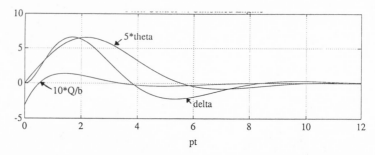

Figure 7.5. Spacecraft with gimbaled engine; response to initial angular velocity $\omega(0) = p$.

$$
\begin{bmatrix} \dot{x}_I \\ \dot{z}_I \end{bmatrix} = \begin{bmatrix} \cos\theta & \sin\theta \\ -\sin\theta & \cos\theta \end{bmatrix} \begin{bmatrix} u \\ w \end{bmatrix}. \tag{7.28}
$$

For $|\delta| \ll 1, \omega^2 \ll b/J_c, \omega_e^2 \ll b/J_c$, (26)–(28) may be approximated by

$$
\dot{u} \cong a_x \cong T/(m + m_e), \tag{7.29}
$$

$$
\dot{w} - \omega u = a_z \cong \frac{1}{m + m_e}(m\ell\dot{\omega} - m_e\ell_e\dot{\omega}_e - T\delta), \tag{7.30}
$$

$$
\dot{x}_I \cong u, \tag{7.31}
$$

$$
\dot{z}_I \cong -u\theta + w, \tag{7.32}
$$

which gives

$$
u \cong a_x t, \tag{7.33}
$$

$$
x_I \cong \frac{1}{2}a_x t^2, \tag{7.34}
$$

$$
w \cong (m\ell\omega - m_e\ell_e\omega_e)/(m + m_e) - a_x \int_o^t \delta dt, \tag{7.35}
$$

$$
z_I \cong -a_x \int_o^t t\theta dt + \int_o^t w dt, \tag{7.36}
$$

where a_x is a constant given in (29).

7.3 Using Off-Modulation or Spin

Instead of a gimbaled engine, one can use an engine with three nozzles, where each of the nozzles can be turned off or throttling for controlled periods of time ("off-modulation"). If the nozzle axes form an equilateral triangle (looking from the rear), then positive and negative pitch and yaw torques can be produced by turning off one or or throttling two nozzles.

A passive method of attitude control during thrust maneuvers is to spin the spacecraft about the thrust axis. Then the thrust misalignment torque rotates with the spacecraft and the spacecraft moves on a helix. If the helix angle is sufficiently small, the thrust maneuver is accomplished with satisfactory accuracy. The helix angle is given by

$$\frac{T\epsilon}{h\omega_s} \tag{7.37}$$

For example, if $T\epsilon = 100$ ft lb, the spin angular momentum h is 1000 ft lb sec, the spin angular velocity ω_s is 10 rad/sec, then the helix angle is .01 rad (.6 deg).

Spin is attractive for unmanned spacecraft since it avoids gimbaling the engine and avoids an active control system.

Problem 7.3.1

Verify the expression for the helix angle in Equation (37).

8

Control of Translational Motions

8.1 Fast Control

The translational motions of the center of mass of a spacecraft are predicted by Newton's equations (see Section 1.1).

The external forces acting on the spacecraft may be divided into control forces $\vec{F}_c$ and other forces $\vec{F}_d$. The control force is usually a reaction force obtained by *throwing mass overboard*. This force is equal to $-c\dot{m}$ where $-\dot{m}$ is the rate at which mass is being thrown overboard, and c is the velocity of the mass particles with respect to the spacecraft (called "specific impulse"). If $|\vec{F}_c| \gg |\vec{F}_d|$, then $\vec{F}_d$ may be regarded as a small disturbance and we shall call this "fast" control. If $|\vec{F}_c|$ is comparable to $|\vec{F}_d|$, then we shall call this "slow" control (see Sections 8.2 and 8.3).

Changes in velocity are usually obtained using a constant thrust force in the direction of desired change. The thrust duration is such that the desired magnitude of velocity change is attained. For fast control

$$\Delta V = \frac{F \Delta t}{m} = \frac{c \frac{\Delta m}{\Delta t} \cdot \Delta t}{m}$$

$$= c \frac{\Delta m}{m}$$

$$\Delta V = c \log \frac{m_o}{m_f}, \tag{8.1}$$

where

$$m_f = m_o - \frac{T}{c}\Delta t,$$

thrust
$$T = c \frac{\Delta m}{\Delta t}$$

and ΔV is the desired velocity change, T is the thrust, (m_f, m_o) are (final, initial) mass, and Δt is the thrust duration.

Rendezvous of one spacecraft with another spacecraft requires sensing relative position and velocity, and the use of either two orthogonal thrusting systems or a single thruster whose thrust direction can be changed.

Fig. 8.1 shows a coordinate system useful for analyzing rendezvous maneuvers; the origin is at the passive spacecraft, which is assumed here to have no acceleration with respect to inertial space. The relative velocity $\vec{V}$ and the line

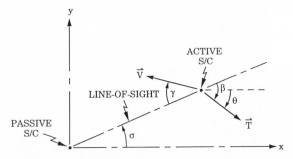

Figure 8.1. Coordinates for analyzing a fast rendezvous maneuver.

of sight determine a plane that we take as the (x, y) plane, and the (x, y) axes are fixed with respect to inertial space. The equations of motion are simply

$$m\dot{u} = T_x \equiv T\cos\theta, \tag{8.2}$$

$$m\dot{v} = T_y \equiv -T\sin\theta, \tag{8.3}$$

$$\dot{x} = u, \tag{8.4}$$

$$\dot{y} = v, \tag{8.5}$$

where $m(t)$ = mass of spacecraft, $T = -c\dot{m}$ = thrust magnitude, θ = thrust direction angle relative to the x-axis, (u, v) = velocity components along the (x, y) axes, and c = specific impulse.

If two orthogonal thrusting systems (T_x, T_y) are used, the spacecraft model reduces to two uncoupled "double-integrator" systems, and control design is quite straightforward (see Problems 8.1.1 and 8.1.2).

We consider here the case with a single on-off thruster and variable thrust direction (Ref. BR-1). We assume a nonzero initial velocity, $V(0)$, and choose the x-axis to be in the direction of $-V(0)$ (Fig. 8.1 shows the situation at a later time when V may *not* be parallel to x). If $\gamma(0) < \pi/2$ and no thrust were used, the initial y-coordinate, $y(0)$, of the active spacecraft would be the distance of closest approach, when $x = 0$. For given values of $y(0), T, V(0)$, and $m(0)$, rendezvous may be achieved using *three constant thrust-direction periods*. There are only two different types of paths, as shown in Fig. 8.2, which correspond to initiating thrust before or after reaching the optimum initial value of $x = (x_o)_{opt}$. If thrust is begun at $(x_o)_{opt}$, only two constant thrust-direction periods are required and the thruster is turned on for the minimum length of time, so this is a *minimum-fuel maneuver*. During the first segment, which takes half the time, $\dot{u} > 0, \dot{v} < 0$; during the second (and final) segment, $\dot{u} > 0, \dot{v} > 0$,

constant deceleration in x
acceleration then deceleration in y

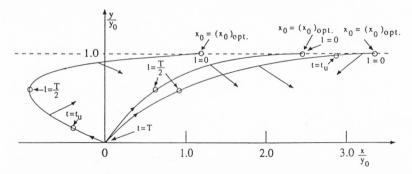

Figure 8.2. Minimum time rendezvous paths using three constant thrust direction periods with $V_o^2/2ay_o = 2$.

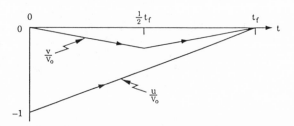

Figure 8.3. Velocity histories for minimum fuel rendezvous with $T/m =$ constant.

and the deceleration is such that $u = v = 0$ when $x = y = 0$. Fig. 8.3 shows the velocity histories for this case when $T/m =$ constant.

For $T/m =$ constant, the spacecraft path consists of a parabola and a straight line. For $x_o > (x_o)_{opt}$, a third parabolic segment is added at the beginning with $\dot{u} < 0$. For $x_o < (x_o)_{opt}$, a third parabolic segment is added in the middle with $\dot{v} > 0$ (see Problem 8.1.1).

Problem 8.1.1 – *Fast Rendezvous with One On-Off Thruster*

Using Figs. 8.1 and 8.2 and assuming $T/m = a =$ constant, show that x and y are brought to zero by the following thrust-direction logic:

$$
\theta = \left\{
\begin{array}{ll}
\theta_o, & 0 < t < \frac{1}{2}t_f \\[2mm]
-\theta_o, & \frac{1}{2}t_f < t < t_f - t_u \\[2mm]
\pi + \theta_o, & t_f - t_u < t < t_f
\end{array}
\right\} x_o < (x_o)_{opt} \Rightarrow \theta_o < \frac{\pi}{2},
$$

$$
\theta = \left\{
\begin{array}{ll}
\theta_o, & 0 < t < t_u \\[2mm]
\pi - \theta_o, & t_u < t < \frac{1}{2}t_f \\[2mm]
\theta_o - \pi, & \frac{1}{2}t_f < t < t_f
\end{array}
\right\} \quad x_o > (x_o)_{opt} \Rightarrow \theta_o > \frac{\pi}{2},
$$

where

$$
a(x_o)_{opt}/V_o^2 = \frac{1}{2\sqrt{2}} \left[1 + \sqrt{1 + \left(\frac{8ay_o}{V_o^2} \right)^2} \right]^{\frac{1}{2}},
$$

and θ_o, t_f, t_u are determined by

$$
y_o = \frac{1}{2}t_f \sin\theta_o,
$$

$$
t_u = \frac{1}{2}(t_f - \sec\theta_o),
$$

$$
x_o = \cos\theta_o(t_f^2 - 2t_u^2), x_o > (x_o)_{opt},
$$

$$
x_o = \cos\theta_o(t_f^2 - 4t_u t_f + 2t_u^2), x_o < (x_o)_{opt},
$$

and (x_o, y_o) are in units of $V_o^2/2a$, (t_f, t_u) are in units of V_o/a. (b) Plot contours of constant t_f in the (x_o, y_o) plane.

Problem 8.1.2 – *Fast Intercept with One On-Off Thruster*

This problem differs from Problem 8.1.1 only in the terminal conditions. Here we want $x(t_f) = y(t_f) = 0$, but the final velocity components are not specified.

(a) Show that minimum time intercept is obtained with one constant direction period (a parabolic path that intercepts the origin).

(b) Determine the minimum-time thrust-direction for arbitrary (x_o, y_o). This solution has a "barrier" in the sense that the minimum time is *not* a continuous function of x_o; if the maneuver is not begun soon enough the S/C must overshoot and come back.

(c) Plot contours of constant t_f on the (x_o, y_o) plane (they are circles with centers at $x_o = V_o t_f$ and radii $at_f^2/2$).

Problem 8.1.3 – *Fast Rendezvous with Two Proportional Thrusters*

For this case, Equations (2)–(5) decouple into two "double-integrator" systems. Show that a proportional plus derivative feedback law will bring x and y to zero, that is,

$$
T_x = -K(x + \tau u),
$$

$$
T_y = -K(y + \tau v),
$$

where τ and K are positive constants.

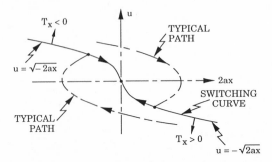

Figure 8.4. Phase-plane paths for rendezvous with an on-off thruster.

Problem 8.1.4 – *Minimum-Time Fast Rendezvous with Two On-Off Thrusting Systems*

For this case, Equations (2)–(5) decouple into two identical "double- integrator" systems. For $T/m = a$ = constant, show that x is brought to zero with only one switch by the logic:

$$
a_x = \begin{cases}
-a, & u > \sqrt{-2ax} \ \ x < 0, \ \ u > -\sqrt{2ax} \ \ x > 0, \\[2mm]
a, & u < \sqrt{-2ax} \ \ x < 0, \ \ u < -\sqrt{2ax} \ \ x > 0,
\end{cases}
$$

which is shown graphically in the phase-space sketch, Fig. 8.4 (Problem 3.3.2 is the corresponding problem for attitude control.)

Problem 8.1.5 – *Position Hold with Two On-Off Thrusters; Bang-Off-Bang Logic*

As in the previous problem, we have two identical double-integrator systems. For T/m = constant, show by means of an $(\dot{x}, x)$ sketch that the spacecraft is brought near to $x = \dot{x} = 0$ using deadband and hysteresis (Schmitt trigger), ending with a low-frequency limit cycle. (See Section 3.3 for discussion of deadband and hysteresis.)

8.2 Slow Control in Nearly Circular Orbit; Cross-Track

The linearized equations of translational motion in circular orbit were derived in Section 1.2. They decouple into a set for cross-track motions and a set for in-track/radial motions. Control of cross-track motions is treated in this section, and control of in-track/radial motions is treated in Section 8.3.

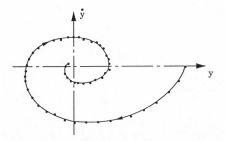

Figure 8.5. Slow control of cross-track position with two opposing proportional thrusters.

Slow Control with Proportional Thrusters

With two proportional thrusters, one thrusting in the $(+y)$ direction and one in the $(-y)$ direction, stabilization of cross-track position may be obtained by using proportional plus derivative feedback:

$$T_y = -K(y + \tau \dot{y}). \tag{8.6}$$

Substituting (6) into (1.12), we have

$$\ddot{y} + \frac{K\tau}{m}\dot{y} + \left(\frac{K}{m} + n^2\right)y = 0, \tag{8.7}$$

which represents stable motion for positive τ and K. Fig. 8.5 shows the damped motion in the $(\dot{y}, y)$ phase plane.

If $K/m \gg n^2$, the problem is essentially the same as the free-space rendezvous problem discussed in Section 8.1.

Slow Control of Cross-Track Position with On-Off Thrusters

With two opposing on-off thrusters, stabilization of cross-track position may be accomplished by using a switching function that is a linear combination of cross-track position error, y, and cross-track velocity, $\dot{y}$ (cf. Section 3.3):

$$T_y = -T_o \text{sgn}(y + \tau \dot{y}). \tag{8.8}$$

A typical $(\dot{y}, y)$ phase-plane trajectory is shown in Fig. 8.6. The segments of the trajectories are circular arcs with centers at $\pm T_o/mn^2$, where n = orbital angular velocity. The system "chatters" (slides down the switching line) in the final segment, and ends up in a limit cycle around $y = \dot{y} = 0$, similar to the one

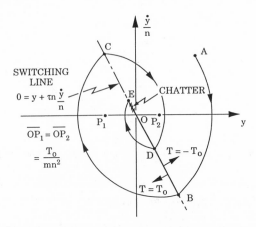

Figure 8.6. Phase-plane path of cross-track position control in circular orbit with a linear switching function, $T = -T_o \text{ sgn } (y + \tau \dot{y})$.

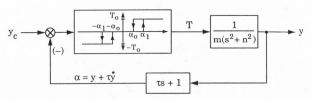

Figure 8.7. Block diagram of cross-track position control in circular orbit with a linear switching function, deadband, and hysteresis (Schmitt trigger).

shown in Fig. 3.6. This waste of fuel can be avoided by using bang-off-bang control, as we showed in Section 3.3 for attitude control.

If the "Schmitt trigger" device is used (deadband with hysteresis) as in Section 3.3, the spacecraft can be brought within a small region around $y = \dot{y} = 0$ with the thrusters off. Fig. 8.7 shows a block diagram of the controlled system.

Fig. 8.8 shows a typical $(\dot{y}, y)$ phase-plane trajectory for the system where $T_o/mn^2 = .2$ km, $\tau = 1/n$, $\alpha_1 = .1$ km, $\alpha_o = .05$ km. On $\bar{AB}$ and $\bar{EF}$, $T = T_o$; on $\bar{CD}$, $T = -T_o$; on $\bar{BC}$, $\bar{DE}$, and after F, $T = 0$. Thus, the final trajectory is *not* a "limit cycle" but a natural motion with $|y| < .04$ km and $T = 0$.

Problem 8.2.1 – *Minimum-Time Bang-Bang Control of Cross-Track Position*

Minimum-time bang-bang control of cross-track position is obtained using the nonlinear switching function shown in Fig. 8.9 (Ref. AT).

With $y(0) = 1$ km, $\dot{y}(0) = 0$, $T_o/mn^2 = .2$ km, calculate the time (in units of $1/n$)

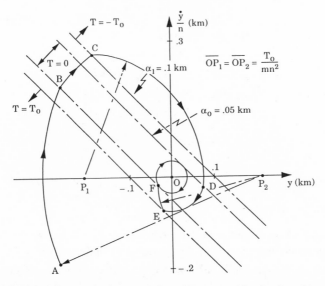

Figure 8.8. Phase plane path of cross-track position control in circular orbit with a linear switching function, deadband, and hysteresis; $T_o/mn^2 = .2$ km, $\tau = 1/n$.

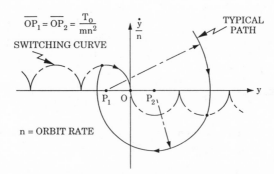

Figure 8.9. Phase-plane path of minimum-time cross-track position control with on-off thrusters (Bushaw).

to $y = \dot{y} = 0$ with the minimum-time switching function and compare it to the time required using the linear switching function with $\tau = 1/n$.

Problem 8.2.2 – *Slow Control of Cross-Track Position Using Only One Thruster*

It is not necessary to use two opposing thrusters, since stabilization can be accomplished with only one.

Show that control can be accomplished using one thruster and sketch a typical $\dot{y}/n$ vs. y trajectory for the following cases:

(a) A proportional thruster.

(b) An on-off thruster with a linear switching function.

(c) An on-off thruster with deadband and hysteresis.

(d) An on-off thruster with the minimum-time switching function.

8.3 Slow Control in Circular Orbit; In-Track/Radial

Equations of Motion

The equations of motion were derived in Section 1.2. It is convenient to put them into dimensionless form by using the following units: time in $1/n$, $(\delta u, \delta w)$ in nR, $(\delta x, \delta z)$ in R, (T_x, T_z) in mg, where n = orbital rate, R = orbit radius, m = mass of S/C, and g = gravitational force per unit mass at that orbit radius:

$$
\begin{bmatrix} \delta \dot{u} \\ \delta \dot{w} \\ \delta \dot{x} \\ \delta \dot{z} \end{bmatrix} = \begin{bmatrix} 0 & 1 & -1 & 0 \\ -1 & 0 & 0 & 2 \\ 1 & 0 & 0 & 1 \\ 0 & 1 & -1 & 0 \end{bmatrix} \begin{bmatrix} \delta u \\ \delta w \\ \delta x \\ \delta z \end{bmatrix} + \begin{bmatrix} 1 & 0 \\ 0 & 1 \\ 0 & 0 \\ 0 & 0 \end{bmatrix} \begin{bmatrix} T_x \\ T_z \end{bmatrix}. \tag{8.9}
$$

The transfer functions from in-track/radial thrust to in-track/radial position deviations are readily deduced from these equations:

$$
\begin{bmatrix} \delta x(s) \\ \delta z(s) \end{bmatrix} = \frac{1}{s^2(s^2+1)} \begin{bmatrix} s^2 - 3 & 2s \\ -2s & s^2 \end{bmatrix} \begin{bmatrix} T_x(s) \\ T_z(s) \end{bmatrix}. \tag{8.10}
$$

There are pole-zero cancellations in three of the four transfer functions, indicating a lack of observability or controllability. Thus we shall choose δx as output (and measurement), and T_x as control, since there are no pole-zero cancellations in that transfer function.

Stabilization of In-Track/Radial Motions

Since all three modes are controllable with T_x, consider the LQ synthesis of a regulator using full state feedback to T_x using proportional thrusters. We take as performance index

$$J = \int_0^\infty \left\{ A[(\delta x)^2 + (\delta z)^2] + B(T_{cx})^2 \right\} dt. \tag{8.11}$$

It follows that the symmetric root characteristic equation (SRCE) is

$$Y^T(-s)AY(s) + B = 0, \tag{8.12}$$

where

$$\begin{bmatrix} \delta x(s) \\ \delta z(s) \end{bmatrix} = Y(s)T_{cx}(s). \tag{8.13}$$

In this case, the transfer function matrix $Y(s)$ is given by

$$Y(s) = \frac{1}{s^2(s^2 + 1)} \begin{bmatrix} s^2 - 3 \\ -2s \end{bmatrix}. \tag{8.14}$$

Thus the SRCE is

$$[s^2 - 3, 2s] \begin{bmatrix} s^2 - 3 \\ -2s \end{bmatrix} + \frac{B}{A}s^4(s^2 + 1)^2 = 0, \tag{8.15}$$

or, in Evans's form:

$$-\frac{B}{A} = \frac{(s^2 - 1)(s^2 - 9)}{s^4(s^2 + 1)^2}. \tag{8.16}$$

Note the "compromise" zeros at $s = \pm 1, \pm 3$. These arise from trying to control both δx and δz with only one control, T_x.

One quadrant of the symmetric root locus (SRL) versus A/B is shown in Fig. 8.10. For $A/B = 1$, the closed-loop regulator eigenvalues are

$$s = -.667 \pm 1.536j, -.996 \pm .279j, \tag{8.17}$$

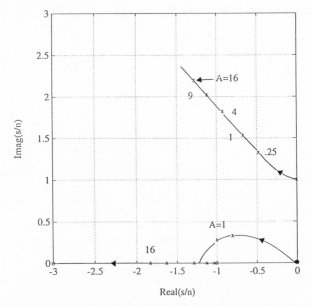

Figure 8.10. Stabilization of in-track/radial motions using proportional tangential thrusters; locus of LQ regulator poles vs. A/B.

and the corresponding state feedback is

$$T_{cx} = [-3.32, 3.27, -2.27, 5.17] \begin{bmatrix} \delta u \\ \delta w \\ \delta x \\ \delta z \end{bmatrix}. \qquad (8.18)$$

Fig. 8.11 shows the *time response*, using the control law (18), where the spacecraft is initially in circular orbit but has an in-track error, namely $\delta x(0)/R = -.001, \delta z(0) = 0, \delta u(0) = 0, \delta w(0) = 0$. The error is eliminated in roughly one orbital period ($nt = 2\pi$); the in-track error overshoots by 70% and returns to zero; the radial displacement is first down ($\delta z > 0$) and then up; the thrust is positive to start with $2.265 \times 10^{-3} \times mg$, and it alternates sign about three times in the first orbit.

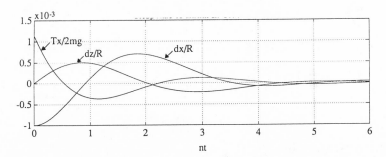

Figure 8.11. Response of controlled spacecraft to an initial in-track error $\delta x(0) = -.001 \times R$.

Problem 8.3.1 – *Stabilization Using Both Tangential and Radial Thrusters*

(a) Using (9) and the performance index

$$J = \int_{o}^{\infty} \left\{ A[(\delta x)^2 + (\delta z)^2] + B(T_{cx}^2 + T_{cz}^2) \right\} dt,$$

plot the closed-loop root locus versus A/B as in Fig. 10.

(b) Using $A/B = 1$, determine the feedback gain matrix and plot the time response as in Fig. 8.11. Answer for K is

$$K = \begin{bmatrix} 1.64 & -.74 & .88 & -1.49 \\ -.74 & 1.85 & -.86 & 2.72 \end{bmatrix}.$$

(c) Compare this two-control system with the one-control system in the text.

9

Flexibility and Fuel Slosh

9.1 Introduction

Flexibility or fuel slosh must be considered by the control designer if the frequency of the lowest vibration or slosh mode is less than about six times the desired control bandwidth. Otherwise there is a strong possibility that this mode will be destabilized by his control system, a phenomenon called "spillover" (because the control effort spills over outside the control bandwidth and destabilizes the vibration mode).

Most spacecraft designed before 1980 were relatively stiff and did not use liquid fuels that could slosh; hence the control designer could approximate them as rigid bodies. However, spacecraft boosters are quite flexible and use liquid fuels that slosh. Booster stabilization systems were designed from the beginning taking into account flexibility. One example is the control system for the "Vanguard" booster, the first successful U.S. satellite (1958, Ref. BLX). However the first attempt by the United States (also 1958) to launch an artificial satellite (the "Jupiter") failed due to an instability caused by liquid fuel slosh in the booster.

Like boosters, many missiles have bending modes that can be excited by control motions. In the 1950s, one of the first air-to-air missile designs (the "Lark") exhibited a servo-elastic instability on a test bench when a feedback loop was closed from a gyro in the nose to the control surfaces at the tail. The inertia forces associated with deflecting the controls excited the first bending mode of the fuselage, and the sensor feedback, designed for a rigid missile, had the wrong sign. This was a classic case of a sensor separated from the actuator by a compliance that we shall discuss in this chapter.

The Skylab solar telescope was stabilized taking into account the flexibility of the large, thin-walled vehicle (excitation by motions of the astronauts).

The Orbiting Solar Observatory (OSO-8) used a high bandwidth raster-scan of its solar telescope, which excited bending modes of the spacecraft that had to be considered in the control design.

Designers of the attitude stabilization system of the Hubble Space Telescope (1990) took into account the vibration of the solar panels caused by thermal stress deformations each time the spacecraft went into or out of the earth's

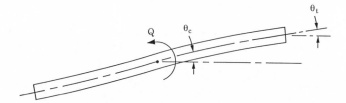

Figure 9.1. Flexible spacecraft modeled as a free-free beam with a torquer at the center.

shadow. However, one of the panels vibrated more than was predicted, seriously reducing the time available for use of the telescope.

9.2 Control Synthesis Using One Vibration Mode

As a relatively simple example, we first consider attitude control of a beamlike spacecraft with a high bandwidth torque actuator (such as a control moment gyro or reaction wheel) at the center (see Fig. 9.1).

Appendix F develops the transfer functions from the central torque $Q_c(s)$ to the slope of the beam at the center $\theta_c(s)$, the slope at the tip $\theta_t(s)$, and the displacement at the tip $u_t(s)$. All of these transfer functions have the same poles on the $j\omega$ axis (an infinite number), which represent the natural frequencies of vibration. The transfer function to $\theta_c(s)$, which is a co-located output, has zeros on the $j\omega$ axis that alternate with the poles. The transfer functions to $\theta_t(s)$ and $u_t(s)$, which are separated (non-co-located) outputs, have zeros on the real axis that occur in positive and negative pairs. The positive real zeros make these "non-minimum-phase" transfer functions and represent an inherent limitation on the speed of response (or bandwidth) using feedback control.

The poles and zeros well outside the control bandwidth (with magnitudes that are greater than about six times the control bandwidth) have little influence on the controlled system. Thus, if the desired bandwidth is less than about one-sixth of the first vibration frequency, the spacecraft is well approximated as a rigid body, which is the approximation we used in the previous chapters.

Let us now consider the case where the desired closed-loop bandwidth is slightly higher, say about one-third of the first vibration frequency. (By this, we mean we desire the real part of the dominant closed-loop eigenvalue to be about one-third of the open-loop natural frequency.) For control synthesis, the spacecraft is then well approximated as a system with a rigid body mode and one vibration mode.

9.2.1 Co-Located Output

From Appendix F, the transfer function from the central torque Q_c to the slope of the beam at the center θ_c (output co-located with the control torque) is

$$\frac{\theta_c(s)}{Q_c(s)} \approx \frac{1 + (s/.228)^2}{s^2(s^2 + 1)}, \tag{9.1}$$

where s is in units of ω_o = the vibration natural frequency,

$$\omega_o = 15.42(EI/\sigma\ell^4)^{1/2},$$

ℓ = the beam half-length, σ = the mass per unit length of the beam, EI = beam stiffness (see Appendix F), Q_c is in units of $EI/J\ell$, and J = moment of inertia of the beam about the center. The zero is on the imaginary axis between the rigid-body poles at $s = 0$ and the vibration poles at $s = \pm j$.

We synthesize a LQR with the performance index

$$J = \int_0^\infty (A\theta_c^2 + Q_c^2)dt. \tag{9.2}$$

Fig. 9.2 shows a locus of the closed-loop poles vs. the weighting parameter A. As A increases, the damping of the modified vibration mode is increased and the modified rigid-body poles move toward the zeros.

From Fig. 9.2, we see that $A = .001$ gives the desired bandwidth, since the dominant response will come from the modified vibration mode. The poles of the modified rigid-body mode are close to the zeros and hence this mode will have smaller residues than the modified vibration mode. For this value of A, the state feedback law is

$$Q = -.020\theta_c - .094\dot{\theta}_c - .012\theta_t - .110\dot{\theta}_t. \tag{9.3}$$

The response to a unit output command is shown in Fig. 9.3 (see Appendix B for a Matlab.m file that calculates response to step output commands). The output response occurs in two stages; the center angle first moves in the right direction (about halfway), then reverses its angular velocity to keep the tip from rotating too far; the input torque is then made negative, and the center angle brings the tip angle into alignment with a small out-of-phase torque supplied for damping.

9.2.2 Separated Output

Now suppose there is a telescope at the tip (output separated from the control torque by compliance), and we wish to control the direction it is pointing in

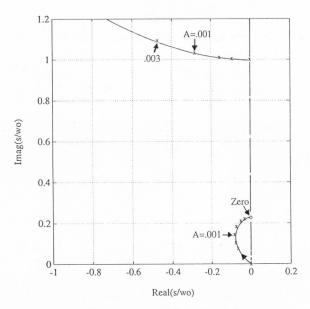

Figure 9.2. Locus of closed-loop poles vs. A for co-located output.

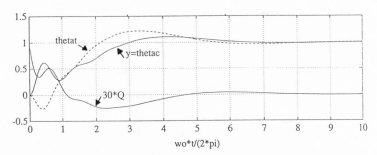

Figure 9.3. Response to output command with co-located output.

space. From Appendix F, the transfer function from the central torque Q_c to the slope of the beam at the tip θ_t is

$$\frac{\theta_t(s)}{Q_c(s)} \approx \frac{1 - (s/.320)^2}{s^2(s^2 + 1)}, \tag{9.4}$$

where the nomenclature was explained below (1).

This system has a right half-plane zero. *Systems with right half-plane zeros are fundamentally deficient with respect to control, in that the bandwidth of the closed-loop system can never exceed the bandwidth of the first right half-plane zero (Ref. KW)*. Physically this is explained by the fact that the tip is controlled by the slope at the center and it takes a finite length of time to produce a finite value of θ_c with a finite torque. Also, in the first vibration mode, the slope at the tip has the opposite sign of the slope at the center (see Fig. F.3); this explains why *the output starts in the wrong direction for an output command* (see Fig. 9.5).

The speed of response (bandwidth) limitation can be seen (Ref. KW) by plotting a locus of closed-loop poles versus the weighting parameter A in the performance index

$$J = \int_0^\infty (A\theta_t^2 + Q^2)dt. \tag{9.5}$$

What does adjusting A correspond to in practice?

Fig. 9.4 shows this root locus; as A is increased, one set of complex poles moves toward the two real zeros. However, the system actually has only one zero there; the other one is a "reflected" zero in this symmetric root locus. Thus one pole is nearly canceled for large A but the other pole is not; since the other poles tend toward infinity, the dominant response of the closed-loop system for large A is produced by the uncancelled pole at the reflected right half-plane zero. This is the best one can do, even with very large control magnitudes.

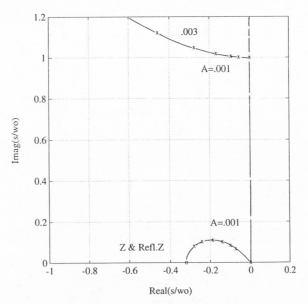

Figure 9.4. Locus of closed-loop poles vs. A for separated output.

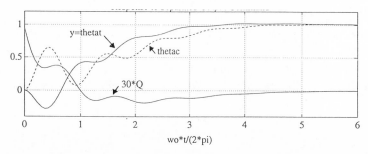

Figure 9.5. Response to output command with separated output.

We chose $A = .001$ so we could compare results with the co-located case with the same control magnitude. The state feedback law (from Matlab) is

$$Q = -.017\theta_c - .125\dot{\theta}_c - .015\theta_t - .183\dot{\theta}_t \qquad (9.6)$$

Fig. 9.5 shows the response of the closed-loop system for a step command in θ_t using this feedback law. Note the response starts in the opposite direction as predicted and settles in a shorter time than in the co-located case. The modified rigid-body mode dominates with its poles near the zero.

9.2.3 Roll-Off Filters

Synthesizing control logic using only one vibration mode means we are consciously neglecting the higher-order vibration modes. When doing this, it is a good idea to insert "roll-off" into the control logic—some all-pole filters near the edge of the bandwidth—so that the loop-transfer gain decreases rapidly ("rolls off") with frequency beyond the control bandwidth. This reduces the possibility of destabilizing the unmodeled higher frequency dynamics ("spillover"), and also reduces the possibility of "aliasing" when using digital control (see Ref. FPW).

A simple way to generate an n-pole filter near the edge of the bandwidth is to replace the control by its nth time derivative in synthesizing a linear-quadratic regulator. LQR synthesis will yield feedbacks on the state, the control, and the first $n - 1$ time derivatives of the control; by choosing the LQR output weight, the bandwidth of these extra poles is easily selected (in a symmetric root locus, the poles will radiate outward from $s = 0$ as the weight is increased). By weighting the nth derivative of the control, we make the control itself very smooth, that is, it contains very little high frequency content.

Example

Consider the example above of synthesizing control logic for a separated output. Let us insert a second-order roll-off filter and still try to achieve step response comparable to that shown in Fig. 9.5. We augment the plant model with two additional state variables, the control Q and its first derivative $\dot{Q}$:

$$\dot{Q} = Q_1, \tag{9.7}$$

$$\dot{Q}_1 = u, \tag{9.8}$$

where $u \equiv \ddot{Q}$ is now the new control. The performance index is

$$J = \int_0^\infty (A\theta_t^2 + u^2)dt. \tag{9.9}$$

Fig. 9.6 shows a locus of closed-loop roots vs. A. For $A = 0$ there are eight poles at $s = 0$ in this symmetric root locus. As A is increased, they radiate outward, four of them going to the zeros and reflected zeros on the real axis. Recall that one pole is nearly canceled for large A but the other pole is not. Since the other poles tend toward infinity, the dominant response of the closed-loop system for large A is produced by the uncancelled pole at the reflected

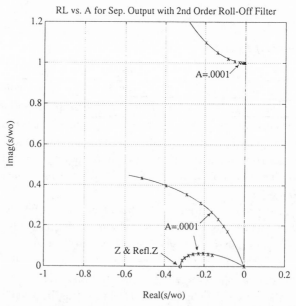

Figure 9.6. Locus of closed-loop poles vs. A for separated output with a second-order roll-off filter.

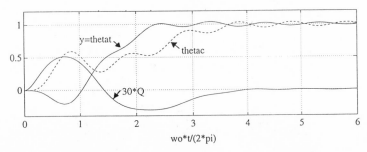

Figure 9.7. Response to output command with separated output and a second-order roll-off filter.

right half-plane zero. This is the best one can do, even with very large control magnitudes.

The modified vibration mode is not as well damped as it was without the roll-off filter. Of course, this means that the next (unmodeled) vibration mode will be affected even less, thus reducing the possibility of spillover, and this was the reason for inserting the roll-off filter.

We chose $A = .0001$, for which the real parts of the poles are comparable to those corresponding to Fig. 9.5. This gives the feedback law

$$\ddot{Q} = -1.048\dot{Q} - .549Q - .0377\dot{\theta}_c - .000348\theta_c - .0681\dot{\theta}_t - .00965\theta_t, \quad (9.10)$$

or, taking the Laplace transform,

$$100Q(s) = -\frac{6.86\dot{\theta}_c(s) + .063\theta_c(s) + 12.40\dot{\theta}_t(s) + 1.758\theta_t}{1 + 2(.707)(s/.741) + (s/.741)^2}, \quad (9.11)$$

which shows explicitly the decrease in gain with frequency.

Fig. 9.7 shows the response to a step output command. The control history $Q(t)$ is very smooth, and the speed of response is comparable to Fig. 9.5. However, there is a small amplitude residual oscillation that takes a long time to settle, associated with the small damping of the vibration mode; this is the price one must pay for the roll-off.

Of course, torquer actuator dynamics (neglected here) introduce some natural roll-off, so we could probably have accomplished the same response with just a first-order roll-off filter.

Problem 9.2.1 – *Two-Finite-Element Approximation*

A two-finite-element approximation of a flexible beamlike spacecraft is two identical rigid bodies connected to each other by a torsion spring at a pinned joint (see Fig. 9.8).

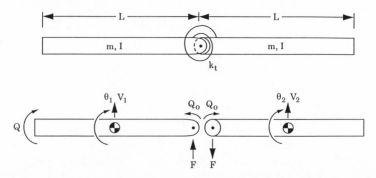

Figure 9.8. Two-finite-element approximation of a free-free beam with a torquer at one end; $Q_o = k_t(\theta_1 - \theta_2)$.

Suppose there is a high bandwidth torque actuator (such as a control moment gyro or reaction wheel) at one end of the spacecraft and a displacement gyro at either the same or the other end, and we wish to control the attitude and the vibration mode of this system. Free-body diagrams of the two bodies are shown in Fig. 9.8.

(a) Show that the equations of motion are

$$
\begin{aligned}
I\ddot{\theta}_1 &= -k_t(\theta_1 - \theta_2) - L \cdot F/2 + Q, \\
I\ddot{\theta}_2 &= -k_t(\theta_2 - \theta_1) - L \cdot F/2, \\
m\dot{V}_1 &= F, \\
m\dot{V}_2 &= -F, \\
\dot{V}_1 &- L \cdot \ddot{\theta}_1/2 = \dot{V}_2 + L \cdot \ddot{\theta}_2/2,
\end{aligned}
$$

where

$$
\begin{aligned}
I &= \text{moment of inertia of each body about its c.m.,} \\
\theta_i &= \text{angular deflection of body } (i = 1, 2), \\
k_t &= \text{torsion spring constant,} \\
F &= \text{shear force between two bodies,} \\
L &= \text{length of each body,} \\
m &= \text{mass of each body,} \\
Q &= \text{torque applied to body 1,} \\
V_i &= \text{transverse velocity of c.m. of body } i.
\end{aligned}
$$

(b) These are five linear equations for $(\ddot{\theta}_1, \ddot{\theta}_2, \dot{V}_1, \dot{V}_2, F)$. Eliminating $\dot{V}_1, \dot{V}_2$, and F, and using $I = mL^2/12$ (moment of inertia of a rod), show that

$$
\begin{bmatrix} \ddot{\theta}_1 \\ \ddot{\theta}_2 \end{bmatrix} = \begin{bmatrix} -.5 & .5 \\ .5 & -.5 \end{bmatrix} \begin{bmatrix} \theta_1 \\ \theta_2 \end{bmatrix} + \begin{bmatrix} 5 \\ -3 \end{bmatrix} Q,
$$

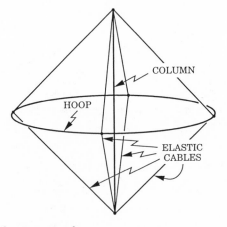

Figure 9.9. Hoop-column spacecraft.

and
$$F = 9.6Q,$$

where time is in units of $1/w_o$, $w_o^2 = 24k_t/mL^2$, Q is in units of $16k_t$, F in k_t/L..

(c) Show that the transfer functions from Q to θ_1 and θ_2 are

$$\begin{bmatrix} \theta_1(s) \\ \theta_2(s) \end{bmatrix} = \frac{1}{s^2(s^2 + 1)} \begin{bmatrix} 5s^2 + 1 \\ -3s^2 + 1 \end{bmatrix} Q(s).$$

Note the positions of the zeros relative to the pole are not very accurate compared to the continuum solution (infinite number of infinitesimal elements) given in Appendix F.

Problem 9.2.2 – *Control of a Hoop-Column Spacecraft*

Fig. 9.9 shows a model of a spacecraft consisting of a circular hoop (which might be the outer perimeter of an antenna) connected to a center column by pretensioned elastic cables. Reaction wheel torquers are placed on the hoop since it is heavier than the column.

The pitch, roll, and yaw motions are uncoupled, and the pitch and roll motions are identical (from symmetry). The pitch equations of motion, including a reaction wheel torquer, are

$$\begin{aligned} I_h \ddot{\theta}_h &= -k(\theta_h - \theta_c) - Ni - c(\dot{\theta}_h - q_w), \\ I_c \ddot{\theta}_c &= -k(\theta_c - \theta_h), \\ I_w \dot{q}_w &= Ni - c(q_w - \dot{\theta}_h), \end{aligned}$$

where

$$iR = e - N(q_w - \dot\theta_h).$$

Total angular momentum H is conserved, where

$$H = I_h\dot\theta_h + I_c\dot\theta_c + I_w q_w.$$

Hence q_w can be replaced by H as the fifth state variable and

$$\dot H = 0.$$

Thus, if $H(0) = 0$, it is always zero and the system can be reduced to fourth order.

(a) Assuming $H = 0$, show that the equations of motion may be written as

$$\ddot\theta_h = -\frac{\epsilon}{1+\epsilon}(\theta_h - \theta_c) - \sigma(1 + \epsilon_w)\dot\theta_h - \sigma\epsilon\dot\theta_c - e,$$

$$\ddot\theta_c = -\frac{1}{1+\epsilon}(\theta_c - \theta_h),$$

where time is in units of $1/\omega_o$, and

$$\omega_o = \sqrt{k(1/I_h + 1/I_c)}$$

is the undamped natural frequency of oscillations of the hoop-column S/C. Also

$$\sigma = (N^2/R + c)/I_w\omega_o,$$
$$\epsilon = I_c/I_h,$$
$$\epsilon_w = I_w/I_h,$$

and e is in units of $RI_h\omega_o^2/N$.

(b) If the diameter of the hoop equals the length of the column and they have the same mass per unit length, then

$$\epsilon = 1/3\pi.$$

Using this value and taking $\sigma = 1, \epsilon_w = .0001$, plot the locus of closed-loop poles versus A using LQR where

$$J = \int_0^\infty (A\theta_h^2 + e^2)dt.$$

(c) For the parameters in (b) and $A = 1$ show that the LQR gains are

$$k = -[1.707, .025, 1.101, .012$$

and plot the response to a unit output command.

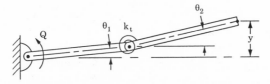

Figure 9.10. Two-finite-element approximation of a flexible robot arm.

Problem 9.2.3 – *Control of a Flexible Robot Arm*

The example of this section is a one-vibration-mode approximation of a *free-free beam*. A flexible robot arm can be modeled as half of a free-free beam, that is, a *pinned-free beam* (see Fig. 9.10).

From Appendix F, the transfer function from torque at the shoulder to tip deflection is

$$\frac{u_t(s)}{Q_s(s)} \approx \frac{1 - (s/.726)^2}{s^2(s^2 + 1)},$$

where u_t = tip deflection in units of ℓ = length of the arm, s is in units of ω_o, Q_s is in units of $EI/J\ell$, J = moment of intertia of the arm about the shoulder, and

$$\omega_o = 15.42(EI/\sigma\ell^4)^{1/2}.$$

(a) Plot a locus of closed-loop poles vs. A for the performance index

$$J = \int_0^\infty (Au_t^2 + Q_s^2)dt.$$

(b) Choose A so that the dominant eigenvalue has a real part that is about $-.5\omega_o$. Calculate and plot the response $u_t(t)$ of the flexible robot arm to a unit output command. Also plot Q_s.

9.3 Parasitic Modes

In the example of the previous section, we assumed that the torquer was located at the center of a uniform beam. In such a model only the antisymmetric beam vibration modes are excited by the torquer. In practice, the spacecraft will not be a uniform beam and the torquer will not be located precisely at the center. This means that the symmetric vibration modes will also be excited but with small amplitudes. We shall call these "parasitic" vibration modes.

The *Orbiting Solar Observatory-8* (OSO-8), launched in 1975 (Ref. YO), is a case in point. It had a de-spun section pointed at the sun, which performed a raster-scan of a body-fixed telescope across the face of the sun (see Fig. 9.11).

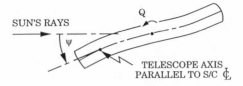

SUN'S RAYS

ψ

Q

TELESCOPE AXIS
PARALLEL TO S/C ₵

Figure 9.11. De-spun section of Orbiting-Solar Observatory 8 (OSO-8).

The telescope was physically displaced from the azimuth torque motor (which was the de-spin motor). The spinning part of the spacecraft acted as a large biased momentum wheel for the de-spun part.

The first symmetric and antisymmetric vibration modes of the de-spun section were within six times the bandwidth needed for the fast raster scan of the telescope, at 93.2 rad/sec and 255 rad/sec respectively. If the torque motor had been placed precisely at the center of mass of the de-spun section, it would not have excited the symmetric mode, that is, there would have been a perfect pole-zero cancellation of the symmetric mode pole for the transfer function from de-spin motor torque to telescope angle. However, this was not the case, so instead there were zeros that were quite close (at $\pm 87.7j$ rad/sec) to the symmetric mode poles.

The actuator bandwidth was about 1500 rad/sec, and the sensor bandwidth, about 600 rad/sec, so that their dynamics were negligible in the design of the control system.

An approximate model of OSO-8 can be obtained from the uniform beam model of Appendix F with the torquer displaced from the center. The torquer can then excite the symmetric as well as the antisymmetric bending modes of the beam. The first antisymmetric mode is the mode of primary concern in designing the controller; the symmetric mode enters as a parasitic oscillation of small amplitude and small damping superimposed on the main response.

If the torquer is placed ahead of the center of the uniform beam by 1.2 percent of the overall length, we match the relative magnitudes of the pole-zero pair corresponding to the symmetric mode given in Ref. YO. Also the ratio of the first symmetric mode frequency to the first antisymmetric mode frequency from the uniform beam model agrees with the data from Ref. YO within 1 percent. This gives the following normalized transfer function from azimuth torque $Q(s)$ to azimuth angle $\psi(s)$ at the telescope (forward tip of the beam):

$$\frac{\psi(s)}{Q(s)} = \frac{[1 + (s/.344)^2][1 - (s/.336)^2]\cdots}{s^2[1 + (s/.365)^2][1 + s^2]\cdots}, \tag{9.12}$$

where time is in units of $1/\omega_o, \omega_o = 255$ rad/sec, Q is in units of $J\omega_o^2$, and $J =$ the moment of inertia of the de-spun section.

The dominant poles of this transfer function are at $s = 0$ and $s = \pm 1.000j$, which represent the rigid-body and first antisymmetric vibration modes. The dominant zeros are at $s = \pm.336$, reflecting the fact that the system is clearly non-minimum-phase (a sudden torque at the center causes the tip angle to start in the opposite direction of the torque). *The right half-plane zero limits the possible bandwidth of the controlled system.*

The first symmetric vibration mode is at $s = \pm.365j$, with nearly cancelling zeros at $s = \pm.336j$; this is a parasitic mode that would not have been there if the torquer had been at the exact center of the beam.

LQ Regulator for the OSO-8

Fig. 9.12 shows a locus of closed-loop roots versus A using the performance index

$$J = \int_0^\infty (A\psi^2 + Q^2)dt, \tag{9.13}$$

where Q is the almost-central torque and ψ is the slope of the uniform beam at the forward tip.

The locus is very similar to Fig. 9.4 except for the parasitic first symmetric mode. The effect of this mode on a step-output command is to introduce a small amplitude oscillation at the first symmetric mode frequency with very small damping, so that the system "rings" for a long time after the command.

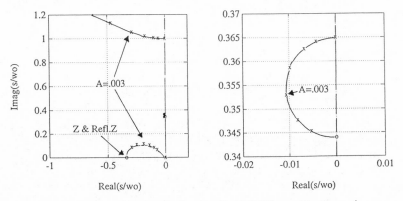

Figure 9.12. OSO-8 azimuth attitude control with LQR—root locus vs. A.

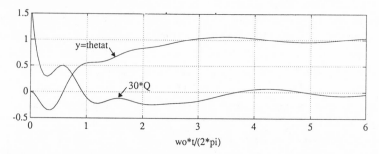

Figure 9.13. OSO-8 with LQR—response to telescope angle command.

This is shown in Fig. 9.13, which is a simulation of a step command in ψ using the LQR corresponding to $A = .003$.

Problem 9.3.1 – *LQ Regulator for OSO-8*

(a) Make a partial fraction expansion of the transfer function in (12).

(b) Using the result in (a), show that a modal-state representation of the transfer function is

$$\begin{bmatrix} \dot{x} \\ \psi \end{bmatrix} = \begin{bmatrix} F & G \\ H & L \end{bmatrix} \begin{bmatrix} x \\ Q \end{bmatrix},$$

where F is block-diagonal with 2 by 2 blocks

$$F_1 = \begin{bmatrix} 0 & 1 \\ 0 & 0 \end{bmatrix}, \quad F_2 = \begin{bmatrix} 0 & 1 \\ -1 & 0 \end{bmatrix}, \quad F_3 = \begin{bmatrix} 0 & 1 \\ -.1336 & 0 \end{bmatrix},$$

and

$$G = [\, 0 \ 1 \ 0 \ 1 \ 0 \ 1\,]^T,$$

$$H = [\, 1 \ 0 \ -11.289 \ 0 \ .3165 \ 0\,],$$

$$L = 0.$$

(c) Verify the locus of closed-loop roots versus the parameter A given in Fig. 12.

(d) From Fig. 12, choosing $A = .003$ gives real parts of the closed-loop eigenvalues about 20 percent of the main vibration frequency. Determine the modal-state feedback gains for $A = .003$ and verify Fig. 9.13, the ψ response for a unit step-output command.

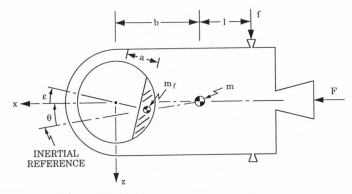

Figure 9.14. Spacecraft with spherical fuel tank during a thrust maneuver.

9.4 Control with Fuel Slosh

During thrust maneuvers the sloshing of fuel in partially filled tanks can interact with the controlled system in such a way as to cause the overall system to be unstable. A fuel slosh instability in the booster caused the failure of the first U. S. attempt to place a spacecraft into orbit in 1958.

This type of instability is possible if the frequency of the first sloshing mode is less than about six times the attitude control bandwidth. Hence this mode (and possibly one or two more) should be included in the dynamic model of the spacecraft used for control synthesis.

To give some insight into this "servo-slosh" problem, we shall consider a spacecraft with a spherical fuel tank and include the lowest frequency sloshing mode in the dynamic model (see Fig. 9.14).

The first sloshing mode in a spherical tank is well approximated by considering the fuel to be "frozen" into a lump that slides on the sides of the tank. The fuel mass acts like a pendulum with its support point at the center of the tank. We shall consider reaction jet control as we did for the rigid spacecraft approximation in Section 7.1.

A free-body diagram of the fuel mass, treated as a pendulum bob suspended by a weightless rod from the center of the tank, is shown in Fig. 9.15.

A free-body diagram of the spacecraft is shown in Fig. 9.16.

In the figures, (a_x, a_z) are components of acceleration of the center of the fuel tank, (ω, ω_f) are angular velocities of (spacecraft, fuel lump), (m, m_f) are masses of (spacecraft, fuel lump), and (I, I_f) are moments of inertia of (spacecraft, fuel lump) about their respective mass centers.

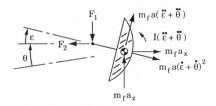

Figure 9.15. Free-body diagram of fuel mass.

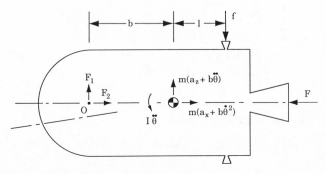

Figure 9.16. Free-body diagram of spacecraft.

The equations of motion of the spacecraft are

$$\sum F_x = 0 = -F_2 - m(a_x + b\omega^2) + F,\qquad(9.14)$$

$$\sum F_z = 0 = -F_1 - m(a_z + b\dot{\omega}) + f,\qquad(9.15)$$

$$\sum M_o = 0 = -I\dot{\omega} - mb(a_z + b\dot{\omega}) + (b+\ell)f.\qquad(9.16)$$

The equations of motion for the fuel lump are

$$\sum F_x = 0 = F_2 - m_f a_x - m_f a\omega_f^2 \cos\epsilon - m_f a\dot{\omega}_f \sin\epsilon,\qquad(9.17)$$

$$\sum F_z = 0 = F_1 - m_f a_z + m_f a\omega_f^2 \sin\epsilon - m_f a\dot{\omega}_f \cos\epsilon,\qquad(9.18)$$

$$\sum M_o = 0 = -I_f\dot{\omega}_f - m_f a^2\dot{\omega}_f - m_f aa_x \sin\epsilon - m_f aa_z \cos\epsilon.\ (9.19)$$

The kinematic equations are

$$\dot{\epsilon} = \omega_f - w,\qquad(9.20)$$

$$\dot{\theta} = \omega.\qquad(9.21)$$

Equations (14)–(19) are six linear equations for the six unknowns F_1, F_2, a_x, a_z, $\dot{\omega}$, and $\dot{\omega}_f$.

Interaction through F_1, F_2

Eliminating F_1, F_2, a_x, and a_z, the equations of motion become (20), (21), and

$$
\begin{bmatrix} I^* & -I_c^* \\ -I_c^* & I_f^* \end{bmatrix} \begin{bmatrix} \dot{\omega} \\ \dot{\omega}_f \end{bmatrix} = \begin{bmatrix} (\ell + b^*)f - m^* ab\omega_f^2 \sin \epsilon \\ -a^* f \cos \epsilon - a^* F \sin \epsilon - m^* ab\omega^2 \sin \epsilon \end{bmatrix},
$$

(9.22)

where

$$
I^* \triangleq I + m^* b^2, \quad m^* \triangleq \frac{mm_f}{m + m_f},
$$

$$
I_f^* \triangleq I_f + m^* a^2, \quad a^* = \frac{m_f}{m + m_f},
$$

$$
I_c^* \triangleq m^* ab \cos \epsilon, \quad b^* = \frac{m_f}{m + m_f}.
$$

For $|\epsilon| \ll 1$, and small deviations from equilibrium, the Laplace transform of the linearized equations of motion is

$$
\begin{bmatrix} I^* s^2 & -I_c^* s^2 & 0 \\ -I_c^* s & I_f^* s & a^* F \\ 1 & -1 & s \end{bmatrix} \begin{bmatrix} \theta(s) \\ \omega_f(s) \\ \epsilon(s) \end{bmatrix} \cong \begin{bmatrix} \ell + b^* \\ -a^* \\ 0 \end{bmatrix} f.
$$

(9.23)

The transfer function from $f(s)$ to $\theta(s)$ may be deduced from these relations:

$$
\frac{\theta(s)}{f(s)} = \frac{A}{s^2} \frac{s^2 + z^2}{s^2 + p^2},
$$

(9.24)

where

$$
A = \frac{(\ell + b^*)I_f^* - a^* I_c^*}{I^* I_f^* - (I_c^*)^2},
$$

$$
z^2 = \frac{a^* F}{I_f^* - a^* I_c^*/(\ell + b^*)},
$$

$$
p^2 = a^* F I_f^* - (I_c^2)/I^*.
$$

The zeros of this transfer function are *above* the poles if

$$
r^2 > b\ell,
$$

(9.25)

where $r^2 = I/m$, which means this would be an unstably interacting mode (see Section 9.5.1).

9.5 Robust Compensator Synthesis

In most of this book we emphasize the best possible response with full state feedback, implying that LQG synthesis using feedback of estimated states will produce almost the same good response as LQR. This is largely true for rigid spacecraft or aircraft, that is, systems with control system bandwidths that are well below the frequency of the first vibration or fuel slosh mode. However, it may not be true for systems with higher control system bandwidths, even when one or more vibration/slosh modes are included in the control design model. The reason for this is that LQG synthesis produces tuned notch compensators, which may not be robust to variations in the vibration frequencies of the spacecraft.

Section 9.2.3 treated the case of modifying LQR synthesis to provide robustness to unmodeled higher frequency dynamics, that is, to avoid spillover. This section treats methods for synthesizing low-order compensators and for modifying LQG compensators to provide robustness to plant parameter changes. However, we first define an important characteristic of flexible systems with non-co-located sensors and actuators, namely unstably interacting modes.

9.5.1 Unstably Interacting Modes

If a rate sensor is co-located with an actuator on a flexible body, and its signal is fed back to the actuator, all vibration modes are stabilized (see a simple proof in the next subsection). If a rate sensor is not co-located with an actuator on a flexible body, and its signal is fed back to the actuator, some vibration modes are stabilized and others are destabilized, depending on the location of the sensor relative to the actuator. The damped modes are called "stably interacting" while the de-stabilized modes are called "unstably interacting."

For the flexible beamlike spacecraft of Section 9.2, the first vibration mode is stably interacting when the rate sensor is co-located with the torquer at the center of the beam, and unstably interacting when the rate sensor is separated from the torquer at the beam tip. The unstable interaction can be understood directly from the mode shape; the slope of the beam at the tip always has the opposite sign of the slope at the actuator so the rate feedback has the wrong sign for the usually stabilizing negative feedback, that is, it produces positive feedback. This is an example of *spillover*, and corresponds exactly to the Lark missile test mentioned in Section 9.1.

Another (equivalent) way of seeing the unstable interaction is to put the transfer function from actuator to rate sensor signal into pole-residue form. Neglecting damping (which is always present in real structures but is usually quite small), the transfer function from an actuator torque Q to a sensed angular rate $\dot{\theta}$ can always be put into the form

$$\frac{\dot{\theta}(s)}{Q(s)} = \frac{1}{Is} + \sum_{i=1}^{\infty} \frac{R_i s}{s^2 + p_i^2}, \tag{9.26}$$

where R_i is the residue of the ith mode. Note that

$$\frac{s}{s^2 + p^2} \equiv \frac{1/2}{s + jp} + \frac{1/2}{s - jp}.$$

Those modes with positive residues are stably interacting modes and those with negative residues are unstably interacting modes. The reason for unstable interaction is quite simple, namely positive instead of negative feedback; the rate sensor is reading the negative of the rate occurring at the actuator.

If all of the vibration modes are stably interacting, it can be shown that the *zeros of the transfer function alternate with the poles* (Ref. MAR). Conversely, if there are any unstably interacting vibration modes, then the poles do *not* alternate with the zeros.

Example
The co-located transfer function of (1) may be written as

$$\frac{\dot{\theta}_c(s)}{Q_c(s)} \approx \frac{1}{s} + \frac{18.24s}{s^2 + 1}, \tag{9.27}$$

which has positive residues at the vibation poles $s = \pm j$, indicating stable interaction.

The separated transfer function of (4) may be written as

$$\frac{\dot{\theta}_c(s)}{Q_c(s)} \approx \frac{1}{s} - \frac{10.77s}{s^2 + 1}, \tag{9.28}$$

which has negative residues at the vibation poles $s = \pm j$, indicating unstable interaction.

Co-located Rate Sensor

The rate at which energy is fed into a system by a torquer is

$$\dot{E} = Q \cdot \Omega, \tag{9.29}$$

where Q is the torque and Ω is the angular velocity of the body at the point where the torquer is located. If Ω is measured by a angular rate sensor (such as a rate gyro) and feedback with a negative gain

$$Q = -k \cdot \Omega, \qquad (9.30)$$

then the energy rate is

$$\dot{E} = -k \cdot \Omega^2, \qquad (9.31)$$

which is negative, proving that the energy of the system decreases with time. Thus none of the modes can be unstably interacting, and the transfer function from Q to Ω must have alternating poles and zeros (Ref. MAR).

The same argument obviously applies for feedback of linear velocity to a co-located force actuator. This fact was recognized by Wykes about 1965 in connection with improving the cockpit ride quality in the B-70 [Ref. WY]; he placed integrating accelerometers, which give vertical velocity, near the hinge line of the elevators, which produce vertical aerodynamic force; this damps the fuselage bending modes. He called this ILAF for "Identically Located Accelerometer and Forcer." Additional conventional sensors were used to stabilize the "rigid-body" motions of the aircraft. Aubrun rediscovered this combination about 1978 in connection with attitude control of flexible spacecraft [Ref. AU] and gave it the name "Low-Authority Control/High-Authority Control" or "LAC/HAC."

9.5.2 *Low-Authority Control/High-Authority Control*

Fig. 17 shows the concept of low-authority control/high-authority control (LAC/HAC) in the s-plane. LAC uses a co-located rate sensor to add damping to all the vibratory modes (but not to the rigid-body mode). HAC uses a separated dispacement sensor to stabilize the rigid body mode, which slightly decreases the damping of the vibratory modes but not enough to produce instability (called "spillover").

Example

Consider the flexible spacecraft of Section 9.2 with a central torquer, a co-located angular rate sensor, and an angle sensor at the tip (a separated sensor). We first close the loop from the rate sensor to the torquer (low-authority control):

$$Q_c(s) = -k_c \dot{\theta}_c. \qquad (9.32)$$

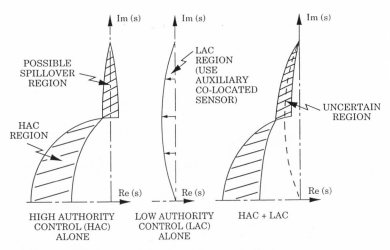

Figure 9.17. The high-authority/low-authority control concept.

Fig. 9.18 shows a root locus vs. the gain k_c. We chose $k_c = .030$, which significantly stabilizes the vibration mode and only slightly stabilizes one of the poles of the rigid-body mode. With this loop closed, the transfer function from incremental torque ΔQ_c to θ_t is

$$\frac{\theta_t(s)}{\Delta Q_c(s)} = \frac{1 - (s/.320)^2}{s(s + .0305)[(s + .2733)^2 + (.9532)^2]}. \tag{9.33}$$

Next we close the second loop from the tip angle sensor to the torquer with the usual lead compensation (the high-authority control) in order to stabilize the rigid-body mode:

$$\Delta Q_c(s) = -k_t \frac{s + .12}{s + .320} \theta_t(s). \tag{9.34}$$

We have placed the lead zero at $s = -.12$ and the pole at $s = -.320$ to cancel the plant zero at that point. A root locus vs. k_t is shown on the right of Fig. 9.18 with the first loop already closed. The rigid-body mode is stabilized and the damping of the vibration mode is decreased somewhat.

We chose $k_t = .05$, since this is where the real parts of the poles corresponding to the two modes are equal. Fig. 9.19 shows the response of this control system to a command for changing the angular position of the flexible body using this HAC/LAC control law:

$$Q_c(s) = -.03 \cdot \dot{\theta}_c - .05 \frac{s + .12}{s + .320} \theta_t. \tag{9.35}$$

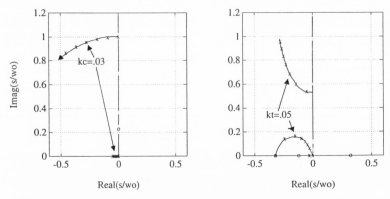

Figure 9.18. Example of a co-located rate sensor (low-authority control) combined with separated angle sensor (high-authority control).

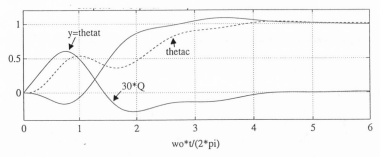

Figure 9.19. Response to output command using a high-authority/low-authority control compensator.

what does this mean

The response is quite comparable to the response with full state feedback in Fig. 9.5. This is not surprising, since we are essentially feeding back all the states except $\dot{\theta}_c$, which had a small gain in the LQR (Eqn. (6)); the lead network approximates $\dot{\theta}_t$. The advantage is that we are using only two instead of four sensors, and not depending on the model to estimate states. This latter dependence is what causes lack of robustness to plant parameter variations.

9.5.3 *LQG Compensator Using a Separated Sensor*

LAC/HAC is usually insensitive to small deviations of the plant dynamics away from the design values, that is, it is "robust" to plant parameter changes. However, it requires two sensors, which is expensive and adds to reliability problems.

It is straightforward to design a compensator using only the separated angle sensor. However, there is a trade-off between performance and robustness; a high performance compensator will be nonrobust to plant parameter changes, and vice-versa. We illustrate this here by again using the flexible spacecraft example of Section 9.2.

Standard LQG Compensator

An LQG compensator for a system with a lightly damped vibration mode and a rigid body mode combines "notch" and lead compensation. The term "notch" comes from the frequency response curve where there are notches near the vibration frequencies indicating that sinusoidal inputs at these frequencies are not passed through the compensator. The lead compensation stabilizes the rigid-body mode.

Fig. 9.20 shows a root locus vs. overall gain (gain=1 corresponds to the LQG compensator) using an LQG compensator for the beamlike spacecraft example of Section 9.2 with a tip angle sensor and a central torquer (see Appendix C for LQG synthesis). The compensator consists of an estimator and feedback of the estimated state as though it were the measured state ("certainty equivalence"). The estimator was designed assuming additive white noise in the control and in the tip angle sensor with spectral density ratio .05. The resulting compensator is

$$\frac{Q_c(s)}{\theta_t(s)} = \frac{.0447(s + .059)[(s - .117)^2 + (1.634)^2]}{(s + 1.086)(s + .322)[(s + .728)^2 + (1.474)^2]}. \tag{9.36}$$

As mentioned previously, this compensator may be interpreted as consisting of two parts:

(1) A lead compensator with poles at $s = -1.086$ and $-.322$, and a zero at $s = -.059$.
(2) A notch compensator with poles at $s = -.728 \pm 1.474j$ and zeros at $s = .117 \pm 1.634j$.

If the plant poles and zeros are exactly where we thought they were, the initial estimate error is zero, and there is negligible noise, then *the response of the closed-loop system to a step output command is exactly the same as if we used state feedback* (cf. Fig. 9.5), since the estimate error dynamics are then not excited.

However, if the plant poles and zeros are not where we thought they were, then the closed loop poles are different from our predictions, since the estimator

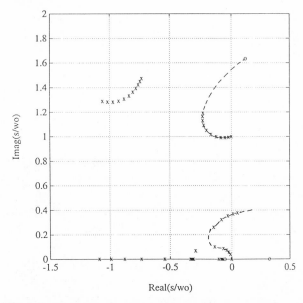

Figure 9.20. LQG compensator for separated sensor/actuator—root locus vs. compensator gain.

uses the plant model for estimating the states from the tip angle measurement. Fig. 9.21 shows a locus of closed-loop poles vs. ϵ where the poles and zeros of the plant are changed by $(1 + \epsilon)$.

The *stability robustness to ϵ is quite good*, in that it takes a 50% increase in the natural frequency to produce instability. However, the *performance degrades rapidly with ϵ*; a 10% decrease causes the real parts of the modified rigid-body poles to decrease by a factor of two, so that the rise time for a step command would nearly double.

Thus, if the vibration frequency is not well known, or if it changes during the operation of the system (for example, use of fuel changes the mass and mass distribution of the spacecraft, which changes the frequency of the bending modes), the performance of the closed-loop system will be substantially poorer than predicted using the nominal plant.

Modifying LQG Compensators for Robustness

It turns out that relatively small modifications of the LQG compensator can make it significantly less sensitive to changes in the vibration frequency. The

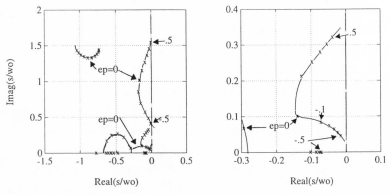

Figure 9.21. LQG compensator—root locus vs. ϵ where plant natural frequency is changed by $(1 + \epsilon)$.

price for this is slightly poorer response at the nominal values of the plant parameters.

Robust compensator design is a subject for another book. However, one method developed recently (Refs. ECB and MB) is to modify the compensator successively to improve its robustness to plant parameter changes Δp. To do this, the references consider plant parameter changes having a quadratic norm σ_1, that is,

$$(\Delta p)^T \Sigma^{-2} \Delta p = \sigma_1^2,$$

where Σ = the standard deviation matrix for parameter changes. σ_1 is chosen to represent the largest expected change (or uncertainty) in the plant parameters. The successive improvement algorithm goes like this (Refs. ECB and MB):

(1) Find the worst set of parameter changes Δp for the given σ_1, in the sense that they cause the largest increase in the quadratic performance index J. This can be done using the σ-analysis algorithm of Ref. ECB.

(2) Redesign the LQG compensator to minimize $W_o J_o + W_1 J_1$, where J_o = performance index with nominal parameters, J_1 = performance index with worst-case parameters having norm σ_1, and W_o, W_1 are weighting factors such that $W_o + W_1 = 1$. This can be done using the algorithm SANDY of Ref. LY.

(3) With the redesigned compensator, repeat (1) and (2) until the process converges, that is, the new compensator is the same as the old compensator. The performance with worst-case parameters having norm σ_1

should be substantially improved, while the performance with nominal parameters should be only slightly degraded.

(4) For better performance robustness over the whole range of parameter changes, repeat the process with one or more intermediate norms on parameter changes (Ref. MB).

Problem 9.5.1 – *HAC/LAC Compensator for Two-Finite-Element S/C Model*

Design a LAC/HAC compensator using successive loop-closure for the two-finite-element S/C model of Problem 9.2.1, using a co-located angular rate sensor and a separated angle sensor.

Problem 9.5.2 – *LQG Compensator for Two-Finite-Element S/C Model*

Design an LQG compensator for the two-finite-element S/C model of Problem 9.2.1, using only a separated angle sensor.

Problem 9.5.3 – *LQG Compensator for OSO-8*

(a) Using a modal model based on (12), plot the estimate-error root locus vs. R_w/R_v where the measurement is z with

$$z = \psi + v,$$

the process noise w is additive to torque Q, and (v, w) are independent white noise processes with spectral densities (R_w, R_v) respectively.

(b) Choose values for A in Problem 9.3.1 and R_w/R_v in (a) and determine the corresponding LQ regulator and estimator gains.

(c) Determine the transfer function of the LQG compensator created by feeding back estimated state with the gains of (b). Where are the compensator poles and zeros?

Problem 9.5.4 – *LQG Compensation with Roll-Off for OSO-8*

Same as Problem 3 but add a second order roll-off filter.

Problem 9.5.5 – *LQG Compensation for Fuel Slosh*

Synthesize an LQG compensator for the system of Eqn. (24) with $z = 1.1p$, using a noisy measurement of pitch angle θ, assuming noisy implementation of the control.

(a) Using the performance index

$$J = \int_0^\infty (\theta^2 + u^2)dt,$$

where $u = Af$, determine the closed-loop eigensystem and corresponding regulator gains.

Hint: A modal state variable form, with time in units of $1/p$, is

$$
\begin{bmatrix} \dot{x}_1 \\ \dot{x}_2 \\ \dot{x}_3 \\ \dot{x}_4 \\ \theta \end{bmatrix} = \begin{bmatrix} 0 & 1 & 0 & 0 & 0 \\ 0 & 0 & 0 & 0 & z^2 \\ 0 & 0 & 0 & 1 & 0 \\ 0 & 0 & -1 & 0 & 1 - z^2 \\ 1 & 0 & 1 & 0 & 0 \end{bmatrix} \begin{bmatrix} x_1 \\ x_2 \\ x_3 \\ x_4 \\ u \end{bmatrix}.
$$

(b) Assume control is implemented with additive white noise, and that there is additive white noise in the measurement. Synthesize an estimator using a noisy measurement of θ, taking $Q/R = 10$, where (Q, R) are spectral densities of the additive white noise in the control implementation and in the measurement respectively.

(c) Plot a root locus against overall gain of the compensator K, such that, when $K = 1$, the roots are the estimator and regulator poles.

10

Natural Motions of Rigid Aircraft

10.1 Equations of Motion

Many aircraft are well approximated as rigid bodies with gravitational, aerodynamic, and propulsive forces acting on them. Their motions may then be described by giving the position and velocity of the center of mass and the orientation and angular velocity of a set of body-fixed axes with respect to a set of reference axes. Usually it is convenient to choose reference axes fixed with respect to the earth. Unless stated otherwise, we shall assume that aircraft velocity is small compared to orbital velocity, so the earth-fixed reference axes may be approximated as inertial axes.

The dynamic equations of motion are

$$\overset{I}{\vec{V}} \equiv \overset{B}{\vec{V}}_B + \vec{\omega}_B \times \vec{V}_B = \vec{g} + \vec{F}_{\text{aero}}/m + \vec{F}_{\text{prop}}/m, \tag{10.1}$$

$$\overset{I}{\vec{H}} \equiv \overset{B}{\vec{H}}_B + \vec{\omega}_B \times \vec{H}_B = \vec{Q}_{\text{aero}} + \vec{Q}_{\text{prop}}, \tag{10.2}$$

where

$$\vec{V}_B = [u, v, w] = \text{body-axis components of the velocity of the center of mass with respect to the reference axes,}$$

$$\vec{\omega}_B = [p, q, r] = \text{body-axis components of the angular velocity of the aircraft with respect to the reference axes,}$$

$$\vec{H}_B \overset{\Delta}{=} \vec{I}_B \cdot \vec{\omega}_B = \text{the body-axis components of the angular momentum of the aircraft with respect to the reference axes,}$$

$$\vec{I}_B = \text{moment of inertia dyadic of the aircraft in body-axes,}$$

$$\overset{I}{(\vec{\ })} = \text{time derivative as seen in inertial axes,}$$

$$\overset{B}{(\vec{\ })} = \text{time derivative as seen in body axes,}$$

$(\vec{F}_{\text{aero}}, \vec{Q}_{\text{aero}})$ = (aerodynamic force, torque about the center of mass),

$(\vec{F}_{\text{prop}}, \vec{Q}_{\text{prop}})$ = (propulsive force, torque about the center of mass),

m = mass of the aircraft,

$\vec{g}$ = gravitational force per unit mass.

The kinematic equations of motion may be written as

$$
\begin{bmatrix} \dot{x} \\ \dot{y} \\ \dot{z} \end{bmatrix} = T_\psi T_\theta T_\phi \begin{bmatrix} u \\ v \\ w \end{bmatrix}, \tag{10.3}
$$

$$
\begin{bmatrix} \dot{\phi} \\ \dot{\theta} \\ \dot{\psi} \end{bmatrix} = (1/c\theta) \begin{bmatrix} c\theta & s\theta s\phi & s\theta c\phi \\ 0 & c\theta c\phi & -c\theta s\phi \\ 0 & s\phi & c\phi \end{bmatrix} \begin{bmatrix} p \\ q \\ r \end{bmatrix}, \tag{10.4}
$$

where

$$
T_\psi \triangleq \begin{bmatrix} c\psi & -s\psi & 0 \\ s\psi & c\psi & 0 \\ 0 & 0 & 1 \end{bmatrix}, \quad T_\theta \triangleq \begin{bmatrix} c\theta & 0 & s\theta \\ 0 & 1 & 0 \\ -s\theta & 0 & c\theta \end{bmatrix}, \quad T_\phi \triangleq \begin{bmatrix} 1 & 0 & 0 \\ 0 & c\phi & -s\phi \\ 0 & s\phi & c\phi \end{bmatrix},
$$
$$\tag{10.5}$$

and

$[x, y, z]$ = location of the aircraft c.m. in the reference axis system,

$[\psi, \theta, \phi]$ = Euler angles of the aircraft body axes with respect to the reference axes (see Figs. 10.1 and 10.2),

$[c(\ ), s(\ )]$ = $[\cos(\ ), \sin(\ )]$.

Almost all aircraft have a right-left plane of geometric symmetry, and, in the NASA standard coordinate system, this is taken as the body axis $x - z$ plane, with the x-axis nominally forward, the z-axis nominally down, and the

y-axis perpendicular to this plane (see Figs. 10.1 and 10.2). If we assume that the $x - z$ plane is also a plane of mass symmetry, then the products of inertia $I_{yz} = I_{yx} = 0$. Writing out Equations (1) and (2) in body-axis components yields

$$m(\dot{u} + qw - rv) = X - mgs\theta + Tc\epsilon,$$
$$m(\dot{v} + ru - pw) = Y + mgc\theta s\phi, \tag{10.6}$$
$$m(\dot{w} + pv - qu) = Z + mgc\theta c\phi - Ts\epsilon,$$

$$I_x\dot{p} + I_{xz}\dot{r} + (I_z - I_y)qr + I_{xz}qp = L,$$
$$I_y\dot{q} + (I_x - I_z)pr + I_{xz}(r^2 - p^2) = M, \tag{10.7}$$
$$I_z\dot{r} + I_{xz}\dot{p} + (I_y - I_x)qp - I_{xz}qr = N,$$

where

$$\vec{F}_{\text{aero}} = [X, Y, Z],$$
$$\vec{F}_{\text{prop}} = [Tc\epsilon, 0, -Ts\epsilon],$$
$$T = \text{propulsive thrust resultant,}$$
$$\epsilon = \text{angle between thrust and body } x\text{-axis,}$$
$$\vec{Q}_{\text{aero}} = [L, M, N] \overset{\triangle}{=} \text{roll, pitch, yaw moments,}$$
$$\vec{Q}_{\text{prop}} \cong 0 \text{ (not true for aircraft with engines in pods below the wing),}$$

$$\vec{I} = \begin{bmatrix} I_x & 0 & I_{xz} \\ 0 & I_y & 0 \\ I_{xz} & 0 & I_z \end{bmatrix}.$$

Steady Rectilinear Flight

In steady rectilinear flight $\dot{u} = \dot{v} = \dot{w} = \dot{p} = \dot{q} = \dot{r} = 0$, and usually the wings are level, $\phi = 0$. In that case, Equations (5)–(7) are equations of *equilibrium*:

$$0 = X - mgs\theta + Tc\epsilon, \tag{10.8}$$
$$0 = Y, \tag{10.9}$$
$$0 = Z + mgc\theta - Ts\epsilon, \tag{10.10}$$
$$0 = L = M = N. \tag{10.11}$$

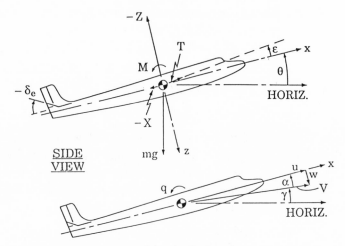

Figure 10.1. Nomenclature for aircraft longitudinal motions.

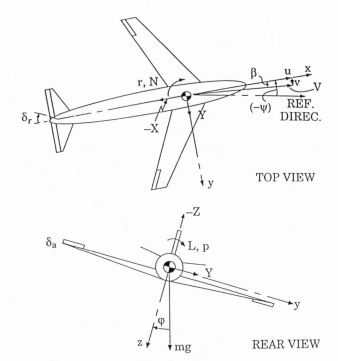

Figure 10.2. Nomenclature for aircraft lateral motions.

Small Deviations from Steady Rectilinear Flight

Assuming that

$$(v^2, w^2) \ll u^2, \tag{10.12}$$

$$(\phi^2, \theta^2, \psi^2, \epsilon^2) \ll 1, \tag{10.13}$$

$$(p^2, q^2, r^2) \ll u^2/b^2, \tag{10.14}$$

where b = wing span, the nonlinear terms in (5)–(7) are negligible.

If the external shape of the aircraft is symmetric about the $x - z$ plane, then (u, w, q, θ) do not produce any (Y, L, N), and (v, p, r, ϕ, ψ) do not produce any (X, Z, M), at least to first order when $p^2 \ll u^2/b^2$. For a given altitude (more precisely, air density), then

$$(Y, L, N) \text{ are functions of } (v, p, r; \delta_a, \delta_r),$$

$$(X, Z, M) \text{ are functions of } (u, w, q; \delta_e),$$

$$T \text{ is a function of } (u, w; \delta_t),$$

where (δ_a, δ_r) =(aileron, rudder) deflections, (δ_e, δ_t) = (elevator, throttle) deflections.

Experience has shown that the aerodynamic forces and moments, for motion frequencies small compared to u/b, are well approximated by *linear functions* with the exception that M should also include a linear term in $\dot{w}$ to account for *downwash lag* between the wings and the empennage and apparent additional mass effects. With these assumptions the equations of motion split into two linear, uncoupled sets for the longitudinal motions involving $(x, z, u, w, q, \theta; \delta_e, \delta_t)$ and the lateral motions involving $(y, v, p, r, \phi, \psi; \delta_a, \delta_r)$.

10.1.1 Longitudinal Equations of Motion

The *longitudinal set* may be written as

$$
\begin{bmatrix}
\delta \dot{u} \\
\delta \dot{w} \\
\dot{q} \\
\delta \dot{\theta}
\end{bmatrix}
=
\begin{bmatrix}
\bar{X}_u & \bar{X}_w & -u_o s\theta_o & -gc\theta_o \\
Z_u & Z_w & u_o c\theta_o & -gs\theta_o \\
\bar{M}_u & \bar{M}_w & \bar{M}_q & 0 \\
0 & 0 & 1 & 0
\end{bmatrix}
\begin{bmatrix}
\delta u \\
\delta w \\
q \\
\delta \theta
\end{bmatrix}
$$

$$+ \begin{bmatrix} X_{\delta_e} & T_{\delta_t} c\epsilon \\ Z_{\delta_e} & -T_{\delta_t} s\epsilon \\ \bar{M}_{\delta_e} & 0 \\ 0 & 0 \end{bmatrix} \begin{bmatrix} \delta(\delta_e) \\ \delta(\delta_t) \end{bmatrix} \tag{10.15}$$

where

$$\bar{M}_{(\cdot)} \triangleq (M_{(\cdot)} + M_{\dot{w}} Z_{(\cdot)})/I_y,$$
$$\bar{X}_{(\cdot)} \triangleq (X_{(\cdot)} + T_{(\cdot)})/m,$$
$$u_o \triangleq \text{steady value of } u,$$
$$\theta_o \triangleq \text{steady value of } \theta.$$

$\delta(\ ) \triangleq$ deviation of ( ) from the steady equilibrium value, and $X_{(\cdot)}, Z_{(\cdot)}, M_{(\cdot)}$, $T_{(\cdot)}$ are the stability derivatives that are estimated from wind tunnel tests, aerodynamic and propulsion theory, and flight tests; they depend on air density and (u_o, w_o). The $M_{(\cdot)}$s also depend on the location of the center of mass. Note that $1/m$ has been absorbed into the $X_{(\cdot)}, T_{(\cdot)}, Z_{(\cdot)}$ stability derivatives, and $1/I_y$ has been absorbed into the $M_{(\cdot)}$ stability derivatives. The $\bar{M}_{(\cdot)}$ terms come from eliminating $\delta\dot{w}$ from the $\delta\dot{q}$ equation.

Horizontal and vertical components of velocity are given by

$$\begin{bmatrix} \dot{x} \\ \dot{z} \end{bmatrix} = \begin{bmatrix} \cos\theta & \sin\theta \\ -\sin\theta & \cos\theta \end{bmatrix} \begin{bmatrix} u \\ w \end{bmatrix},$$

where $-\dot{z} \equiv \dot{h}$ = rate of climb.

For $\theta^2 \ll 1, w^2 \ll u^2, \dot{x} = u$, and $\dot{h} \cong -w + u_o\theta$, which implies

$$\delta\dot{x} \cong \delta u, \tag{10.16}$$
$$\delta\dot{h} \cong -\delta w + u_o\delta\theta. \tag{10.17}$$

10.1.2 *Lateral Equations of Motion*

The *lateral set* may be written as

$$
\begin{bmatrix} \dot{v} \\ \dot{r} \\ \dot{p} \\ \dot{\phi} \end{bmatrix} \cong \begin{bmatrix} Y_v & -u_o c\theta_o & u_o s\theta_o & gc\theta_o \\ \bar{N}_v & \bar{N}_r & \bar{N}_p & 0 \\ \bar{L}_v & \bar{L}_r & \bar{L}_p & 0 \\ 0 & \tan\theta_o & 1 & 0 \end{bmatrix} \begin{bmatrix} v \\ r \\ p \\ \phi \end{bmatrix} + \begin{bmatrix} Y_{\delta_a} & Y_{\delta_r} \\ \bar{N}_{\delta_a} & \bar{N}_{\delta_r} \\ \bar{L}_{\delta_a} & \bar{L}_{\delta_r} \\ 0 & 0 \end{bmatrix} \begin{bmatrix} \delta_a \\ \delta_r \end{bmatrix} \quad (10.18)
$$

$$\dot{\psi} \cong r/c\theta_o \quad (10.19)$$

$$\dot{y} \cong u_o\psi + v, \quad (10.20)$$

where

$$
\begin{bmatrix} \bar{L}_{(\cdot)} \\ \bar{N}_{(\cdot)} \end{bmatrix} \triangleq \begin{bmatrix} I_x & I_{xz} \\ I_{xz} & I_z \end{bmatrix}^{-1} \begin{bmatrix} L_{(\cdot)} \\ N_{(\cdot)} \end{bmatrix}.
$$

As in the longitudinal case, m has been absorbed into the $Y_{(\cdot)}$ stability derivatives.

10.1.3 *Angle of Attack, Sideslip Angle, and Flight Path Angles*

Instead of rectangular components of velocity (u, v, w), polar coordinates (V, α, β) are often used in analyzing aircraft motions, where V is velocity magnitude, α is angle of attack, and β is sideslip angle. The most common definitions of α and β (see Figs. 10.1 and 10.2) are

$$u = V \cos\alpha \cos\beta, \quad (10.21)$$

$$v = V \cos\alpha \sin\beta, \quad (10.22)$$

$$w = V \sin\alpha, \quad (10.23)$$

which imply that

$$V = \sqrt{u^2 + v^2 + w^2}, \quad (10.24)$$

$$\alpha = \tan^{-1}\left(w/\sqrt{u^2 + v^2}\right), \quad (10.25)$$

$$\beta = \tan^{-1}(v/u). \quad (10.26)$$

For $v^2, w^2 \ll u^2$, these relations are well approximated as

$$V \cong u, \tag{10.27}$$

$$\alpha \cong w/u, \tag{10.28}$$

$$\beta \cong v/u. \tag{10.29}$$

Instead of $-\dot{h} \equiv \dot{z}$ and $\dot{y}$, vertical and horizontal flight path angles γ and λ are occasionally used in analyzing aircraft motions. The most common definitions of γ and λ are

$$\dot{x}_I = V \cos\gamma \cos\lambda, \tag{10.30}$$

$$\dot{y}_I = V \cos\gamma \sin\lambda, \tag{10.31}$$

$$-\dot{z}_I \equiv \dot{h} = V \sin\gamma, \tag{10.32}$$

where (x_I, y_I, z_I) are the coordinates of the center of mass of the aircraft with respect to a set of earth-fixed axes.

Equations (30)–(32) imply that

$$V = \sqrt{\dot{x}_I^2 + \dot{y}_I^2 + \dot{z}_I^2}, \tag{10.33}$$

$$\gamma = \tan^{-1}\left(\dot{h}/\sqrt{\dot{x}_I^2 + \dot{y}_I^2}\right), \tag{10.34}$$

$$\lambda = \tan^{-1}\left(\dot{y}_I/\dot{x}_I\right). \tag{10.35}$$

For $\dot{y}_I^2, \dot{h}^2 \ll \dot{x}_I^2$, these relations are well approximated as

$$V \cong \dot{x}_I, \tag{10.36}$$

$$\gamma \cong \dot{h}/\dot{x}_I, \tag{10.37}$$

$$\lambda \cong \dot{y}_I/\dot{x}_I. \tag{10.38}$$

10.2 Natural Longitudinal Motions

The natural motions are best described by an example. Hence we consider a small general-aviation aircraft, the Navion, whose stability derivatives are given by Teper (Ref. TE, see Fig. 10.3):

$$
\begin{bmatrix} \delta\dot{u} \\ \delta\dot{w} \\ \dot{q} \\ \delta\dot{\theta} \end{bmatrix} = \begin{bmatrix} -.045 & .036 & 0 & -.322 \\ -.370 & -2.02 & 1.76 & 0 \\ .191 & -3.96 & -2.98 & 0 \\ 0 & 0 & 1.0 & 0 \end{bmatrix} \begin{bmatrix} \delta u \\ \delta w \\ q \\ \delta\theta \end{bmatrix}, \tag{10.39}
$$

where the units are (u, w) in ft/sec, q in crad/sec, θ in crad.

The equilibrium ("trim") condition is at $u_o = 176$ ft/sec, $w_o = 1.84$ ft/sec $(\alpha_o \cong w_o/u_o$ $=0.6$ deg), $\gamma_o = 0(\theta_o = 0.6$ deg), and sea level altitude.

The normalization of the variables by factors of 10 was chosen so that one unit of each variable is approximately of comparable significance. If we do this, the relative magnitude of elements in the coefficient matrices has some significance; if some elements are very small compared to others, this usually means that the element may be neglected—replaced by zero—without any important change in the mathematical model. It also has advantages in numerical calculations, in that the matrices are approximately "balanced."

10.2.1 *The Phugoid and Short-Period Modes*

If we look for exponential solutions, that is, let $( ) = ( )_o \exp(st)$, then (1) becomes

$$
\begin{bmatrix} s+.045 & -.036 & 0 & .322 \\ .370 & s+2.02 & -1.76 & 0 \\ -.191 & 3.96 & s+2.98 & 0 \\ 0 & 0 & -1.0 & s \end{bmatrix} \begin{bmatrix} \delta u_o \\ \delta w_o \\ q_o \\ \delta\theta_o \end{bmatrix} = 0. \tag{10.40}
$$

The *characteristic equation* of this system is the determinant of the 4×4 matrix set equal to zero. This yields a fourth-order polynomial in s, which, when factored, is

$$
0 = \left[(s+2.51)^2 + (2.59)^2\right]\left[(s+.017)^2 + (.213)^2\right]. \tag{10.41}
$$

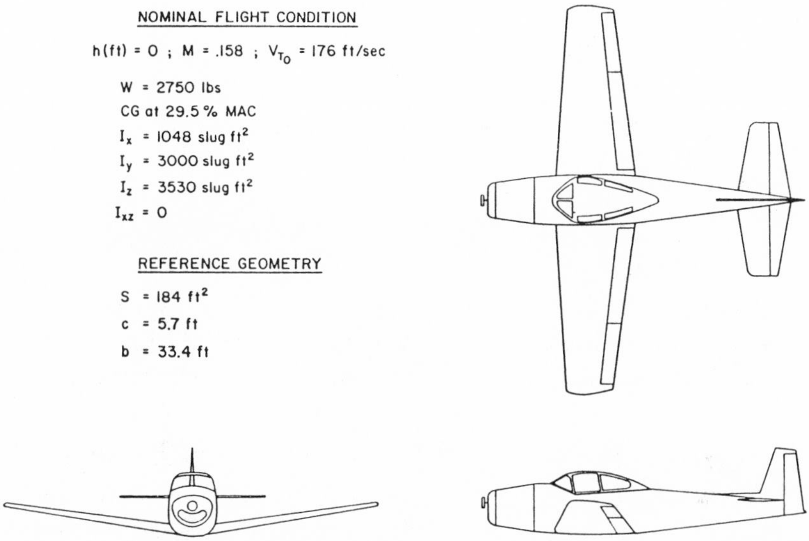

NOMINAL FLIGHT CONDITION

$h(ft) = 0$; $M = .158$; $V_{T_0} = 176$ ft/sec

W = 2750 lbs
CG at 29.5 % MAC
I_x = 1048 slug ft^2
I_y = 3000 slug ft^2
I_z = 3530 slug ft^2
I_{xz} = 0

REFERENCE GEOMETRY

S = 184 ft^2
c = 5.7 ft
b = 33.4 ft

Figure 10.3. Three-view of Navion aircraft.

The mode corresponding to the higher frequency ($s = -2.51 \pm 2.59j$ rad/sec) is called the *short-period mode*. The other complex mode ($s = -.017 \pm .213j$ rad/sec) is called the *phugoid mode* (the name was coined by Bryant, whose Greek was apparently a little shaky; "phugos" means "flight" as in "fugitive").

The *eigenvectors* (or "mode shapes") are obtained by substituting each of the eigenvalues back into (2). We *normalize* these complex eigenvectors by choosing the largest element to be unity. The eigenvalues and eigenvectors are listed in Table 10.1.

Table 10.1

```
% Eigensystem of Navion natural longitudinal motions.
% x = [u,w,q,theta]'; units ft, sec, crad.

F        =
-0.045    0.036    0.000   -0.322
-0.370   -2.020    1.760    0.000
 0.191   -3.960   -2.980    0.000
 0.000    0.000    1.000    0.000


% Eigenvalues and eigenvectors:
```

```
[T,EV]=eig(F)
EV          =
-0.0171+/-.2134i   -2.5054+/-2.5949i
T           =
 1.0000+/-.0000i   -0.0029-/+.0195i
-0.0592+/-.0013i   -0.1200-/+.6562i
 0.1430+/-.0086i    1.0000+/-.0000i
-0.0934+/-.6627i   -0.1926-/+.1994i
```

Figs. 10.4 and 10.5 show state variable histories for pure phugoid motion and pure short period motion. These were obtained by using the corresponding real parts of the eigenvectors as initial conditions with $\dot{x} = Fx$. (One can use the real part of $e^{j\theta}$ times the eigenvector for any θ—the initial phase is arbitrary.) Note the small damping of the phugoid motion and the large damping of the short-period motion.

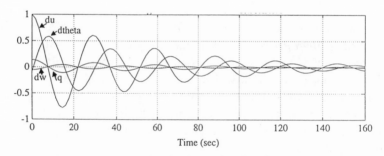

Figure 10.4. Pure phugoid motion for the Navion aircraft.

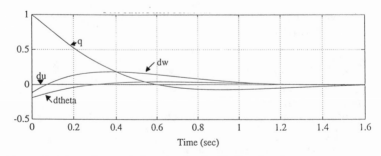

Figure 10.5. Pure short period motion for the Navion aircraft.

10.2.2 General Natural Motion

The *general natural motion* is a superposition of these two modes of motion, that is,

$$x(t) = \sum_{k=0}^{4} c_k V(:, k) * exp(s_k t),\qquad(10.42)$$

where $c_2 = \tilde{c}_1$ and $c_4 = \tilde{c}_3$ are arbitrary complex constants and $\tilde{(\,)}$ indicates complex conjugate. Since $V(:, 2)$ and s_2 are complex conjugates of $V(:, 1)$ and s_1 and $V(:, 4)$ and s_4 are complex conjugates of $V(:, 3)$ and s_3, the expression for $x(t)$ is real. Thus, there are four arbitrary real constants for the fourth-order system (1).

10.2.3 Phasor Interpretation of Complex Modes

The complex modes may be thought of as superpositions of contra-rotating, attenuating (or amplifying) vectors, called "phasors." The real part of the eigenvalue gives the attenuation (if it is negative) or amplification (if it is positive), and the imaginary part gives the frequency of rotation of the phasor. Each element of the eigenvector can be written in the polar form $ae^{i\theta}$ where a is the magnitude and θ is the phase angle. The mode is represented by two phasors, one rotating clockwise and the other counterclockwise; in this way, superposition always gives real elements in the solution. Fig. 10.6 shows phasor diagrams for the Navion longitudinal motions.

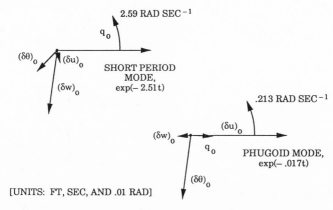

Figure 10.6. Phasor diagrams of longitudinal modes of Navion aircraft.

The short-period phasor rotates with frequency 2.59 rad sec^{-1}, and all the magnitudes attenuate like $\exp(-2.51t)$ where t is in seconds. The phugoid phasor rotates with frequency .213 rad sec^{-1}, and all the magnitudes attenuate like $\exp(-.017t)$.

Note δu_o is negligible in the short-period mode and δw_o (angle-of-attack change) is negligible in the phugoid mode. This is the basis of approximations commonly made in analyzing longitudinal motions.

Problem 10.2.1 – *Eigensystem for 747 Longitudinal Motions at h = 40 kft*

From Heffley and Jewell (Ref. HE-1) the perturbation equations of longitudinal motion for the 747 airplane in level flight at an altitude of 40 kft and velocity = 774 ft/sec (M=0.80) are

$$
\begin{bmatrix} \delta \dot{u} \\ \delta \dot{w} \\ \dot{q} \\ \delta \dot{\theta} \end{bmatrix} = \begin{bmatrix} -.003 & .039 & 0 & -.322 \\ -.065 & -.319 & 7.74 & 0 \\ .020 & -.101 & -.429 & 0 \\ 0 & 0 & 1 & 0 \end{bmatrix} \begin{bmatrix} \delta u \\ \delta w \\ q \\ \delta \theta \end{bmatrix},
$$

where the units are ft, sec, and crad (see Fig. 12.2).

(a) Find the eigenvalues and eigenvectors of the natural motion.

(b) Discuss the quantities that should be sensed for feedback to modify each of the modes.

Problem 10.2.2 – *Eigensystem for 747 Longitudinal Motions in Landing Configuration*

From Teper (Ref. TE) the perturbation equations of longitudinal motion for the 747 airplane at sea level in the landing configuration with velocity = 221 ft/sec are

$$
\begin{bmatrix} \delta \dot{u} \\ \delta \dot{w} \\ \dot{q} \\ \delta \dot{\theta} \end{bmatrix} = \begin{bmatrix} -.021 & .122 & 0 & -.322 \\ -.209 & -.530 & 2.22 & 0 \\ .017 & -1.64 & -.412 & 0 \\ 0 & 0 & 1 & 0 \end{bmatrix} \begin{bmatrix} \delta u \\ \delta w \\ q \\ \delta \theta \end{bmatrix},
$$

where the units are ft, sec, and crad (see Fig. 12.2).

(a) Find the eigenvalues and eigenvectors of the natural motion.

(b) Discuss the quantities that should be sensed for feedback to modify each of the modes.

10.3 Natural Lateral Motions

As in the previous section, we describe the natural lateral motions by using the Navion airplane as an example. Using data from Teper (Ref. TE), we see that the equations of motion at sea level and a speed of 176 ft/sec are

$$
\begin{bmatrix} \dot{v} \\ \dot{r} \\ \dot{p} \\ \dot{\phi} \end{bmatrix} = \begin{bmatrix} -.254 & -1.76 & 0 & .322 \\ 2.55 & -.76 & -.35 & 0 \\ -9.08 & 2.19 & -8.40 & 0 \\ 0 & 0 & 1.0 & 0 \end{bmatrix} \begin{bmatrix} v \\ r \\ p \\ \phi \end{bmatrix}, \tag{10.43}
$$

where the units are ft, sec, and crad.

10.3.1 *The Roll, Spiral, and Dutch Roll Modes*

If we look for exponential solutions, that is, let $(\cdot) = (\cdot)_o \exp(st)$, then (43) becomes

$$
\begin{bmatrix} s+.254 & 1.76 & 0 & -.322 \\ -2.55 & s+.76 & .35 & 0 \\ 9.08 & -2.19 & s+8.40 & 0 \\ 0 & 0 & -1.0 & s \end{bmatrix} \begin{bmatrix} v_o \\ r_o \\ p_o \\ \phi_o \end{bmatrix} = 0. \tag{10.44}
$$

The characteristic equation of this system is the determinant of the 4×4 coefficient matrix set equal to zero. This yields a fourth-order polynomial in s that, when factored, is

$$
0 = (s+8.433)\left[(s+.486)^2 + (2.334)^2\right](s+.0088). \tag{10.45}
$$

The mode corresponding to the larger real eigenvalue (s = -8.433 rad/sec) is called the *roll mode*. The mode corresponding to the smaller real eigenvalue (s = -.0088 rad/sec) is called the *spiral mode*. The mode corresponding to the complex eigenvalues (s = -.486±2.334j rad/sec) is called the *dutch roll mode* (motion in this mode looks like the motion that a Dutch skater makes on the ice).

The *eigenvectors* (or "mode shapes") are shown in Table 10.2.

Table 10.2

```
% Eigensystem of the Navion natural lateral motions.
% x = [v,r,p,phi]'; units ft, sec, crad.

F          =
   -0.254   -1.760        0     0.322
    2.550   -0.760    -0.350        0
   -9.080    2.190    -8.400        0
        0        0     1.000        0


% Eigenvalues and eigenvectors:

[T,ev]=eig(F)
ev            =
   -0.0088   -0.4862+/-2.3335i   -8.4327
T             =
    0.0506    0.0947+/-.7938i    0.0135
    0.1759    1.0000             0.0411
   -0.0088   -0.0924-/+.8835i    1.0000
    1.0000   -0.3550+/-.1136i   -0.1186
```

Figs. 10.7 to 10.9 show state variable histories for pure roll motion, pure dutch roll motion, and pure spiral motion. These were obtained by using the corresponding real parts of the eigenvectors as initial conditions with $\dot{x} = Fx$. Note the long time duration (small damping) of the spiral mode, the short time duration (large damping) of the roll mode, and the lightly damped oscillatory motion of the dutch roll mode.

The *general natural motion* is a superposition of these three modes (see Section 10.2).

As in Section 10.2, *phasor diagrams* are helpful in thinking about the modes. Fig. 10.10 shows phasor diagrams for this example.

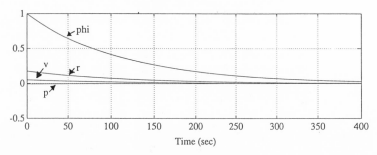

Figure 10.7. Pure spiral motion for the Navion aircraft.

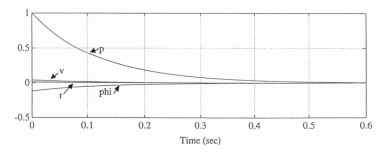

Figure 10.8. Pure roll motion for the Navion aircraft.

Problem 10.3.1 – *Eigensystem for 747 Lateral Motions at h = 40 kft*

From Heffley and Jewell (Ref. HE-1) the perturbation equations of lateral motion for the 747 airplane in level flight at an altitude of 40 kft and velocity = 774 ft/sec (M = 0.80) are

$$
\begin{bmatrix} \dot{v} \\ \dot{r} \\ \dot{p} \\ \dot{\phi} \end{bmatrix} = \begin{bmatrix} -.0558 & -.774 & 0 & .322 \\ .773 & -.115 & -.318 & 0 \\ -.394 & .039 & -.465 & 0 \\ 0 & 0 & 1.0 & 0 \end{bmatrix} \begin{bmatrix} v \\ r \\ p \\ \phi \end{bmatrix},
$$

where the units are ft, sec, and crad (see Fig. 12.2).

(a) Find the eigenvalues and eigenvectors of the natural motion.

(b) Discuss the quantities that should be sensed for feedback to modify each of the modes.

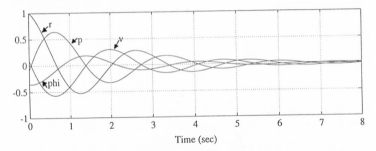

Figure 10.9. Pure dutch roll motion for the Navion aircraft.

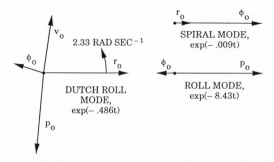

Figure 10.10. Phasor diagrams of lateral modes of Navion aircraft.

Problem 10.3.2 – *Eigensystem for 747 Lateral Motions in Landing Configuration*

From Teper (Ref. TE) the perturbation equations of lateral motion for the 747 airplane at sea level in the landing configuration with velocity = 221 ft/sec are:

$$
\begin{bmatrix} \dot{v} \\ \dot{r} \\ \dot{p} \\ \dot{\phi} \end{bmatrix}
=
\begin{bmatrix}
-.089 & -2.19 & .328 & .319 \\
.076 & -.217 & -.166 & 0 \\
-.602 & .327 & -.975 & 0 \\
0 & 0 & 1.0 & 0
\end{bmatrix}
\begin{bmatrix} v \\ r \\ p \\ \phi \end{bmatrix},
$$

where the units are ft, sec, and crad (see Fig. 12.2).

(a) Find the eigenvalues and eigenvectors of the natural motion.

(b) Discuss the quantities that should be sensed for feedback to modify each of the modes.

10.4 Wind Disturbances

10.4.1 *Description of Winds*

The wind velocity at any point in the atmosphere varies randomly. It is often approximated as a constant mean wind plus a fluctuating gust component whose average value over time is zero. If the gust magnitudes are negligibly small (that is, they have negligible effect on the motions of the aircraft), the effects of a constant wind are easily analyzed by using reference axes that move with the air mass, since they will be inertial axes. The velocity with respect to the ground is found by adding (vectorially) the aircraft velocity with respect to the air mass (air speed) and the velocity of the air mass with respect to the ground (wind velocity). With a constant horizontal wind the aircraft, in general, flies "crabbed" with respect to the ground, that is, its horizontal velocity vector is not parallel to its plane of symmetry.

10.4.2 *Gust Velocity Components*

The components of the gust velocity vary randomly with time. Nonetheless, from many years of measurements of gusts and of aircraft responses to gusts, we know that each component is *correlated* for a short time with itself and nearly uncorrelated with the other two components. With these correlations, we can make estimates of the root-mean-square (RMS) response of an aircraft to gusts, given the RMS gust velocity, σ (see Appendix E).

Also available from the measurements are estimates of the probability of encountering gusty air (turbulence) at various altitudes and, having encountered it, the probability distribution of σ. Chalk et al. (Ref. CH) concluded that the probability of encountering turbulence decreases from 80% at sea level to 7% at an altitude of 40,000 ft. They also conclude that the distribution of σ is nearly gaussian with a standard deviation of 2.3 ft/sec.

10.4.3 *Wind as a Disturbance*

If we use reference axes fixed with respect to the earth, then the velocity perturbations in the aerodynamic force and moment terms on the right-hand sides of the equations of motion (10.15)–(10.20) should be replaced by *perturbations in airspeed*. Airspeed perturbations are $(\delta u - u_w, \delta v - v_w, \delta w - w_w)$ where $(\delta u, \delta v, \delta w)$ are ground-speed perturbations and (u_w, v_w, w_w) are wind-velocity components in body axes.

The effects of gradients in wind velocity over the dimensions of the aircraft can also be approximated by thinking of the local air mass as rotating like a rigid body (as well as translating). This means that the angular velocities (p, q, r) in the aerodynamic force and moment terms on the right-hand sides of (10.15)–(10.20) should be replaced by $(p - p_w, q - q_w, r - r_w)$, where (p_w, q_w, r_w) are local air mass angular velocities due to wind. Of the three components, only p_w has a significant effect on most aircraft due to its effect on the wings.

Summarizing, the rigid body equations of motion for small perturbations from steady rectilinear flight are

$$
\begin{bmatrix} \delta\dot{u} \\ \delta\dot{w} \\ \dot{q} \\ \delta\dot{\theta} \end{bmatrix}
=
\begin{bmatrix}
\bar{X}_u & \bar{X}_w & -u_o s\theta_o & -gc\theta_o \\
Z_u & Z_w & u_o c\theta_o & -gs\theta_o \\
\bar{M}_u & \bar{M}_w & \bar{M}_q & 0 \\
0 & 0 & 1 & 0
\end{bmatrix}
\begin{bmatrix} \delta u - u_w \\ \delta w - w_w \\ q \\ \delta\theta \end{bmatrix}
-
\begin{bmatrix} 0 \\ 0 \\ \bar{M}_q \\ 0 \end{bmatrix}
q_w, \quad (10.46)
$$

$$\delta\dot{h} = -\delta w + u_o\delta\theta, \tag{10.47}$$

$$\delta\dot{x} = \delta u, \tag{10.48}$$

$$
\begin{bmatrix} \dot{v} \\ \dot{r} \\ \dot{p} \\ \dot{\phi} \end{bmatrix}
=
\begin{bmatrix}
Y_v & -u_o c\theta_o & u_o s\theta_o & -gc\theta_o \\
\bar{N}_v & \bar{N}_r & \bar{N}_p & 0 \\
\bar{L}_v & \bar{L}_r & \bar{L}_p & 0 \\
0 & t\theta_o & 1 & 0
\end{bmatrix}
\begin{bmatrix} v - v_w \\ r \\ p \\ \phi \end{bmatrix}
\tag{10.49}
$$

$$
-
\begin{bmatrix}
0 & 0 \\
\bar{N}_r & \bar{N}_p \\
\bar{L}_r & \bar{L}_p \\
0 & 0
\end{bmatrix}
\begin{bmatrix} r_w \\ p_w \end{bmatrix},
\tag{10.50}
$$

$$\delta\dot{y} = v + u_o\psi, \tag{10.51}$$

$$\dot{\psi} = r, \tag{10.52}$$

where $s\theta_o = \sin\theta_o$, $c\theta_o = \cos\theta_o$, $t\theta_o = \tan\theta_o$, and (q_w, r_w) are usually negligible.

10.4.4 *Stochastic Models for Gusts*

Appendix E discusses models of random dynamic processes. From empirical data, Holley (Ref. HO) gives the following approximate dynamic models of the (u_w, v_w, w_w) and p_w gusts.

Forward Gust Velocity, u_w

$$\frac{u_w(s)}{\eta_u(s)} = \frac{c_1}{sL/V + c_1},\qquad (10.53)$$

where

$\eta_u(t)$ is zero mean white noise with spectral density,

$$Q_u = \frac{2\sigma^2 L}{c_1 V}, \Rightarrow \text{RMS } u_w = \sigma,$$

$$c_1 \triangleq \frac{(1 + 3\beta/2)^{2/3}}{1 + 3\beta},$$

$$\beta \triangleq b/2L,$$

$$L \cong L_\infty \frac{h}{h + h_o} \triangleq \text{turbulence integral scale},$$

$$h = \text{altitude},$$

$$L_\infty = 2{,}000\text{ft}, \ h_o = 2{,}500\text{ft},$$

$$b = \text{wing span},$$

$$V = \text{aircraft velocity},$$

$$\sigma = \text{RMS gust velocity}.$$

Vertical Gust Velocity, w_w

$$\frac{w_w(s)}{\eta_w(s)} = \frac{c_3}{(sL/V)^2 + c_2(sL/V) + c_3},\qquad (10.54)$$

where

$\eta_w(t)$ is zero-mean white noise with spectral density,

$$Q_w = 2\frac{c_2}{c_3}\frac{\sigma^2 L}{V} \Rightarrow \text{RMS } w_w = \sigma,$$

$$c_2 = \frac{1+3\beta}{2\beta^{4/3}}, \quad c_3 = \frac{(1+\beta)^{2/3}}{\beta^{4/3}}.$$

A plot of the poles of (53) versus β is given in Fig. 10.11. As $\beta \to 0, c_3/c_2 \to 2, c_3 \to \infty$, so that (53) becomes

$$\frac{w_w(s)}{\eta_w(s)} \cong \frac{2}{(2\beta^{4/3}sL/V + 1)(sL/V + 2)}. \tag{10.55}$$

Side Gust Velocity, v_w

$$\frac{v_w(s)}{\eta_w(s)} = \frac{c_3}{(sL/V)^2 + c_2(sL/V) + c_3}, \tag{10.56}$$

where

$\eta_w(t)$ is zero-mean white noise with spectral density,

$$Q_w = 2\frac{c_2}{c_3}\frac{\sigma^2 L}{V} \Rightarrow \text{RMS } v_w = \sigma.$$

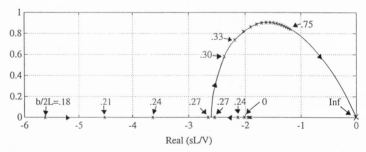

Figure 10.11. Vertical gust eigenvalues vs. $b/2L$, where b = wing span, L = gust correlation distance, V = aircraft velocity.

Roll Gust Velocity, p_w

$$\frac{p_w(s)}{\eta_p(s)} = \frac{1}{sb/(1.2V) + 1},\tag{10.57}$$

where

$\eta_p(t)$ is zero-mean white noise with spectral density,

$$Q_p = \left(\frac{5}{3}\right)^4 \sigma^2/LV \quad \Rightarrow \quad \text{RMS } p_w = 2.15\sigma/\sqrt{Lb}.$$

Example DC-8 in cruise at 33,000 ft

$$V = 824 \text{ ft/sec}, \quad b = 142.3 \text{ ft}, \quad L = 1860 \text{ ft}, \quad \sigma = 2.3 \text{ ft/sec}$$

$$\Rightarrow \beta = .0383.$$

Hence

$$c_1 = .931, \quad c_2 = 43.2, \quad c_3 = 79.6.$$

Finally, then,

$$\frac{u_w(s)}{\eta_u(s)} = \frac{1}{s/.413 + 1},$$

$$\frac{v_w(s)}{\eta_w(s)} = \frac{w_w(s)}{\eta_w(s)} = \frac{1}{(s/.853 + 1)(s/18.3 + 1)},$$

$$\cong \frac{1}{s/.853 + 1},$$

$$\frac{p_w(s)}{\eta_p(s)} = \frac{1}{s/6.95 + 1},$$

$$Q_u = 25.7 \text{ ft}^2/\sec,$$

$$Q_v = Q_w = 13.0 \text{ ft}^2/\sec,$$

$$Q_p = 2.66 \times 10^{-5} \sec^{-1},$$

$$\text{RMS } p_w = 9.61 \times 10^{-3} \text{ rad/sec},$$

$$= .551 \text{ deg/sec}.$$

11

Aircraft Sensors

11.1 Introduction

This book is concerned with the synthesis of control logic, but this is obviously only one (usually small) part of designing a control system. Control requires actuators, closed-loop control requires sensors, and there must be a hardware realization of the control logic. Ideally, the frequency response of the sensors and actuators should be relatively flat and have small phase-shift over the closed-loop bandwidth of the control system. The hardware should be accurate, reliable, and of reasonable volume, mass, and cost.

The knowledge about hardware design resides primarily in industry, in the minds of experienced engineers and, to some extent, in company handbooks and design codes. In this chapter, we briefly cover the operating principles of vertical and directional gyros and inertial measurement units, which combine gyros and specific force sensors. The concept of complementary filtering is discussed. The operating principles of sensors such as air-data sensors (yielding airspeed, barometric altitude, Mach number, and sometimes angle of attack and sideslip angle), magnetic compasses, rate gyros, specific force sensors (accelerometers), and radio position sensors are covered elsewhere (for example in Refs. KAY and WR).

11.2 Vertical Gyros

Concept (Ref. Ch. 11 of KAY by W. G. Wing)

Vertical gyros (VGs) or artificial horizons have been standard sensors on aircraft for many decades. A VG consists of a two-degree-of-freedom gyro whose spin axis is nominally vertical, with two nominally horizontal specific force sensors mounted on the inner gimbal (see Fig. 11.1).

If gimbal bearing torques and mass unbalance were negligible, the gyro spin axis would stay fixed with respect to inertial space, and, as the aircraft attitude changed, roll and pitch could be read from the two angle sensors mounted on the inner and outer gimbals. Of course, gimbal bearing torques and mass unbalance

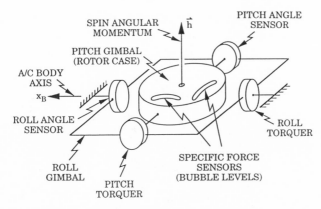

Figure 11.1. Schematic drawing of a typical vertical gyro.

are not negligible, and if no corrective action were taken the spin axis would gradually drift away from the vertical. Typical uncorrected drift rates are about 10 deg/hr; inertial quality gyros have much lower drift rates but are much more expensive.

To correct this drift, signals from the two specific force sensors (usually electrolytic bubble levels) are used to torque the gimbals, with a smoothing time of about one minute, in such a way that the two specific force signals are kept close to zero (gyro erection). If the average horizontal acceleration of the aircraft over the last minute is zero, this should keep the gyro spin axis near vertical. This is an example of *complementary filtering*, where two sensors, one with good short-time response (the gyro) and one with good long-time response (the time-averaged specific force), are combined to produce an instrument with good short-time and long-time response.

Erection Cutoff

If the aircraft is in a steady turn for a minute or longer, the vertical gyro will develop a large error since the specific force sensors read the apparent vertical, which is perpendicular to the wings in a coordinated turn, and the gyro axis is torqued toward this apparent vertical. For this reason most VGs have an erection cutoff device, which stops the torquing signals when the signals from the specific force sensors reach a certain magnitude.

Dynamics of the Averaging Vertical Gyro

Consider the aircraft with roll angle ϕ as shown in Fig. 11.2. The indicated roll

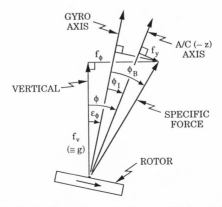

Figure 11.2. Nomenclature for vertical gyro with aircraft in roll.

angle, ϕ_I, is equal to

$$\phi_I = \phi - \epsilon_\phi, \tag{11.1}$$

where ϵ_ϕ = gyro drift angle in roll. The roll bubble level, which is mounted on the gyro rotor case, reads an angle ϕ_B where

$$\phi_B = \tan^{-1} f_\phi / f_V - \epsilon_\phi \cong f_\phi / g - \epsilon_\phi, \tag{11.2}$$

where f_V = vertical specific force $\cong g$, and f_ϕ = side component of horizontal specific force.

A similar figure could be drawn for the aircraft in pitch, where

$$\theta_I = \theta - \epsilon_\theta, \tag{11.3}$$

$$\theta_B = \tan^{-1} f_\theta / f_V - \epsilon_\theta \cong f_\theta / g - \epsilon_\theta, \tag{11.4}$$

where θ_I = indicates pitch angle, ϵ_θ = gyro drift angle in pitch, θ_B = reading of pitch bubble level, and f_θ = backward component of horizontal specific force.

The angular momentum of the gyro, in components parallel to the inner gimbal axes, is

$$\vec{H}_G = \left[J_x \dot{\epsilon}_\phi, J_y \dot{\epsilon}_\theta, -h \right], \tag{11.5}$$

where (J_x, J_y) = transverse moments of inertia of rotor + rotor case + gimbals, and h = spin angular momentum of gyro.

The angular velocity of the inner gimbal axes is

$$\vec{\omega}_G = \left[\dot{\epsilon}_\phi, \dot{\epsilon}_\theta, 0 \right], \tag{11.6}$$

so the equations of motion of the gyro are

$$\overset{G}{\vec{H}_G} + \vec{\omega}_G \times \vec{H}_G = \vec{Q} \tag{11.7}$$

or

$$J_x \ddot{\epsilon}_\phi - h\dot{\epsilon}_\theta \cong Q_{cx} + Q_{dx}, \qquad (11.8)$$

$$J_y \ddot{\epsilon}_\theta + h\dot{\epsilon}_\phi \cong Q_{cy} + Q_{dy}, \qquad (11.9)$$

$$-\dot{h} \cong 0, \qquad (11.10)$$

where

$$\left(Q_{cx}, Q_{cy}\right) = \text{(roll, pitch) control torques applied to the gimbals,}$$

$$\left(Q_{dx}, Q_{dy}\right) = \text{(roll, pitch) disturbance torques due to friction, etc.}$$

The control torques in the averaging vertical gyro are made proportional to the readings of the bubble levels, ϕ_B and θ_B:

$$Q_{cx} = -k\theta_B \cong -k(f_\theta/g - \epsilon_\theta), \qquad (11.11)$$

$$Q_{cy} = k\phi_B \cong k(f_\phi/g - \epsilon_\phi). \qquad (11.12)$$

The disturbance torques may be approximated as viscous damping torques due to gimbal bearing friction and other random torques:

$$Q_{dx} = -D\left(\dot{\epsilon}_\phi - \dot{\phi}\right) + Q'_{dx}, \qquad (11.13)$$

$$Q_{dy} = -D\left(\dot{\epsilon}_\theta - \dot{\theta}\right) + Q'_{dy}. \qquad (11.14)$$

Thus, the Laplace transform of the VG equations of motion is

$$\begin{bmatrix} J_x s^2 + Ds & -hs - k \\ \\ hs + k & J_y s^2 + Ds \end{bmatrix} \begin{bmatrix} \epsilon_\phi(s) \\ \\ \epsilon_\theta(s) \end{bmatrix}$$

$$\cong \begin{bmatrix} (-k/g)f_\theta(s) + Ds\phi(s) + Q'_{dx}(s) \\ \\ (k/g)f_\phi(s) + Ds\theta(s) + Q'_{d_y}(s) \end{bmatrix}. \qquad (11.15)$$

The characteristic equation of (15), with $J_x \cong J_y \overset{\Delta}{=} J$, is

$$(Js + D)^2 s^2 + (hs + k)^2 = 0. \qquad (11.16)$$

Now $h/k \cong 1$ minute, whereas typically J/h is on the order of milliseconds, so the roots of (16) are given approximately by

$$s \cong -\frac{D}{J} \pm \frac{h}{J}j \, , \quad -\frac{kh}{h^2 + D^2}\left(1 \pm \frac{D}{h}j\right). \qquad (11.17)$$

The fast roots correspond to a damped nutation mode whereas the slow roots correspond to the erection (or precession) mode. Thus, the erection mode may be approximated in (15) by putting $J_x \cong J_y \cong 0$; in addition, D/h is usually $\ll 1$, so that we may place $D \cong 0$ in (15). This decouples the system into

$$\tau \dot{\epsilon}_\theta + \epsilon_\theta \cong \frac{f_\theta}{g} - \frac{Q'_{dx}}{k}, \tag{11.18}$$

$$\tau \dot{\epsilon}_\phi + \epsilon_\phi \cong \frac{f_\phi}{g} + \frac{Q'_{dx}}{k}, \tag{11.19}$$

where $\tau = h/k$. For $\tau \cong 1$ minute, vertical gyros are accurate to about one degree.

Finally, to use (18)–(19) with the aircraft equations of motion, we must relate (f_θ, f_ϕ) to aircraft body-axis components of specific force (f_x, f_y):

$$f_\theta \cong - f_x + g\theta,$$
$$f_\phi \cong f_y + g\phi.$$

Substituting (18) and (19) into (1) and (3), we obtain

$$\phi_I(s) \cong \frac{\tau s}{\tau s + 1}\phi(s) - \frac{1}{\tau s + 1}\left[\frac{f_y(s)}{g} + \frac{Q'_{dy}(s)}{k}\right],$$

$$\theta_I(s) \cong \frac{\tau s}{\tau s + 1}\theta(s) + \frac{1}{\tau s + 1}\left[\frac{f_x(s)}{g} + \frac{Q'_{dx}(s)}{k}\right].$$

Thus,

$$\phi_I \cong \begin{cases} \phi & \omega \gg 1/\tau \\ \\ -f_y/g & \omega \ll 1/\tau \end{cases}, \tag{11.20}$$

$$\theta_I \cong \begin{cases} \theta & \omega \gg 1/\tau \\ \\ f_x/g & \omega \ll 1/\tau \end{cases}, \tag{11.21}$$

as we commented earlier in connection with the complementary filter interpretation of the vertical gyro.

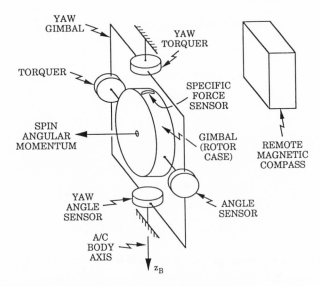

Figure 11.3. Schematic drawing of a typical directional gyro.

11.3 Directional Gyros

Concept (Ref. Ch. 11 of KAY by W. G. Wing)

Directional gyros (DGs) have also been standard sensors on aircraft for many decades. A DG consists of a two-degree-of-freedom gyro whose spin axis is nominally horizontal pointing toward magnetic north, with a specific force sensor on the inner gimbal (see Fig. 11.3).

If gimbal bearing torques and mass unbalance were negligible, the gyro spin axis would stay fixed with respect to inertial space and, as the aircraft heading changed, the yaw angle could be read from the sensor mounted on the outer gimbal. Of course, gimbal bearing torques and mass unbalance are not negligible, and, if no corrective action were taken, the spin axis would gradually drift away from its horizontal, north-pointing position.

To correct this drift, signals from a specific force sensor on the rotor case and a remotely located magnetic compass are used to torque the gimbals, with a smoothing time of about one minute, in such a way that the specific force sensor signal is kept close to zero and the yaw angle is close to the yaw angle read by the magnetic compass (gyro erection). This is another example of complementary filtering (see previous section on the vertical gyro).

Dynamics of the Averaging Directional Gyro

The equations of motion are nearly identical to those of the averaging vertical gyro:

$$\psi_I = \psi - \epsilon_\psi, \tag{11.22}$$

$$J_y \ddot{\epsilon}_\theta + h \dot{\epsilon}_\psi = Q_{cy} + Q_{dy}, \tag{11.23}$$

$$J_z \ddot{\epsilon}_\psi - h \dot{\epsilon}_\theta = Q_{cz} + Q_{dz}, \tag{11.24}$$

$$Q_{cy} = k(\epsilon_M - \epsilon_\psi), \tag{11.25}$$

$$Q_{cz} = -k\theta_B \cong k(-f_\theta/g + \epsilon_\theta), \tag{11.26}$$

$$Q_{dy} = -D(\dot{\epsilon}_\theta - \dot{\theta}) + Q'_{dy}, \tag{11.27}$$

$$Q_{dz} = -D(\dot{\epsilon}_\psi - \dot{\psi}) + Q'_{dz}, \tag{11.28}$$

where the nomenclature is the same as in the section on the vertical gyro (VG), except for the following (see Fig. 11.4):

$$\psi_I = \text{indicated heading angle,}$$
$$\epsilon_M = \text{heading angle error of magnetic compass,}$$
$$\epsilon_\psi = \text{heading angle error of gyro axis (drift angle).}$$

Using the same frequency-separation argument that was used in the VG section, the yaw erection motion is well described by the approximate equation:

$$\frac{h}{k}\dot{\epsilon}_\psi + \epsilon_\psi \cong \epsilon_M + \frac{1}{k}Q'_{dy}. \tag{11.29}$$

Again, a typical value of the smoothing time constant, h/k, is one minute; thus $\epsilon_M + Q_{dy}/k$ is well approximated as a random bias (RMS about one degree of arc) plus white noise.

Other Versions

A cheaper version of the DG uses manual instead of automatic slaving in heading, that is, the pilot occasionally adjusts the DG so that it agrees with the magnetic compass.

The DG display can be made to read heading relative to true north if a device for manually entering the magnetic deviation is added.

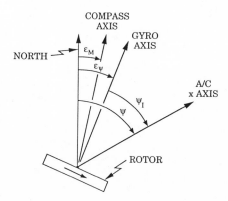

Figure 11.4. Nomenclature for directional gyro from above.

11.4 Inertial Measurement Units

Introduction

Inertial measurement units (IMUs) have been on transoceanic civil transports since the early 1970s. These IMUs have RMS latitude errors that are less than three or four miles and RMS longitude errors that grow at a rate of about one mile per hour. They have *gimballed stable platforms* that are torqued to stay locally horizontal. Some are also torqued to stay north pointing, while others rotate slowly about the local vertical to average out horizontal errors, while "remembering" where north is.

IMUs were made possible by the development[1] of:

(a) *Rate gyros* having RMS *drift rates*, ϵ, on the order of one-thousandth of the earth's rotation rate (15 deg/hr), which is three orders of magnitude smaller than the drift rates of gyros used in VGs and DGs.

(b) *Specific force sensors* (accelerometers) having RMS errors, b, on the order of one-thousandth of the earth's gravitational force per unit mass (32.2 ft sec^{-2}).

Concept for a Stationary IMU

Three gyros of inertial quality on a stationary vehicle can determine the *direction of the earth's polar axis within .001 rad $\cong$ .06 deg.*

---

[1]By C. S. Draper and colleagues at the MIT Instrumentation Laboratory in the early 1940s.

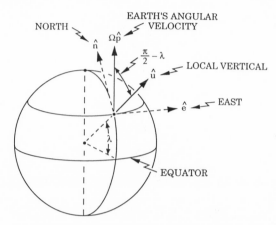

Figure 11.5. Determination of east ($\hat{e}$) and north ($\hat{n}$) directions, given $\hat{u}$, the local vertical, and $\hat{p}$, a unit vector parallel to the earth's axis.

Three specific force sensors of inertial quality on a stationary vehicle can determine the *direction of the local vertical within .001 rad* $\cong$ *.06 deg.*

The angle between the local vertical $\hat{u}$ and the earth's polar axis $\hat{p}$ (see Fig. 11.5) is equal to 90° minus the latitude, λ, so a stationary IMU can determine the latitude of the observer to within .001 $\sqrt{2}$ rad, that is, a *latitude position error* of $(.001)(3960)\sqrt{2} \cong 5.6$ miles.

A vector perpendicular to the local vertical and the earth's polar axis, $\hat{e} = \hat{p} \times \hat{u} / \cos \lambda$, (see Fig. 11.5) is horizontal and parallel to the east-west direction within .001 $\sqrt{1 + 2\tan^2 \lambda}$ rad. A vector perpendicular to the local vertical and the east-west direction, $\hat{n} = \hat{u} \times \hat{e}$, is horizontal and parallel to the *north-south direction within* .001 $\sqrt{1 + 2\tan^2 \lambda}$ *rad.* This error is an order of magnitude smaller than the errors of the best magnetic compasses and is about the same as the error of the best shipboard gyrocompasses.

A gyro with sensitive axis parallel to $\hat{p}$, on a stationary vehicle, should read 15 deg/hr. If the gyro has a drift rate ϵ, observers in a closed, stationary vehicle would infer that they are moving eastward at a velocity

$$\epsilon R \cos \lambda, \tag{11.30}$$

where R = radius of the earth. Thus the *longitude position error* of an IMU grows at this rate. For ϵ = .015 deg/hr, R=3960 mi, λ =40°, this rate is $(.015/57.3)(3960)\cos 40° \cong 0.8$ mi/hr.

Concept for a Moving IMU

When the IMU is on a moving vehicle, it is impossible to measure the local vertical precisely since the specific forces on the vehicle are no longer precisely vertical and equal to g. It is also impossible to measure the earth's polar axis precisely since the vehicle's attitude is not fixed with respect to the earth.

However, if the IMU was *aligned*—if the platform was made locally horizontal and north pointing, while the vehicle was stationary with respect to the earth at a known location on the earth—then it can be maintained locally level and north pointing very accurately by *computing* the rates of change of the direction of the local vertical and the direction of north, and torquing the platform so that it rotates at these rates. Note the earth's polar axis is fixed in inertial space.

The *computation* involves integrating the outputs of the two horizontal specific force sensors to estimate the two components of horizontal velocity with respect to the earth; knowing initial latitude and longitude, this gives an estimate of the current rate of change of latitude and longitude. Taking into account the earth's rotation, the rate of change of the directions of the local vertical and north at this current position can then be computed. The gyros on the platform are used with a torquer-servo loop to command the desired platform rotation rates.

With one more integration, position change can be estimated, so that a continuous estimate of current position is available.

The Longitude Channel at the Equator

Fig. 11.6 shows a stable platform on a vehicle moving eastward at the equator with a small level-angle error ϕ. A rate gyro on the platform reads the angular velocity of the platform, ω, with an error ϵ (the gyro drift rate), that is,

$$z_\omega = \omega + \epsilon. \tag{11.31}$$

A sensor on the platform reads the specific force parallel to the platform. We assume that the vertical component of specific force is approximately equal to g, and we denote the eastward component of specific force as a. Thus the sensor reads

$$z_a = a + g\phi + b, \tag{11.32}$$

where b = error in specific force sensor.

The equations of motion of the vehicle and the platform attitude are

$$\dot{x} = v, \tag{11.33}$$

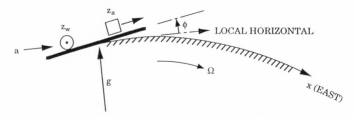

Figure 11.6. IMU stable platform on a vehicle moving eastward at the equator.

$$\dot{v} = a, \tag{11.34}$$
$$\dot{\phi} = \omega + \Omega + v/R, \tag{11.35}$$

where

v = eastward velocity relative to the earth,

ω = platform rotation rate with respect to inertial space,

Ω = earth rotation rate,

R = radius of earth.

The torquer servo loop commands platform rotation rate

$$\omega = -\frac{1}{t}(z_\omega - \omega_c)\omega_c = -\Omega - \hat{v}/R, \tag{11.36}$$

where

$$\dot{\hat{v}} = z_a = a + g\phi + b,$$
$$\hat{v} \overset{\Delta}{=} \text{ estimated eastward velocity.}$$

Since the commanded platform rotation rate is implemented with feedback from the platform gyro, it is implemented with an error equal to the drift rate, ϵ, so

$$z_\omega \longrightarrow \omega_c, \omega \longrightarrow -\Omega - \hat{v}/R - \epsilon \tag{11.37}$$

Thus

$$\dot{\phi} = -\epsilon - (v - \hat{v})/R, \tag{11.38}$$
$$\dot{\hat{v}} - \dot{v} = g\phi + b. \tag{11.39}$$

Taking the Laplace transform of (38)–(39), we have

$$\begin{bmatrix} s & 1/R \\ -g & s \end{bmatrix} \begin{bmatrix} \phi(s) \\ \delta v(s) \end{bmatrix} = \frac{1}{s} \begin{bmatrix} -\epsilon \\ b \end{bmatrix} + \begin{bmatrix} \phi(0) \\ 0 \end{bmatrix}, \tag{11.40}$$

where

$$\delta v \overset{\Delta}{=} \hat{v} - v = \text{error in velocity estimate,}$$

$$\phi(0) = \text{initial platform level-angle error.}$$

The transfer functions from ϵ and b to ϕ and δv are

$$\begin{bmatrix} \phi(s) \\ \\ \delta v(s) \end{bmatrix} = \frac{1}{s(s^2 + \omega_s^2)} \begin{bmatrix} s & -1/R \\ g & s \end{bmatrix} \begin{bmatrix} -\epsilon + s\phi(0) \\ \\ b \end{bmatrix}, \qquad (11.41)$$

where $\delta v(0) = 0$ and

$$\omega_s \overset{\Delta}{=} \sqrt{g/R} = \text{Schuler frequency.} \qquad (11.42)$$

The alignment procedure is such that $\phi(0) = -b/g$, so

$$\phi(t) = \frac{-\epsilon}{\omega_s} \sin \omega_s t - \frac{b}{g}, \qquad (11.43)$$

$$\delta v(t) = \epsilon R(1 - \cos \omega_s t). \qquad (11.44)$$

Now longitude change is estimated by

$$\dot{\hat{x}} = \hat{v} \qquad (11.45)$$

so

$$\delta \dot{x} = \delta v \qquad (11.46)$$

where $\delta x(0) = 0$ and $\delta x \overset{\Delta}{=} \hat{x} - x =$ estimate error in longitude. From (43)–(44) and (46) then,

$$\delta x(t) = -\epsilon R \left(t - \frac{\sin \omega_s t}{\omega_s} \right). \qquad (11.47)$$

Thus, the platform tilt error, ϕ, and the velocity error are bounded and oscillatory, while the longitude error is unbounded and grows at a rate ϵR, the same as we estimated for the stationary IMU.

The RMS magnitude of the platform tilt error is

$$\left[\frac{1}{2} \left(\frac{\epsilon}{\omega_s} \right)^2 + \left(\frac{b}{g} \right)^2 \right]^{\frac{1}{2}}.$$

For $\epsilon = .015$ deg/hr, $\omega_s = 4.47$ rad/hr=1/83 cycles/min, $b/g = .001$, the RMS tilt error is approximately .06 deg. Thus the platform tilt error stays the same order of magnitude as the error in determining it when the vehicle is stationary!

Conclusion

Angle sensors on the gimbals of an IMU give *roll, pitch, and yaw angles* to an accuracy of about .06 deg. This is a two-order-of-magnitude improvement over VGs and DGs, and there is no problem in a steady turn as there is with the VG.

IMUs give horizontal ground velocity to an accuracy of about one mile per hour. If these measurements are combined with vertical velocity from a baro-inertial altimeter (see next section) and the gimbal angles, components of *ground velocity* in aircraft *body axes* can be determined again to an accuracy of about *one mile per hour*.

11.5 Baro-Inertial Altimeters

Introduction

The vertical channel of an IMU is unstable, as we shall demonstrate. However, it can be stabilized by using it in conjunction with a barometric altimeter. This is another example of complementary filtering; the IMU is accurate at high frequencies, whereas the barometric altimeter is good at low frequencies. In particular, baro-inertial altimeters give very *precise measurements of rate of climb* with essentially no time lag.

Vertical Channel of an IMU

A sensor on the stable platform reads the vertical component of specific force a, with an error b, that is,

$$z_a(t) = a(t) + b_a. \tag{11.48}$$

The equations of motion of the vehicle, and hence the platform, are

$$\begin{aligned} \dot{h} &= v, \\ \dot{v} &= a - g(h), \end{aligned} \tag{11.49}$$

where h = altitude, v = vertical velocity, and $g(h)$ = gravitational force per unit mass.

If one tried to use the specific force measurement, $z(t)$, to estimate altitude, one would have

$$\begin{aligned} \dot{\hat{h}} &= \hat{v}, \\ \dot{\hat{v}} &= z - g(\hat{h}), \end{aligned} \tag{11.50}$$

where $\hat{h}$ = estimated altitude, $\hat{v}$ = estimated vertical velocity.

Let

$$\tilde{h} \triangleq \hat{h} - h = \text{error in estimate of altitude,} \tag{11.51}$$

$$\tilde{v} \triangleq \hat{v} - v = \text{error in estimate of vertical velocity.} \tag{11.52}$$

Using (48)–(50), we have

$$\dot{\tilde{h}} = \tilde{v},$$
$$\dot{\tilde{v}} = b + \left[g(h) - g(\hat{h}) \right]. \tag{11.53}$$

Now $g(h) \cong g(\hat{h}) + \partial g / \partial h (h - \hat{h})$ where

$$\frac{\partial g}{\partial h} \cong -\frac{2g}{R}, \tag{11.54}$$

so (53) may be written as

$$\dot{\tilde{h}} = \tilde{v},$$
$$\dot{\tilde{v}} \cong b + \frac{2g}{R} \tilde{h}. \tag{11.55}$$

For $\tilde{h}(0) = \tilde{v}(0) = 0$, (55) yields

$$\tilde{h}(t) = \frac{b}{2g/R} \left(\cosh \sqrt{\frac{2g}{R}} t - 1 \right), \tag{11.56}$$

which verifies our earlier assertion that the vertical channel of an IMU, by itself, is fundamentally unstable.

To stabilize the vertical channel, we use the pressure altitude signal from the barometric altimeter, $h_B(t)$, in a modification of (50):

$$\dot{\hat{h}} = \hat{v} + k_h(h_B - \hat{h}),$$
$$\dot{\hat{v}} = z - g(\hat{h}) + k_v(h_B - \hat{h}), \tag{11.57}$$

where the gains k_h and k_v are to be selected. We again use (51)–(52) and define

$$\tilde{h}_B \triangleq -h_B + \hat{h}. \tag{11.58}$$

Then subtracting (49) from (57) and using (48) and (50) gives us

$$\dot{\tilde{h}}_B = \tilde{v} + k_h \tilde{h}_B,$$
$$\dot{\tilde{v}} = b_a + \frac{2g}{R} \tilde{h}_B + k_v \tilde{h}_B. \tag{11.59}$$

Taking the Laplace transform of (59) with $b = 0$, we have

$$\begin{bmatrix} s + k_h & -1 \\ -2g/R + k_v & s \end{bmatrix} \begin{bmatrix} \tilde{h}(s) \\ \tilde{v}(s) \end{bmatrix} = 0. \tag{11.60}$$

The characteristic equation of (60) is

$$s^2 + k_h s + k_v - 2g/R = 0. \tag{11.61}$$

If we choose to place the two estimate error poles at $s = -1/\tau$, where $\tau \cong 1$ minute, the *desired* characteristic equation is

$$\left(s + \frac{1}{\tau}\right)^2 \equiv s^2 + \frac{2}{\tau}s + \frac{1}{\tau^2} = 0. \tag{11.62}$$

Comparing coefficients of (61) and (62), we find

$$\begin{aligned} k_h &= 2/\tau, \\ k_v &= 2g/R + 1/\tau^2. \end{aligned} \tag{11.63}$$

With this choice of gains, the baro-inertial altimeter is *stable* and has the good high-frequency response of the IMU and the good low-frequency response of the barometric altimeter.

12

Control of Longitudinal
Motions of Aircraft

12.1 Introduction

There are many types of autopilots, ranging from the relatively simple wings-level autopilots on some small aircraft to the sophisticated multi-mode autopilots using an inertial reference system (IRS) on large aircraft. The history of the development of autopilots goes back to 1914 when Lawrence Sperry demonstrated automatic flight at the Paris airshow (see Refs. DAV, MC, and PE-2 for excellent coverage of this history).

It is convenient to divide autopilot logic into two parts, (a) stability augmentation or hold logic, and (b) command logic.

A *stability augmentation system* (SAS) either stabilizes an inherently unstable aircraft or augments the damping of certain modes of motion that have insufficient natural damping. For example, the spiral mode on many aircraft is slightly unstable or else has very little damping (see Section 10.3), so active control of the ailerons is used, with feedback of roll angle from a vertical gyro, to provide adequate damping of this mode. Another example is active stabilization of the Dutch roll mode by feedback of yaw rate, or lateral acceleration at the tail, to the rudder (many jet aircraft have insufficient natural damping of Dutch roll).

A *command or hold autopilot* has one or more of the following three types of engagement: (a) preselect, (b) synchronous-select, and (c) postselect.

With *preselect logic*, the pilot can enter a desired flight condition into the autopilot, then push an "arm" or "engage" button, and the autopilot will then fly the airplane to the specified flight condition and hold it there. The preselect command is entered into the autopilot with a keyboard or by turning a knob that changes a digital read-out displaying the command. For example, cruise altitude is often preselected, and the command is "armed," that is, the altitude-hold logic will start to function when the actual altitude is within a preselected range (the "window") of the selected altitude.

With *synchronous-select logic*, the pilot sets up the desired flight condition and then pushes a button on the autopilot, which then holds the flight condition as it was when the button was pushed.

If *postselect logic* is available, it is possible for the pilot to change the flight condition by entering commands into the autopilot, but only after the synchronous-select mode has been used. Postselection commands are entered either by knob-dial or by control wheel steering; the latter involves either displacing the wheel by an amount proportional to the command (such as a bank angle command) or by putting a force on the wheel proportional to the command (such as pitchrate).

The usual longitudinal command-hold outputs are (a) velocity or Mach number, (b) climb rate, (c) altitude, (d) pitch attitude, and (e) glide slope for automatic approach to landing.

Examples of command-hold autopilots are (a) autopilots used in remotely piloted vehicles (RPVs), where an operator on the ground radios commands to the aircraft such as heading changes, altitude changes, etc., (b) autopilots in large aircraft, where virtually the entire flight path can be dialed in before or shortly after take-off; the climb rate for ascent, the desired cruise altitude, the desired cruise velocity and heading (or latitude and longitude) of several way points, etc., (c) automatic landing systems where the final flare, de-crab (if needed), touchdown, and thrust reversal are performed by the autopilot.

Autopilot design criteria are inherently subjective. Hence it is important to obtain opinions on desired behavior of the controlled aircraft from the people who will use, maintain, and pay for the autopilot. Some of the criteria that are commonly specified are enumerated here:

(a) The aircraft should return to (or arrive at) the desired flight condition in a reasonable length of time after a disturbance (or command), and without too much overshoot.

(b) Control excursions should be within a specified range (the "control authority"), which is usually a small fraction of the maximum range of the control. This leaves a margin for other functions that may be superimposed and avoids the possibility of saturating the control.

(c) Steady-state errors in the presence of constant disturbances or commands should be acceptably small.

(d) Motions sensed by the passengers must not be excessive or frightening.

(e) Loads on the aircraft structure should be within specified ranges, which again are usually small fractions of the maximum loads that the structure can sustain.

Control law synthesis is also a somewhat subjective area. Every designer has his or her own favorite methods. Historically, *frequency response methods* (Ref. MC) were developed first (1930s to 1940s), and they are still the most commonly used methods (Ref. MI). These methods make use of Bode, Nyquist, or Nichols plots of system response to sinusoidal inputs over the range of relevant frequencies. They use the concepts of phase margin and gain margin to ensure robust closed-loop behavior. Compensator synthesis is based on successive loop closures, selecting certain types of compensators (proportional, integral, derivative, lead, lag, notch, etc.), and adjusting parameters until satisfactory closed-loop response is attained. These methods are described in detail in Ref. MC.

Root locus methods were introduced by Evans (Ref. EV) in the late 1940s. The essential feature is the ability to quickly sketch the locus of closed-loop roots in the complex s-plane versus some parameter of the system, usually a feedback gain. As with frequency response methods, compensator synthesis is based on successive loop closures, adjusting parameters until satisfactory closed-loop response is attained. These methods are described in Refs. BL, MC, and PE-2; an example is given in Section 12.4.

Quadratic synthesis methods were introduced by Kalman in the late 1950s and early 1960s (Refs. KAL-1 and KAL-2). These methods make use of a quadratic performance index to synthesize state-feedback gains using numerical algorithms that minimize the performance index. Compensators may be synthesized by treating disturbances and sensor noise as gaussian random processes and determining gains for a state-variable estimator, so that estimated state is fed back instead of measured state. Hence these methods are also called *linear-quadratic-gaussian* (LQG) methods. A brief summary of these methods is given in Appendixes C and D.

Modal decomposition methods have been known for nearly a century, but they only became practical for systems of high order with complex eigenvalues when digital computers and new algorithms (see Refs. FRA, WI, MA, and PO) made them rapid, accurate, and economical. With these algorithms, it is possible to transform a high-order, stationary linear dynamic system into modal coordinates in only a few minutes (the time to enter the system coefficients into the computer). Having the system in modal coordinates enables the designer to see immediately which modes are *controllable* by which controls, *observable* by which measurements, and *disturbable* by which disturbances. Modal decomposition is a generalization of *partial fraction expansion of transfer functions*. A summary of modal decomposition is given in Appendix A, and we shall use this method to analyze the controllability, observability, and disturbability of aircraft modes.

12.2 Steady-State Control

Equilibrium Points for Linear Systems (cf. Appendix B)

Consider a linear, time-invariant system

$$\dot{x} = Fx + Gu, \tag{12.1}$$

$$y = Hx + Lu, \tag{12.2}$$

where x is the state vector, u is the control vector, and y is the output vector. If $u = 0$, clearly $x = 0$, $u = 0$, $y = 0$ is the set point (or equilibrium point) where $\dot{x} = 0$. If $u = u_s = $ constant $\neq 0$, *nonzero equilibrium points* may exist where $x = x_s = $ constant $\neq 0$, $y = y_s = $ constant $\neq 0$. For the common case where the dimension of $y = $ dimension of u, these nonzero set points must satisfy

$$\begin{bmatrix} 0 \\ y_c \end{bmatrix} = \begin{bmatrix} F & G \\ H & L \end{bmatrix} \begin{bmatrix} x_s \\ u_s \end{bmatrix}. \tag{12.3}$$

If solutions exist, then they are of the form

$$\begin{bmatrix} x_s \\ u_s \end{bmatrix} = \begin{bmatrix} N_{xy} \\ N_{uy} \end{bmatrix} [y_c], \tag{12.4}$$

where

$$\begin{bmatrix} N_{xy} \\ N_{uy} \end{bmatrix} = \begin{bmatrix} F & G \\ H & L \end{bmatrix}^{-1} \begin{bmatrix} 0 \\ I \end{bmatrix}. \tag{12.5}$$

Equilibrium Points for Longitudinal Motions of an Aircraft

The two outputs of most interest in the longitudinal motions of an aircraft are airspeed and climb rate. To control them, we use elevator and throttle. Before discussing stability, we shall discuss equilibrium flight and the elevator/throttle changes that are required to change from one airspeed/climb-rate equilibrium condition to another.

Example 1–Navion Airplane at Sea Level Near $V = 176$ ft/sec

From Teper (Ref. TE), the perturbation equations for longitudinal motion of the Navion airplane are

$$
\begin{bmatrix} \dot{u} \\ \dot{w} \\ \dot{q} \\ \dot{\theta} \end{bmatrix} = \begin{bmatrix} -.045 & .036 & 0 & -.322 \\ -.370 & -2.02 & 1.76 & 0 \\ .191 & -3.96 & -2.98 & 0 \\ 0 & 0 & 1 & 0 \end{bmatrix} \begin{bmatrix} u - u_w \\ w - w_w \\ q \\ \theta \end{bmatrix} + \begin{bmatrix} 0 & 1 \\ -.282 & 0 \\ -11.0 & 0 \\ 0 & 0 \end{bmatrix} \begin{bmatrix} \delta_e \\ \delta_t \end{bmatrix},
$$

(12.6)

where (u_w, w_w) are perturbations in wind velocity components in body axes, and the other symbols were defined in Section 10.2. Recall that (u, w, u_w, w_w) are in ft/sec, q in crad/sec, θ in crad, and note that we have omitted the perturbation symbol $\delta(\ )$ to reduce clutter in the equations. Eqn. (6) is for perturbations away from equilibrium level flight at an airspeed of 176 ft/sec with $u_w = w_w = 0$.

The two outputs of interest here are

$$
\begin{aligned}
u - u_w &= \text{change in airspeed}, \\
\dot{h} = -w + 1.76\theta &= \text{climb rate}.
\end{aligned}
$$

(12.7)

The steady-state changes in elevator and throttle (δ_e and δ_t) required to produce desired small changes in steady-state airspeed and climb rate may be determined from (6) and (7) by putting

$$
\dot{u} = \dot{w} = \dot{q} = \dot{\theta} = 0.
$$

(12.8)

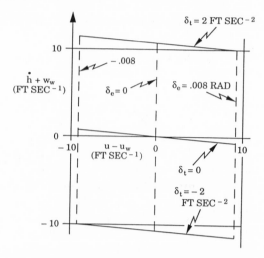

Figure 12.1. Steady-state elevator and throttle settings as functions of airspeed and climb rate; Navion airplane at sea level, $V_o = 176$ ft/sec.

Using a computer code to solve these linear equations, assuming constant values for u_w and w_w, we obtain

$$
\begin{bmatrix} \delta_e \\ \delta_t \end{bmatrix} = \begin{bmatrix} .0877 & 0 \\ .0163 & .1830 \end{bmatrix} \begin{bmatrix} u - u_w \\ \dot{h} + w_w \end{bmatrix} ,
$$

$$
w = w_w - .1954(u - u_w),
$$

$$
\theta = -.1110(u - u_w) + .568(\dot{h} + w_w),
$$

$$
q = 0. \tag{12.9}
$$

Fig. 12.1 shows the relationship between (δ_e, δ_t) and $(u - u_w, \dot{h} + w_w)$. Note that *steady state $\dot{h}$ is controlled mainly by δ_t*, but also slightly by δ_e. Note also the very small value of δ_e (only .009 crad $\cong$ .5 deg) to change airspeed by 10 ft/sec.

Example 2–747 Airplane at $h = 40$ kft, Near $M = 0.8$

From Heffley and Jewell (Ref. HE-1) the perturbation equations of longitudinal motion for the 747 airplane cruising in level flight at an altitude of 40 kft and velocity = 774 ft/sec (Mach number 0.80) are

$$
\begin{bmatrix} \dot{u} \\ \dot{w} \\ \dot{q} \\ \dot{\theta} \end{bmatrix} =
\begin{bmatrix} -.003 & .039 & 0 & -.322 \\ -.065 & -.319 & 7.74 & 0 \\ .0201 & -.101 & -.429 & 0 \\ 0 & 0 & 1 & 0 \end{bmatrix}
\begin{bmatrix} u - u_w \\ w - w_w \\ q \\ \theta \end{bmatrix}
$$

$$
+ \begin{bmatrix} .01 & 1 \\ -.18 & -.04 \\ -1.16 & .598 \\ 0 & 0 \end{bmatrix}
\begin{bmatrix} \delta_e \\ \delta_t \end{bmatrix}, \tag{12.10}
$$

where the units are ft, sec, and crad (see Fig. 12.2).

Note the large pitch-up moment due to thrust increase ($\dot{q}$ due to δ_t), which is caused by the location of the engines in pods well below the center of mass.

Putting $\dot{u} = \dot{w} = \dot{q} = \dot{\theta} = \dot{u}_w = 0$ in (10), and using

$$
\dot{h} = -w + 7.74\theta, \tag{12.11}
$$

we may determine steady-state δ_e and δ_t to give required $u - u_w$ and $\dot{h} + w_w$, and the corresponding values of w and θ:

$$
\begin{bmatrix} \delta_e \\ \delta_t \\ w - w_w \\ \theta \end{bmatrix} =
\begin{bmatrix} .038 & .0229 \\ .0020 & .0413 \\ -.225 & -.0181 \\ -.0291 & .1269 \end{bmatrix}
\begin{bmatrix} u - u_w \\ \dot{h} + w_w \end{bmatrix}. \tag{12.12}
$$

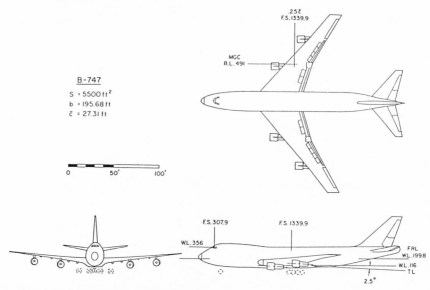

Figure 12.2. Three-view of 747 aircraft.

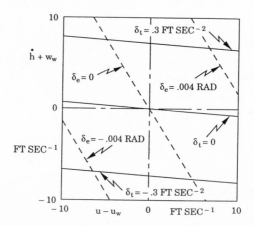

Figure 12.3. Steady-state elevator and throttle settings as functions of airspeed and climb rate; 747 at 40 kft, M = 0.8.

Figure 12.3 shows (δ_e, δ_t) as functions of $(u - u_w, \dot{h} + w_w)$. The lines of constant throttle setting (δ_t = constant) are similar to those of Fig. 12.1 for the Navion, but the lines of constant elevator setting (δ_e = constant) are quite different. This difference occurs largely because the elevators must be used to counteract the pitching moment caused by δ_t.

Problem 12.2.1 – *Steady-State Longitudinal Controls; 747 in Landing Configuration*

From the following equations of motion, sketch a figure of (δ_e, δ_t) versus $(u - u_w, \dot{h} + w_w)$ for the 747 airplane in landing configuration at a weight of 564,000 lb with $V = 221$ ft/sec:

$$
\begin{bmatrix} \dot{u} \\ \dot{w} \\ \dot{q} \\ \dot{\theta} \end{bmatrix}
=
\begin{bmatrix}
-.021 & .122 & 0 & -.322 \\
-.209 & -.530 & 2.21 & 0 \\
.017 & -.164 & -.412 & 0 \\
0 & 0 & 1 & 0
\end{bmatrix}
\begin{bmatrix} u - u_w \\ w - w_w \\ q \\ \theta \end{bmatrix}
+
\begin{bmatrix}
.010 & 1 \\
-.064 & -.044 \\
-.378 & .544 \\
0 & 0
\end{bmatrix}
\begin{bmatrix} \delta_e \\ \delta_t \end{bmatrix},
$$

where the units are ft, sec, and crad (see Fig. 12.2).
Partial answer:

$$
\begin{bmatrix} \delta_e \\ \delta_t \\ w - w_w \\ \theta \end{bmatrix}
=
\begin{bmatrix}
.241 & .222 \\
.0085 & .1426 \\
-.424 & -.0386 \\
-.1919 & .435
\end{bmatrix}
\begin{bmatrix} u - u_w \\ \dot{h} + w_w \end{bmatrix}.
$$

12.3 Observability and Controllability

Introduction

The equations of motion in modal coordinates show controllability, observability, and disturbability explicitly (see Appendix A). If there are modes with insufficient stability, and these modes are observable with one or more sensed outputs, and are controllable with one or more control inputs, then feedback of the appropriate output to the appropriate control, possibly with dynamic compensation, can be found to provide the degree of stability desired.

Example 1–Navion at Sea Level; Fourth Order Model

As an example we consider the Navion airplane in horizontal flight near sea level at a velocity of 176 ft/sec. The perturbation equations of longitudinal motion were given in Equation 12.6. For emphasis, we repeat again that we have *scaled* the variables by factors of 10 so that one unit of each variable is roughly of comparable significance to the autopilot designer. In this case, we chose (u, w, u_w, w_w) in ft/sec, (θ, δ_e) in .01 rad, q in .01 rad/sec, and δ_t in ft sec^{-2}. We also omit the perturbation symbol δ to avoid clutter.

Using an eigenvalue/eigenvector code, the modal form of Equation 12.6 was determined to be

$$\dot{\xi} = F_m \xi + G_m u + \Gamma_m w, \tag{12.13}$$

where

$$F_m = \begin{bmatrix} -2.52 & 2.59 & 0 & 0 \\ -2.59 & -2.51 & 0 & 0 \\ 0 & 0 & -.017 & .213 \\ 0 & 0 & -.213 & -.017 \end{bmatrix}, \tag{12.14}$$

$$G_m = \begin{bmatrix} -11.0 & -.143 \\ 2.42 & -.064 \\ .016 & .998 \\ 2.46 & -.080 \end{bmatrix}, \tag{12.15}$$

$$\Gamma_m = \begin{bmatrix} -.194 & 3.97 \\ -.532 & -3.80 \\ .034 & -.099 \\ .212 & .001 \end{bmatrix}, \tag{12.16}$$

(ξ_1, ξ_2) are the short-period modal coordinates, and (ξ_3, ξ_4) are the phugoid modal coordinates. These coordinates are related to the state variables by the eigenvector (transformation) matrix:

$$
\begin{bmatrix} u \\ w \\ q \\ \theta \end{bmatrix} = \begin{bmatrix} -.003 & -.019 & 1 & 0 \\ -.120 & -.656 & -.059 & -.001 \\ 1 & 0 & .143 & -.009 \\ -.193 & -.199 & -.093 & -.663 \end{bmatrix} \begin{bmatrix} \xi_1 \\ \xi_2 \\ \xi_3 \\ \xi_4 \end{bmatrix}. \tag{12.17}
$$

The inverse transformation is given by

$$
\begin{bmatrix} \xi_1 \\ \xi_2 \\ \xi_3 \\ \xi_4 \end{bmatrix} = \begin{bmatrix} -.143 & .008 & .998 & -.013 \\ -.064 & -1.524 & -.182 & .005 \\ .998 & -.030 & -.001 & 0 \\ -.080 & .460 & -.235 & -1.507 \end{bmatrix} \begin{bmatrix} u \\ w \\ q \\ \theta \end{bmatrix}. \tag{12.18}
$$

Examining (14)–(16), we note that:

(a) The natural damping of the short period mode is quite adequate.

(b) The natural damping of the phugoid mode is inadequate.

(c) The short-period mode is well controlled by elevator, and slightly controllable with throttle.

(d) The phugoid mode is controllable by either elevator or throttle.

(e) The short-period mode is disturbed almost an order of magnitude more by vertical gusts, w_w, than it is by horizontal gusts, u_w, since

$$
\left[(3.97)^2 + (3.80)^2 \right]^{\frac{1}{2}} / \left[(.194)^2 + (.532)^2 \right]^{\frac{1}{2}} \cong 9.70.
$$

(f) The phugoid mode is disturbed about twice as much by horizontal gusts as by vertical gusts.

(g) The phugoid mode is much less disturbed by gusts than the short period mode is.

Examining (18), we find that

(a) The short-period mode is observed strongly in pitch rate q and angle of attack w, and is also seen to a lesser extent in pitch angle θ.

(b) The phugoid mode is strongly observed in forward speed u and in pitch angle θ, and is seen to a lesser extent in pitch rate q.

From the observations above, we conclude that

(a) The short-period mode needs no stability augmentation.

(b) The phugoid mode needs stability augmentation, and this could be accomplished by feeding back forward speed u, or pitch angle θ, to either the elevator δ_e or the throttle δ_t.

(c) The main disturbances are vertical gusts, and they affect the well-damped short-period mode more than the phugoid mode.

Phugoid Approximation

An approximate reduced-order model suitable for analyzing control inputs in the phugoid mode bandwidth can be obtained by treating the *short period mode as quasi-steady*, that is,

$$
\begin{bmatrix} 0 \\ 0 \end{bmatrix} \cong \begin{bmatrix} -2.51 & 2.59 \\ -2.59 & -2.51 \end{bmatrix} \begin{bmatrix} \xi_1 \\ \xi_2 \end{bmatrix} + \begin{bmatrix} -11.0 & -.143 \\ 2.44 & -.064 \end{bmatrix} \begin{bmatrix} \delta_e \\ \delta_t \end{bmatrix}
$$
$$
+ \begin{bmatrix} -.194 & 3.97 \\ -.532 & -3.80 \end{bmatrix} \begin{bmatrix} u_w \\ w_w \end{bmatrix},
$$

or

$$
\begin{bmatrix} \xi_1 \\ \xi_2 \end{bmatrix} \cong \begin{bmatrix} -1.630 & -.0403 \\ 2.658 & .01634 \end{bmatrix} \begin{bmatrix} \delta_e \\ \delta_t \end{bmatrix} + \begin{bmatrix} -.143 & .008 \\ -.064 & -1.524 \end{bmatrix} \begin{bmatrix} u_w \\ w_w \end{bmatrix}.
$$

Replacing (ξ_1, ξ_2) by their quasi-steady-state values, the order of the system is reduced from four to two:

$$
\begin{bmatrix} u \\ w \\ q \\ \theta \end{bmatrix} \cong \begin{bmatrix} 1 & 0 \\ -.059 & -.001 \\ .143 & -.009 \\ -.093 & -.663 \end{bmatrix} \begin{bmatrix} \xi_3 \\ \xi_4 \end{bmatrix} + \begin{bmatrix} -.047 & 0 \\ -1.549 & -.006 \\ -1.630 & -.040 \\ -.216 & .005 \end{bmatrix} \begin{bmatrix} \delta_e \\ \delta_t \end{bmatrix}
$$

$$
+ \begin{bmatrix} .002 & .030 \\ .059 & .999 \\ -.143 & .008 \\ .040 & .302 \end{bmatrix} \begin{bmatrix} u_w \\ w_w \end{bmatrix},
$$

where

$$
\begin{bmatrix} \dot{\xi}_3 \\ \dot{\xi}_4 \end{bmatrix} = \begin{bmatrix} -.017 & .213 \\ -.213 & -.017 \end{bmatrix} \begin{bmatrix} \xi_3 \\ \xi_4 \end{bmatrix} + \begin{bmatrix} .016 & .998 \\ 2.457 & -.080 \end{bmatrix} \begin{bmatrix} \delta_e \\ \delta_t \end{bmatrix}
$$

$$
+ \begin{bmatrix} .034 & -.099 \\ .212 & .001 \end{bmatrix} \begin{bmatrix} u_w \\ w_w \end{bmatrix}.
$$

The same result can be obtained by neglecting appropriate terms in the transfer functions given by Teper (Ref. TE); for example,

$$
\frac{\theta(s)}{\delta_e(s)} = \frac{-11.00(s + .0522)(s + 1.911)}{\left[(s + 2.51)^2 + (2.59)^2\right]\left[(s + .017)^2 + (.213)^2\right]},
$$

$$
\cong \frac{-1.616(s + .0522)}{(s + .017)^2 + (.213)^2},
$$

where s has been neglected compared to 1.911 and 2.59 since $|s| \cong .2$ in the phugoid mode bandwidth.

Short-Period Approximation

A simpler approximate model suitable for analyzing control inputs in the short-period bandwidth can be obtained by *approximating the phugoid eigenvalues as zeros*, that is,

$$
\begin{bmatrix} \dot{\xi}_3 \\ \dot{\xi}_4 \end{bmatrix} \cong \begin{bmatrix} .016 & .998 \\ 2.46 & -.080 \end{bmatrix} \begin{bmatrix} \delta_e \\ \delta_t \end{bmatrix}.
$$

The same result can be obtained by neglecting appropriate terms in the transfer functions given by Teper; for example,

$$
\begin{aligned}
\frac{w(s)}{\delta_e(s)} &= \frac{-.282(s + 72.0)[(s + .022)^2 + (.255)^2]}{[(s + 2.51)^2 + (2.59)^2][(s + .017)^2 + (.213)^2]}, \\
&\cong \frac{-20.3}{(s + 2.51)^2 + (2.59)^2},
\end{aligned}
$$

where .022, .209, .017, and .213 have been neglected compared to s and s has been neglected compared to 72.0 since $|s| \cong 3$ in the short-period response. Note this reduces to a second-order transfer function because of the double pole-zero cancellation near $s = 0$.

Example 2–747 at 40 kft; Fifth-Order Model Including Altitude

To synthesize altitude command-hold logic, we must include altitude change, h, as a state variable in our mathematical model:

$$
\dot{h} = -w + V\theta.
$$

Since this equation is only coupled one-way to the other four equations in $\dot{u}, \dot{w}, \dot{q}$, and $\dot{\theta}$, the fifth eigenvalue of the open-loop system is $s = 0$. We shall call the corresponding mode the *energy mode* since, as we shall demonstrate, the modal coordinate is nearly proportional to the change in kinetic plus potential energy of the aircraft. We shall also show that the energy mode is well controlled with thrust and only slightly controllable with elevator.

The specific energy of the airplane, E, is

$$
E \cong \frac{1}{2}[(V + u)^2 + w^2] + g(H_o + h),
$$

where u and w are perturbations in velocity in level flight at velocity V, $g =$ gravitational force per unit mass, and h is the perturbation in altitude from H_o. Changes in E are thus given approximately by

$$\delta E \cong Vu + gh.$$

The open-loop eigenvectors for the fifth-order system that includes altitude may be determined directly, or, if the eigenvectors are already known for the fourth-order system, the h component may be found from

$$(h_o)_i = \left(\frac{-w_o + V\theta_o}{s}\right)_i,$$

where i = short period or phugoid, (w_o, θ_o) are components of the fourth-order eigenvectors, and s = eigenvalue. The resulting fifth-order eigenvector may then be renormalized.

Using a modal transformation computer code, the modal form of Equation 12.10 plus

$$\dot{h} = w + 7.74\theta$$

is given in Table 12.1.

Table 12.1

```
% 747 at 40 kft; modal form longitudinal EOM; units ft, sec, crad.
% x = [u,w,q,theta,h]'; u = [delta e delta t]'; w = [uw,ww]'.
% xdot = Fx + Gu + Ga*w; mdot = Fm*m + Gm*u + Gam*w; m = Tx.
F =
    -0.0030      0.0390           0    -0.3220           0
    -0.0650     -0.3190      7.7400           0           0
     0.0201     -0.1010     -0.4290           0           0
          0           0      1.0000           0           0
          0     -1.0000           0      7.7400           0
[G Ga] =
     0.0100      1.0000      0.0030     -0.0390
    -0.1800     -.0400      0.0650      0.3190
    -1.1600      .5980     -0.0201      0.1010
          0           0           0           0
          0           0           0           0


[v,d]=eig(F);ev=conj(diag(d)')
          0                 -0.3750 + 0.8817i   -0.3750 - 0.8817i
    -0.0005 + 0.0675i   -0.0005 - 0.0675i
T=[v(:,1),real(v(:,2)),imag(v(:,2)),real(v(:,4)),imag(v(:,4))]
          0      0.0111     -0.0079     -0.0415     -0.0021
```

```
            0    1.0000   -0.0000   -0.0058    0.0002
            0   -0.0071    0.1139   -0.0006   -0.0001
            0    0.1123   -0.0396   -0.0008    0.0087
       1.0000   -0.2411    0.2512    1.0000   -0.0000
Ti=inv(T)
      24.0373   -0.6581    1.4594    5.7200    1.0000
      -0.1392    1.0080   -0.0305   -0.0597         0
      -0.1338    0.0614    8.7827    0.0255         0
     -24.0372    0.8858   -3.6733   -5.7408         0
      -1.0409  -12.5793   39.8637  114.6949         0
Fm=T\F*T
            0    0.0000   -0.0000    0.0000    0.0000 % Energy mode
            0   -0.3750    0.8817    0.0000   -0.0000 % Short Period
            0   -0.8817   -0.3750   -0.0000    0.0000   mode
            0    0.0000   -0.0000   -0.0005    0.0675 % Phugoid
            0   -0.0000   -0.0000   -0.0675   -0.0005   mode
Gm=T\G;Gam=T\Ga;[Gm Gam]
      -1.3340   24.9363   -0.0000   -1.0000
      -0.1475   -0.1978    0.0657    0.3239
     -10.2003    5.1158   -0.1729    0.9119
       3.8612  -26.2693    0.0593    0.8490
     -43.9881   23.3008   -1.6220    0.0540
```

From Table 12.1, we see that

(a) $Gm(1,2)$ is large compared to $Gm(1,1)$, indicating that the energy mode is primarily controlled by thrust and almost uncontrollable with elevator. Deflecting the elevators creates a pitching moment but does not change kinetic energy of translation or potential energy (altitude); the kinetic energy of rotation is negligibly small compared to kinetic energy of translation. The *elevator controls energy distribution* between kinetic and potential; climbing increases potential energy at the expense of kinetic energy and descending increases kinetic energy at the expense of potential energy.

(b) $Gam(1,1)$ is zero, indicating that the energy mode is undisturbed by horizontal gusts.

(c) $Ti(1,5) = 1$ indicates that the energy mode is observed only in altitude.

(e) $Ti(1,[1\ 5])= [24\ 1]$, so the energy modal coordinate is nearly proportional to energy change, since

$$\delta E \cong 774u + 32.2h,$$
$$\cong 32.2(24u + h).$$

Problem 12.3.1 – *747 at 40 kft; Fourth-Order Model*

Consider the 747 airplane in cruise at an altitude of 40 kft and at Mach number 0.8 (velocity equal to 774 ft/sec). The perturbation equations of longitudinal motion were given in Equation 12.10.

(a) Show that the real modal form of (12.10) is

$$
\begin{bmatrix} \dot{\xi}_1 \\ \dot{\xi}_2 \\ \dot{\xi}_3 \\ \dot{\xi}_4 \end{bmatrix}
=
\begin{bmatrix}
-.375 & .882 & 0 & 0 \\
-.882 & -.375 & 0 & 0 \\
0 & 0 & 0 & .068 \\
0 & 0 & -.068 & 0
\end{bmatrix}
\begin{bmatrix} \xi_1 \\ \xi_2 \\ \xi_3 \\ \xi_4 \end{bmatrix}
$$

$$
+
\begin{bmatrix}
.388 & -2.25 \\
-1.151 & .5.66 \\
-.388 & 2.25 \\
.003 & 2.11
\end{bmatrix}
\begin{bmatrix} \delta_e \\ \delta_t \end{bmatrix}
+
\begin{bmatrix}
-.143 & -.002 \\
.168 & -1.152 \\
.143 & .002 \\
-.008 & .075
\end{bmatrix}
\begin{bmatrix} u_w \\ w_w \end{bmatrix},
$$

where (ξ_1, ξ_2) are the short-period modal coordinates and (ξ_3, ξ_4) are the phugoid modal coordinates.

(b) Show that the real modal coordinates are related to the state variables by the eigenvector (transformation) matrix

$$
\begin{bmatrix} u \\ w \\ q \\ \theta \end{bmatrix}
=
\begin{bmatrix}
.011 & -.003 & .021 & .472 \\
.792 & .280 & .009 & .065 \\
-.375 & .882 & 0 & .068 \\
1 & 0 & 1 & 0
\end{bmatrix}
\begin{bmatrix} \xi_1 \\ \xi_2 \\ \xi_3 \\ \xi_4 \end{bmatrix},
$$

and that the inverse transformation is given by

$$
\begin{bmatrix} \xi_1 \\ \xi_2 \\ \xi_3 \\ \xi_4 \end{bmatrix}
=
\begin{bmatrix}
-.103 & 1.107 & -.352 & -.008 \\
-.206 & .469 & .985 & 0 \\
.103 & -1.107 & .352 & 1.008 \\
2.11 & 2.57 & -.001 & -.044
\end{bmatrix}
\begin{bmatrix} u \\ w \\ q \\ \theta \end{bmatrix}.
$$

Problem 12.3.2 – *Navion at Sea Level; Fifth-Order Model*

(a) Using Equations (12.6) and

$$\dot{h} = -w + 1.76\theta,$$

where h is in units of ft, w in ft/sec, θ in crad, *show that* the real modal form of the Navion longitudinal equations of motion is

$$
\begin{bmatrix} \dot{\xi}_1 \\ \dot{\xi}_2 \\ \dot{\xi}_3 \\ \dot{\xi}_4 \\ \dot{\xi}_5 \end{bmatrix}
=
\begin{bmatrix}
-2.51 & 2.59 & 0 & 0 & 0 \\
-2.59 & -2.51 & 0 & 0 & 0 \\
0 & 0 & -.017 & .213 & 0 \\
0 & 0 & -.213 & -.017 & 0 \\
0 & 0 & 0 & 0 & 0
\end{bmatrix}
\begin{bmatrix} \xi_1 \\ \xi_2 \\ \xi_3 \\ \xi_4 \\ \xi_5 \end{bmatrix}
$$

$$
+
\begin{bmatrix}
-10.98 & -.143 \\
2.43 & -.064 \\
2.18 & -5.45 \\
13.25 & -.492 \\
-1.015 & 5.47
\end{bmatrix}
\begin{bmatrix} \delta_e \\ \delta_t \end{bmatrix}
+
\begin{bmatrix}
-.194 & 3.97 \\
-.532 & -3.80 \\
.012 & .533 \\
-1.172 & .083 \\
0 & -1
\end{bmatrix}
\begin{bmatrix} u_w \\ w_w \end{bmatrix},
$$

where (ξ_1, ξ_2) are the short-period modal coordinates, (ξ_3, ξ_4) are the phugoid modal coordinates, and ξ_5 is the energy modal coordinate.

(b) Show that the modal coordinates are related to the state variables by

$$
\begin{bmatrix} u \\ w \\ q \\ \theta \\ h \end{bmatrix}
=
\begin{bmatrix}
-.003 & -.019 & -.180 & -.031 & 0 \\
-.120 & -.657 & .011 & -.002 & 0 \\
1 & 0 & .026 & -.003 & 0 \\
-.193 & -.199 & -.003 & .122 & 0 \\
.103 & -.015 & 1 & 0 & 1
\end{bmatrix}
\begin{bmatrix} \xi_1 \\ \xi_2 \\ \xi_3 \\ \xi_4 \\ \xi_5 \end{bmatrix},
$$

and the inverse transformation is

$$
\begin{bmatrix} \xi_1 \\ \xi_2 \\ \xi_3 \\ \xi_4 \\ \xi_5 \end{bmatrix} = \begin{bmatrix} -.143 & .008 & .998 & -.013 & 0 \\ -.064 & -1.524 & -.182 & .005 & 0 \\ -5.45 & .585 & -.213 & -1.392 & 0 \\ -.492 & -2.452 & 1.267 & 8.117 & 0 \\ 5.466 & -.609 & .108 & 1.393 & 1 \end{bmatrix} \begin{bmatrix} u \\ w \\ q \\ \theta \\ h \end{bmatrix}.
$$

(c) Discuss the observability, controllability, and disturbability of the short-period, phugoid, and energy modes, and show that ξ_5 is nearly proportional to change in energy, δE.

Problem 12.3.3 – *Longitudinal Modal Equations for DC-8 in Cruise at 33 kft*

From Teper (Ref. TE) the perturbation equations of longitudinal motion are

$$
\begin{bmatrix} \dot{u} \\ \dot{w} \\ \dot{q} \\ \dot{\theta} \end{bmatrix} = \begin{bmatrix} -.015 & .004 & 0 & -.322 \\ -.074 & -.806 & 8.24 & 0 \\ -.0749 & -1.07 & -1.344 & 0 \\ 0 & 0 & 1.0 & 0 \end{bmatrix} \begin{bmatrix} u - u_w \\ w - w_w \\ q \\ \theta \end{bmatrix}
$$
$$
+ \begin{bmatrix} 0 & 1 \\ -.3 & 0 \\ -4.57 & 1.02 \\ 0 & 0 \end{bmatrix} \begin{bmatrix} \delta_e \\ \delta_t \end{bmatrix},
$$

where the units are ft, sec, and crad. The $\dot{q}$ due to δ_t term was estimated to be ℓ/r^2 where ℓ = the distance of the resultant thrust line below the center of mass and r = radius of gyration in pitch; here $\ell \cong 5.0$ ft, $r \cong 22.1$ ft.

Determine the modal form of the equations of motion and the transformation between modal coordinates and the original state variables.

Problem 12.3.4 – *Longitudinal Modal Equations for 747 in Landing Configuration*

Using the equations of motion given in Problem 12.2.1, determine the modal form of the equations of motion, and the transformation between modal coordinates and the original state variables.

12.4 Control of Climb Rate and Airspeed

12.4.1 *Climb-Rate/Airspeed Hold*

Suppose a pilot sets up a steady airspeed and climb rate and then keeps the elevator and throttle fixed. If both the short-period and phugoid modes are stable, any small perturbations in climb rate or airspeed due to disturbances over a finite time will eventually damp out.

If one of the modes is unstable, the pilot will have to stabilize it by almost continously moving the controls; this is very tiring since it requires constant attention. Even if both modes are stable, if one of them is lightly damped, the pilot will want to move the controls to damp out the effects of disturbances more quickly, that is, the pilot will augment the stability of the lightly damped mode. Again, this is very tiring.

For aircraft with unstable or lightly damped modes, the pilot's workload is greatly decreased if the aircraft is provided with a "stability augmentation system" (SAS), which, in this case, would be a climb-rate/airspeed hold autopilot. The SAS should do essentially what the pilot would do, or else the pilot will not feel comfortable using the SAS.

In conventional aircraft, the short-period mode is usually well damped, while the phugoid mode is lightly damped or even slightly unstable (cf. the Navion and 747 examples in Section 12.3).

The phugoid mode is observable with pitch angle and controllable with elevator. Thus feedback of pitch angle to elevator could be used to increase phugoid damping. The first autopilots (such as Sperry in 1914) did just that, using simple proportional feedback, and were called "displacement autopilots" (presumably because the elevator is displaced in proportion to the pitch angle displacement). Unfortunately, with proportional feedback, the increased damping of the phugoid mode comes at the expense of decreasing the damping of the short-period mode, and displacement autopilots for the Navion and the 747, while very simple, are not very satisfactory (cf. Problem 12.4.1).

A significantly better SAS is obtained by feeding back all four states to elevator, or by introducing dynamic compensation on a measurement of pitch angle (cf. Problem 12.4.2). Such a SAS can easily match the stability augmentation provided by an expert pilot.

An even better SAS is obtained by feeding back all four states to the *throttle* as well as to the elevator, that is, one can improve on the stability augmentation provided by an expert pilot, since he or she does not usually move the throttle to damp out a disturbance. We shall design such a SAS for both the Navion and the 747, using the linear-quadratic-regulator (LQR) method (see Appendix C).

Nearly the same results can be obtained by sensing only climb rate and airspeed (or pitch angle and airspeed) by using a dynamic compensator. Such a compensator can be designed using classical *successive loop-closure* methods. However, only an experienced designer can achieve coordinated, graceful control with this method (cf. Refs. BL and MC). Compensator synthesis with linear-quadratic-gaussian (LQG) methods or differential-game (DG) methods is much simpler, and easily produces coordinated, graceful controls (see Appendixes C and D).

A Climb-Rate/Airspeed SAS for the Navion

The outputs we wish to control are climb-rate $\dot{h}$ and airspeed deviation u, using elevator perturbation δ_e and throttle perturbation δ_t. Thus, we shall use a quadratic performance index of the form

$$J = \int_0^\infty [A(u^2 + \dot{h}^2) + \delta_e^2 + B_t \delta_t^2] dt, \tag{12.19}$$

where the units are ft, sec, and crad.

By this choice, the autopilot will use approximately one crad of elevator and $1/\sqrt{B_t}$ ft/sec^2 of propulsive specific force for $1/\sqrt{A}$ ft/sec of airspeed error or $1/\sqrt{A}$ ft/sec of climb- rate error. If we assume the elevator authority is about 10 crad and the throttle authority is about 5 ft/sec^2, then $B_t = (10/5)^2 \equiv 4$. A reasonable value of A would be unity. Fig. 12.4 shows a symmetric root locus vs. A with $B_t = 4$; this is *not an Evans root locus*, since A enters quadratically in this two-input, two-output symmetric-root characteristic equation.

Table 12.2 is a MATLAB diary of synthesizing the linear-quadratic-regulator (LQR) using $A = 1$ and $B_t = 4$ in the quadratic performance index.

It is interesting to find the "worst" initial condition vector of unit magnitude (see Appendix C.9), in the sense that it produces the largest value of the performance index J. Since the optimal return function is $J = x^T(0)Sx(0)$, this is easily done by finding the largest eigenvalue of the Riccati matrix S, and

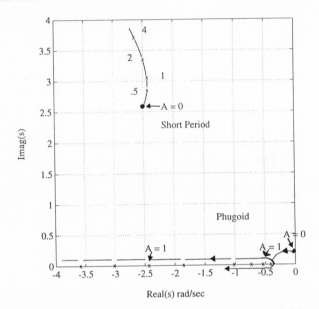

Figure 12.4. Root locus vs. A for Navion LQR climb-rate/airspeed hold autopilot.

using the corresponding eigenvector, normalized to unit length, as the initial condition. This is done in Table 12.2, and the results are plotted in Fig. 12.5.

Table 12.2

```
% Climb-rate/airspeed SAS for Navion; response to worst initial
% condition vector of unit magnitude. x = [u,w,q,theta]';
% u = [delta e,delta t]'; xdot = Fx + Gu; y = Hx + Lu; units ft,.
% sec, crad [F,G,H] from Eqn. 12.6.

A=eye(2);B=diag([1 4]);
[k,S]=lqr(F,G,H'*A*H,B);k
    0.1348     0.4909    -0.2233    -1.7678
    0.4566    -0.0071    -0.0029    -0.0109
[evec,d]=eig(F-G*k);ev=conj(diag(d)')
  -2.4277 + 3.0527i  -2.4277 - 3.0527i  -2.4263  -0.5381
evec
    0.0015 - 0.0172i    0.0015 + 0.0172i  -0.0029   1.0000
   -0.0673 - 0.5698i   -0.0673 + 0.5698i   1.0000  -0.3227
    1.0000 + 0.0000i    1.0000 - 0.0000i  -0.2868  -0.0387
   -0.1596 - 0.2007i   -0.1596 + 0.2007i   0.1182   0.0719
```

```
[vs,ds]=eig(S);
[J,l]=max(diag(ds));J
   2.6136
x0=vs(:,1)/norm(vs(:,1));x0'          % Worst unit initial condition
   -0.0415    -0.3228     0.0725      0.9428
H1=[H;[0 0 0 1];-k];L1=[zeros(5,2)];
t=[0:.03:3]';;
u=zeros(101,2);
y=lsim(F-G*k,G,H1,L1,u,t,x0);
plot(t,y)
```

From Table 12.2 we see that the poles of the short-period mode have been modified only slightly by the feedback, whereas the phugoid mode poles have been modified significantly to two real poles at -.538 and -2.426 rad/sec; this means about $3/.538 \cong 5.6$ seconds to damp out the worst initial condition error, which seems reasonable. The gains are all on the order of unity or less, which also seems reasonable.

The slowest mode (s = -.538 rad/sec) is the only one that involves significant airspeed perturbations. Thus, if the airspeed is not perturbed, only the other three, faster, modes will be excited, and this means only about $3/2.427 \cong 1.2$ seconds to damp out an initial condition error.

From Table 12.2 the worst initial condition is a high rate of climb (negative w and positive pitch angle) with a positive pitch rate and only a small airspeed deviation. A large elevator deflection is needed to pitch the aircraft down, while only a small throttle change is needed.

For comparison, Fig. 12.6 shows the response of the aircraft without control (controls fixed at their trim values for horizontal flight) to the worst initial conditions of unit magnitude (see Appendix C.9). It takes several minutes, instead of several seconds, for the perturbations to damp out (because of the very lightly damped phugoid mode). The worst initial condition is almost the same as in the closed-loop case.

Climb-Rate/Airspeed Command for the Navion

The SAS previously developed stabilizes the Navion airplane in horizontal flight (zero climb rate) at sea level with an airspeed of 176 ft/sec. We can command the airplane to a nonzero climb rate and/or a different airspeed simply by changing the set point (see Appendix B). This is done in Table 12.3, and the response to commands is shown in Figs. 12.7 and 12.8.

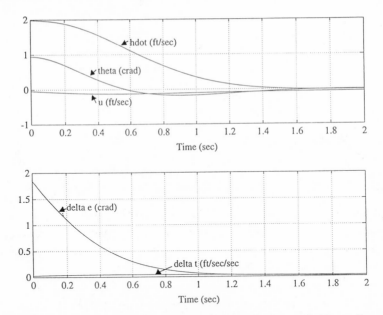

Figure 12.5. Response of Navion with LQR climb-rate/airspeed hold autopilot to the worst initial condition vector of unit magnitude.

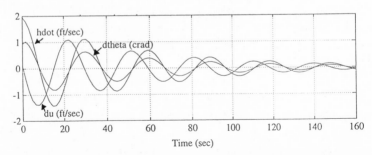

Figure 12.6. Open-loop response of Navion to the worst initial condition vector of unit magnitude.

Table 12.3

```
% Climb-rate/airspeed command for Navion.
% xdot = Fx + Gu; y = Hx + Lu; units ft, sec, crad.
% Use F,G,H,k from Table 12.2
L=zeros(2);
H1=[0 0 0 1];                        % To calculate theta response
```

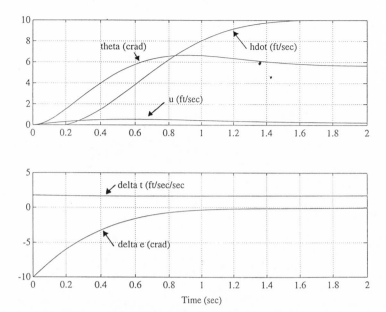

Figure 12.7. Response of Navion with LQR SAS to command climb rate = 10 ft/sec and no velocity change.

```
L1=zeros(1,2);
t=[0:.1:10]';
yc=[0 10]';
[y,y1,u]=stepcmd(F,G,H,L,k,yc,H1,L1,t);
plot(t,[y y1])
plot(t,u)
yc=[10 0]';
[y,y1,u]=stepcmd(F,G,H,L,k,yc,H1,L1,t);
plot(t,[y y1])
plot(t,u)
```

Note the 10 ft/sec climb-rate command takes only about 2 sec and uses reasonable control amplitudes, 10 crad of elevator deflection and 1.5 ft/sec² in specific thrust. The 10 ft/sec velocity command takes 6 to 8 sec and uses 4.5 ft/sec² of specific thrust and about 3 crad of elevator deflection.

A Climb-Rate/Airspeed SAS for the 747 at 40 kft

The 747 airplane has an *inertial reference system*, so good estimates of u, w,

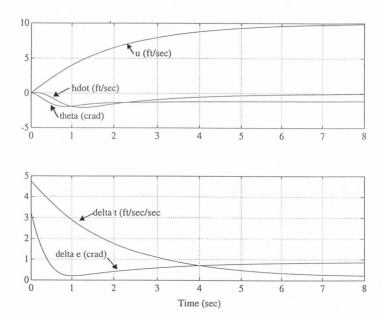

Figure 12.8. Response of Navion with LQR SAS to command for 10 ft/sec increase in velocity and zero climb rate.

and θ are available, and q can be obtained by pseudo-differentiating θ. Thus state feedback can be used without any dynamic compensator.

Unlike the propellor-driven Navion, where blade pitch can be changed very rapidly to change thrust, there is a significant time lag in changing thrust with the turbojet engines on the 747. We model this with a first-order system with a four-second time constant:

$$\dot{\delta}_t = -.25(\delta_t - \delta_{tc}).$$

We synthesize state feedback gains using Equation 12.10 and the preceding thrust lag equation as the dynamic model (see Table 12.3). We assume δ_t can be estimated, and we use the performance index

$$J = \int_0^\infty [A(u^2 + \dot{h}^2) + \delta_e^2 + \delta_{tc}^2]dt,$$

where the units are ft, sec, and crad. This indicates we are willing to use 1 crad of elevator and 1 ft/sec^2 of commanded thrust specific force (δ_{tc}) for an error of $1/\sqrt{(A)}$ ft/sec in either u or $\dot{h}$. Fig. 12.9 is a symmetric root locus vs. A; again this is *not an Evans root locus*, since A enters quadratically in this two-input, two-output symmetric-root characteristic equation.

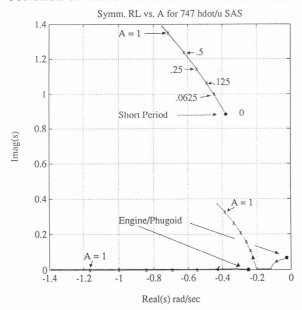

Figure 12.9. Root locus vs. *A* for 747 climb-rate/airspeed hold autopilot.

The closed-loop eigenvalues, the state feedback gains, and the worst initial conditions of unit magnitude are given in Table 12.4 for $A = .25$. The response to the worst initial conditions is shown in Fig. 12.10.

Table 12.4

```
747 climb-rate/airspeed SAS at 40 kft.
State feedback gains, closed-loop eigenvalues, and worst initial
conditions.  x=[u,w,q,theta,delta t]'; u=[delta e,delta tc].
Units ft, sec, crad.
k =
   .034    .363  -1.090  -3.890   -.324
   .487   -.009    .075    .084   1.270
ev =
  -.546 + 1.141j , -.546 - 1.141j , -.292 + .207j ,
  -.292 -  .207j , -.843
xo' =
  -.010  -.101    .135    .986    .015
```

For comparison, Fig. 12.11 shows the response of the 747 without control (controls fixed at their trim values for horizontal flight) to the worst initial

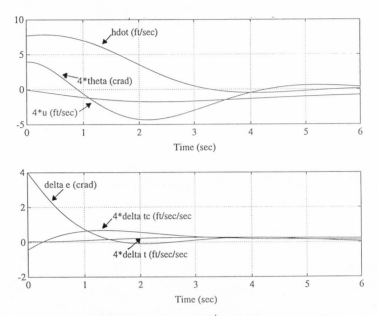

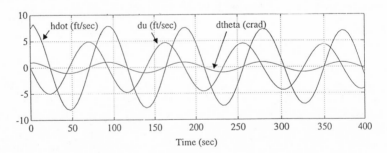

Figure 12.10. Response of 747 at 40 kft with h/u SAS to the worst initial
condition vector of unit magnitude.

Figure 12.11. Open-loop response of 747 at 40 kft to the worst initial condition
vector of unit magnitude.

conditions of unit magnitude. The phugoid mode is almost undamped, so it
dominates the response, which lasts nearly 10 minutes (3 time constants =
3/(60)(.005)). The worst initial condition is almost the same as in the closed-
loop case.

Climb-Rate/Airspeed Command for the 747 at 40 kft

The climb-rate/airspeed hold logic previously developed stabilizes the 747 with zero climb rate at an airspeed of 774 ft/sec at altitudes near 40 kft. To command the airplane to a nonzero climb rate and/or a slightly different airspeed, we simply change the equilibrium set point in the hold logic (Appendix B).

Figs. 12.12 and 12.13 show the responses of the 747 airplane to commands in climb rate and airspeed, respectively.

Note the 10 ft/sec climb-rate command takes only about 4 sec and uses reasonable control amplitudes, 5 crad of elevator deflection, and 1 ft/sec^2 in specific thrust. The 10 ft/sec velocity command takes about 10 sec and uses 1.6 ft/sec^2 of specific thrust and only 1 crad of elevator deflection.

Problem 12.4.1 – *A Displacement Autopilot for the Navion*

In Example 1 of Section 12.3, we showed that the phugoid mode of the Navion airplane could be stabilized by feeding back pitch angle, θ, to elevator, δ_e. In the phugoid mode bandwidth, we also showed that the transfer function from δ_e to θ is well approximated by

$$\frac{\theta(s)}{\delta_e(s)} \cong \frac{-1.626(s + .0523)}{(s + .017)^2 + (.213)^2}.$$

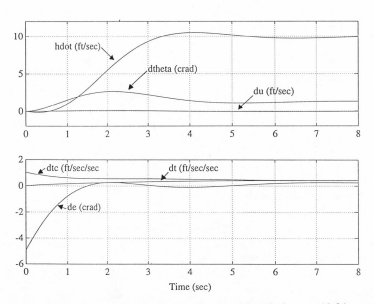

Figure 12.12. Response of 747 at 40 kft to command for climb rate = 10 ft/sec and no velocity change.

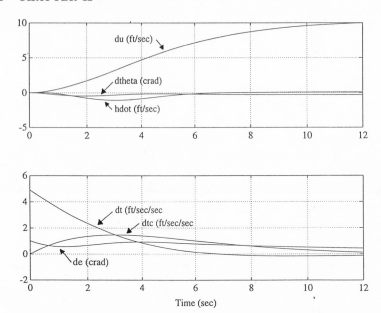

Figure 12.13. Response of 747 at 40 kft to command for 10 ft/sec increase in velocity and zero climb rate.

(a) Using proportional feedback,

$$\delta_e = k\theta,$$

show that the closed-loop characteristic equation in Evans's form is

$$-1.626k \cong \frac{(s + .017)^2 + (.213)^2}{s + .0523}$$

and *plot a root locus* versus k.

(b) Determine k so that there is a double real pole at the break-in point of the root locus of (a). *Show that*, while the phugoid damping has been increased, the short-period damping has been decreased.

(c) Using the value of k from (c), calculate and plot the response of the closed-loop system of (b) to the worst initial conditions of unit magnitude, and compare the response to that of Fig. 12.6, where all four states are fed back to both controls (instead of just one state to one control as here).

Problem 12.4.2 – *Climb-Rate/Airspeed Autopilot for 747 at 40 kft*

Verify Table 12.4 and Figs. 12.9 to 12.13.

Problem 12.4.3 – *Stabilization of 747 at 40 kft Using Lead Compensation on θ to δ_e*

Synthesize a phugoid SAS for the 747 at 40 kft using only θ feedback to δ_e. The transfer function from δ_e to θ is

$$\frac{\theta(s)}{\delta_e(s)} = \frac{-1.16(s + .0113)(s + .295)}{[s^2 + (.0676)^2][(s + .375)^2 + (.882)^2]}.$$

One solution is lead compensation on θ:

$$\frac{\delta_e(s)}{\theta(s)} = 3.50\frac{s + .6}{s + 3.6}.$$

Compare this SAS with the full state feedback SAS developed in this section.

Problem 12.4.4 – *Climb-Rate/Airspeed SAS for 747 in Landing Configuration*

Using the equations of motion given in Problem 12.2.1, synthesize a state feedback SAS for the 747 in its landing configuration.

Problem 12.4.5 – *Climb-Rate/Airspeed SAS for DC-8 in Cruise*

Use the equations of motion given in Problem 12.3.3, but add a first-order system to represent the lag of thrust behind thrust command:

$$\dot{\delta}_t = -.25(\delta_t - \delta_{tc}).$$

(a) Verify that the system has transmission zeros at $s = -10.3$ and $s = 8.98$ rad/sec, which do not limit the speed of response of the closed-loop system.

(b) Plot a locus of closed-loop roots vs. A where the performance index is

$$J = \int_o^\infty [A(u^2 + \dot{h}^2) + 100 * (\delta_e^2 + \delta_{tc}^2)]dt.$$

Note this implies that we are willing to use 1 centirad of δ_e and 1 ft/sec² of $\delta_t c$ when there is a $10/\sqrt{(A)}$ airspeed error or a $10/\sqrt{(A)}$ ft/sec climb-rate error.

(c) For $A = 3$ show that the modified engine-lag/phugoid poles are at $s = -.4681$ and $-.1997 \pm .0744$ rad/sec, and determine the corresponding state feedback gains.

(d) Using nonzero setpoint logic, determine the feedforward gains for commanding airspeed and climb rate.

(e) Plot the responses to commands for 10 ft/sec climb rate and 10 ft/sec increase in airspeed.

Problem 12.4.6 – *Longitudinal Stability Augmentation for DC-8 in Cruise*

From the equations of motion given in Problem 12.3.3, the transfer function from δ_e to θ is

$$\frac{\theta(s)}{\delta_e(s)} = \frac{-4.57(s + .0154)(s + .735)}{[(s + 1.08)^2 + (2.96)^2][(s + .0064)^2 + (.0239)^2]}.$$

(a) Show, by means of a root locus sketch, that proportional feedback of θ to δ_e stabilizes the phugoid mode but reduces damping of the short-period mode.

(b) Show, by means of a root locus plot, that *proportional plus derivative feedback*, $\delta_e = k(\theta + \tau\dot{\theta})$, augments the stability of both modes (take $1/\tau = .8$ sec^{-1}). Note θ and $\dot{\theta}$ could be obtained from a vertical gyro and a rate gyro, respectively. Determine k so that the faster of the two modified phugoid eigenvalues is at $s = -.1$ sec^{-1}, and find the exact closed-loop poles for this k and τ.

(c) Show, by means of a root locus sketch, that *lead compensation* on the θ signal augments the stability of both modes, where

$$\delta_e(s) = k\frac{\tau_1 s + 1}{\tau_2 s + 1}\theta(s).$$

Take $1/\tau_1 = 0.8$ sec^{-1}, $1/\tau_2 = 5.0$ sec^{-1}. Determine k so that the faster of the two modified phugoid eigenvalues is at $s = -.1$ sec^{-1}, and find the exact closed-loop poles for this k, τ_1, and τ_2.

Problem 12.4.7 – *Longitudinal Control of 747 Using Thrust Only ($\delta e = 0$)*

In an emergency when the aerodynamic controls malfunction, many multi-engine aircraft can still be controlled *using thrust only* (e.g. the United Airlines DC-10 incident in the 1980's when the hydraulic lines were severed by turbine blades from the failure of the center engine). See problem 13.5.4 for lateral control using differential thrust on left and right engines.

(a) Synthesize a climb-rate hold autopilot using thrust deviations only, with the elevators fixed at zero deflection, for the 747 at sea level in its landing configuration (cf. Problem 12.4.4).

(b) Plot the response to a command for an increase in climb-rate of 10 ft/sec.

12.5 Control of Altitude and Airspeed

12.5.1 *Altitude/Airspeed Hold*

If zero climb rate is commanded using a climb-rate/airspeed SAS, the airplane will slowly drift in altitude since the climb rate is never measured or estimated

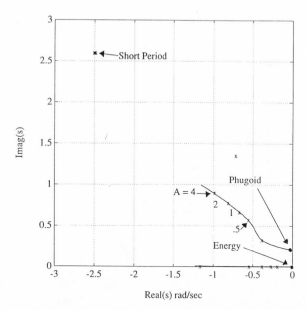

Figure 12.14. Root locus vs. A for Navion altitude/airspeed hold autopilot.

perfectly; to hold or command altitude there must be a feedback of measured altitude.

Thus we must add altitude as a fifth state variable and develop new feedback gains for this augmented system.

An Altitude/Airspeed SAS for the Navion

We use the performance index

$$J = \int_0^\infty [A(u^2 + (h/4)^2) + \delta_e^2 + 9\delta_t^2]dt, \qquad (12.20)$$

where again the units are ft, sec, and crad. By this we indicate that we are willing to use 1 crad of elevator deflection and 1/3 ft/sec^2 of specific thrust if there is a change in airspeed of $1/\sqrt{A}$ ft/sec or a change in altitude of $4/\sqrt{A}$ ft.

Fig. 12.14 shows a root locus vs. A. Note the short-period mode is hardly changed at all over this range of A (0 to 4).

Choosing $A = 1$ stabilizes the energy mode to $s = -.27$ rad/sec, which means that perturbations in energy will be damped out in approximately $3/.27 \cong 12$ sec, which seems reasonable. Table 12.5 shows the feedback gain matrix, the closed-loop eigenvalues, and the worst initial condition vector of unit magnitude.

Table 12.5

Navion altitude/airspeed SAS. State feedback gains, closed-loop eigenvalues, and worst initial conditions. x=[u,w,q,theta,h]';
u=[delta e,delta t]. Units ft, sec, crad.

```
k =
    .254    .286   -.123   -.945   -.230
    .291   -.010   -.002    .006    .032
ev =
  -2.50 + 2.60j , -2.50 - 2.60j , -.668 + .665j ,
   -.668 - .665j , -.270
xo' =
    .990   -.046   -.004    .051    .123
```

Note that only velocity is a significant feedback to throttle, and velocity perturbation is the dominant element in the worst initial condition.

Fig. 12.15 shows the response to the worst initial conditions, which verifies our prediction of about 12 sec to damp out initial perturbations.

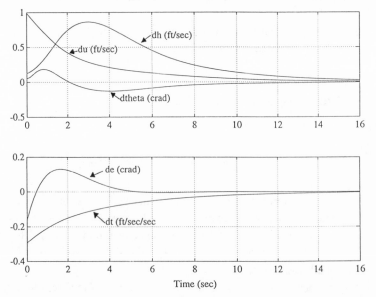

Figure 12.15. Response of Navion with h/u SAS to the worst initial condition vector of unit magnitude.

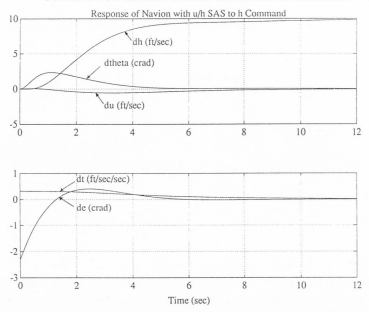

Figure 12.16. Response of Navion with LQR SAS to command for an increase in altitude of 10 ft with no velocity change.

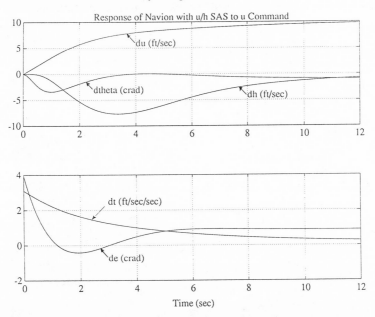

Figure 12.17. Response of Navion with LQR SAS to command for a 10 ft/sec increase in velocity with no change in altitude.

Figs. 12.16 and 12.17 show satisfactory responses to step commands for a 10 ft increase in altitude and a 10 ft/sec increase in velocity, respectively. Reasonable amounts of control are used in both maneuvers.

Note the 10 ft altitude command takes about 10 sec and uses reasonable control amplitudes, 2 crad of elevator deflection, and .3 ft/sec^2 in specific thrust. The 10 ft/sec velocity command takes 10 to 12 sec and uses 3 ft/sec^2 of specific thrust and about 4 crad of elevator deflection.

An Altitude/Airspeed SAS for the 747 at 40 kft

We use the performance index

$$J = \int_0^\infty [A(u^2 + (h/10)^2) + \delta_e^2 + \delta_{tc}^2]dt, \tag{12.21}$$

where again the units are ft, sec, and crad. By this we indicate that we are willing to use 1 crad of elevator deflection and 1 ft/sec^2 of specific thrust if there is a change in airspeed of $1/\sqrt{(A)}$ ft/sec or a change in altitude of $10/\sqrt{(A)}$ ft.

Fig. 12.18 shows a root locus vs. A. The short period mode hardly changes over this range of A (0 to 1). The engine and energy modes combine into a complex eigenvalue pair, and the modified phugoid mode eigenvalues stay complex.

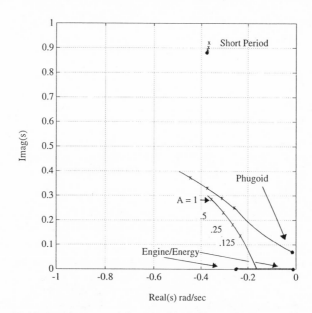

Figure 12.18. Root locus vs. A for 747 altitude/airspeed hold autopilot.

Choosing $A = .25$ stabilizes the modified energy/engine mode to $s = -.264\pm$ $.182j$ sec-1, which means that perturbations in energy will be damped out in approximately $3/.264 \cong 12$ sec, which seems reasonable. Table 12.6 shows the feedback gain matrix, the closed-loop eigenvalues, and the worst initial condition vector of unit magnitude.

Table 12.6

747 altitude/airspeed SAS. State feedback gains, closed-loop
eigenvalues, and worst initial conditions. x=[u,w,q,theta,h,
delta t]'; u=[delta e,delta tc]. Units ft, sec, crad.

```
k =
    .108    .194   -.531  -1.88   -.048  -.153
    .472   -.012    .039   .101    .016  1.223
ev =
   -.371 ± .893j , -.309 ± .289j , -.264 ± .182j
xo' =
   -.018  -.097    .249   .948    .031   .168
```

Note that only velocity and thrust specific force are significant feedbacks to throttle, and the worst initial condition corresponds mostly to climb-rate positive pitch angle and negative w as in the Navion example.

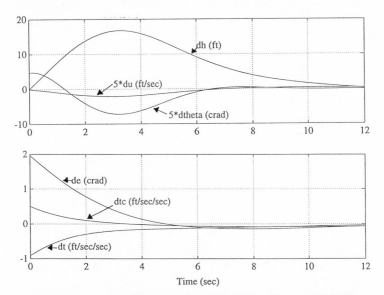

Figure 12.19. Response of 747 at 40 kft with h/u SAS to the worst initial condition vector of unit magnitude.

Fig. 12.19 shows the response to the worst initial conditions, which verifies our prediction of about 12 sec to damp out initial perturbations.

Figs. 12.20 and 12.21 show satisfactory responses to step commands for a 100 ft increase in altitude and a 10 ft/sec increase in velocity, respectively.

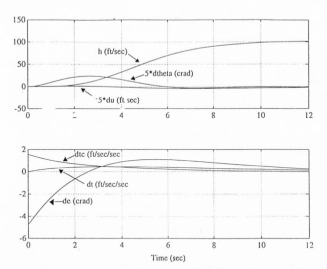

Figure 12.20. Response of 747 at 40 kft to command for an increase in altitude of 100 ft with no velocity change.

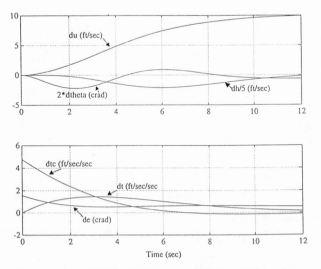

Figure 12.21. Response of 747 at 40 kft to command for a 10 ft/sec increase in velocity with no change in altitude.

Reasonable amounts of control are used in both maneuvers. Note that the 100 ft altitude command takes about 10 sec and uses reasonable control amplitudes, 5 crad of elevator deflection and .6 ft/sec^2 in specific thrust. The 10 ft/sec velocity command also takes about 10 sec and uses 1.5 ft/sec^2 of specific thrust and about 1.5 crad of elevator deflection.

12.5.2 *Altitude Capture*

Altitude capture is implemented by adding logic that switches the autopilot mode from climb-rate/airspeed hold to altitude/airspeed command at an appropriate time τ_{sw} before the desired altitude is reached. The desired cruise altitude is entered into the control computer some time before the desired cruise altitude is reached.

Let the switching altitude be an amount h_{sw} below the desired cruise altitude (see Fig. 12.22). It is reasonable to choose h_{sw} to be proportional to the climb-rate:

$$h_{sw} = \tau_{sw}\dot{h}, \qquad (12.22)$$

where τ_{sw} is the parameter to be selected.

Altitude Capture Logic for the Navion

Using level flight at 176 ft/sec as the reference equilibrium condition the state variables in a steady climb at 176 ft/sec are given by (12.6):

$$u = w = q = 0,$$
$$\theta = .568\dot{h}. \qquad (12.23)$$

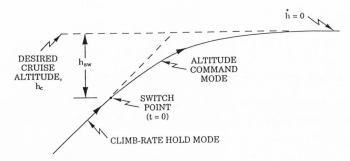

Figure 12.22. Altitude capture nomenclature.

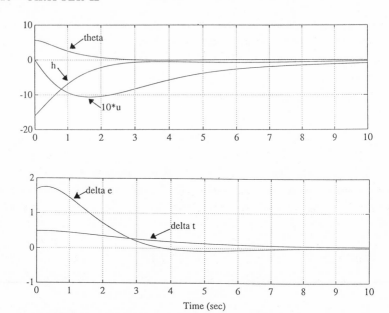

Figure 12.23. Navion altitude capture trajectory for a climb-rate of 10 ft/sec and $h_{sw} = 16$ ft.

Consider an example of altitude capture for the Navion when the steady climb rate is 10 ft/sec. From (23):

$$\theta(0) = 5.68 \quad \text{crad.} \tag{12.24}$$

Using this initial value (and $u(0) = w(0) = q(0) = 0$), with the altitude-hold logic, we chose $h_{sw} = 16$ ft. Fig. 12.23 shows the corresponding altitude capture path; note the drop in throttle δ_t and the elevator pulse (increasing δ_e for about 3 seconds).

Precision Command or Hold Logic Using Integral-Error Feedback

For more precise control of outputs, integral-error feedback should be used. If y_1 is a measured output and y_{1c} is its commanded value, introduce an additional state variable e_1, where

$$\dot{e}_1 = y_1 - y_{1c}.$$

If e_1 is included in the feedback logic and the closed loop is stable, then $\dot{e}_1 \to 0$ in the steady state and

$$y_1 \to y_{1c}.$$

This will hold even when

(a) Unmodeled constant disturbances exist.

(b) The model contains small errors.

(c) Unmodeled small nonlinearities exist.

As an example, altitude command logic may be interpreted as integral-error logic for climb-rate commands; simply replace h (measured altitude change) by e, where $\dot{e} = \dot{h}_m - \dot{h}_c$, where $\dot{h}_m$ is measured climb rate, and $\dot{h}_c$ is commanded climb rate; this will bring $\dot{h}_m$ to $\dot{h}_c$ precisely in spite of modeling errors and constant disturbances. Note, however, that if $\dot{h}_c = 0$, the altitude will still drift since $\dot{h}_m$ is never exactly correct.

Problem 12.5.1 – *Altitude/Airspeed Autopilot for the Navion*

Verify Table 12.5 and Figs. 12.14 to 12.17.

Problem 12.5.2 – *Altitude/Airspeed Autopilot for 747 at 40 kft*

Verify Table 12.6 and Figs. 12.18 to 12.21.

Problem 12.5.3 – *Altitude Capture Logic for the 747 at 40 kft*

(a) Using level flight at 40 kft and 774 ft/sec airspeed as the reference equilibrium condition, show that the equilibrium states in a steady climb with airspeed = 774 ft/sec are

$$\begin{aligned}
\delta_e &= .0229\dot{h}, \\
\delta_{tc} &= \delta_t = .0413\dot{h}, \\
u &= q = 0, \\
w &= -.0181\dot{h}, \\
\theta &= .1269\dot{h}.
\end{aligned}$$

(b) Using the altitude/airspeed hold logic in the text, and the initial conditions from (a) along with $h = -h_{sw}$, verify the altitude capture path for $\dot{h}(0) = 20$ ft/sec and $h_{sw} = 80$ ft shown in Fig. 12.24. Note the drop in throttle command δ_{tc} and the increase in δe at $t = 0 +$.

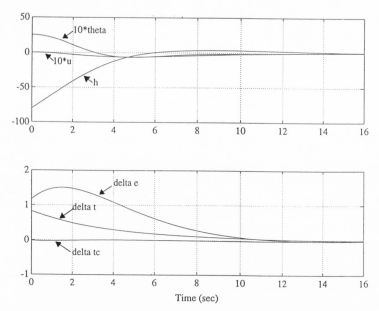

Figure 12.24. 747 altitude capture at 40 kft with a climb rate of 20 ft/sec and $h_{sw} = 80$ ft. Units ft, sec, crad.

12.6 Glide-Slope Capture and Hold

The Instrument Landing System (ILS)

Current automatic landing (autoland) systems use the ILS and a radar altimeter. The ILS consists of two fan beams, a glide-slope beam and a localizer beam (see Fig. 12.25).

The ILS-derived measurements of d and y are inputs to the pilot (or autopilot) who keeps $d \cong 0, y \cong 0$ on the approach. At an altitude of about 50 ft above the runway, d is replaced by h from the radar altimeter and a flare is performed that reduces the vertical descent rate from between 7 to 15 ft/sec to between 1 to 2 ft/sec for touchdown (see Section 12.7). The localizer beam is used during approach, flare, and the landing rollout (deceleration on the runway to taxiing speed). Use of the localizer beam is discussed in Chapter 13 on lateral control.

Glide-Slope Capture and Hold

The glide-slope radio signal received in the airplane gives angular deviation of the aircraft from the center of the beam. This signal is displayed to the pilot, who

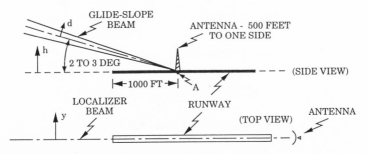

Figure 12.25. ILS beams for landing Approach.

can move the controls to null this signal for a manual ILS approach. The ILS glide-slope signal may also be fed into the autopilot and used with appropriate logic to move the elevators and the throttle automatically so as to keep the airplane on the glide-slope beam with a specified airspeed.

If the glide-slope signal is multiplied by the altitude above the runway and an appropriate scaling factor, the result will be an approximation to d, the linear deviation of the airplane from the center of the beam. Autopilot logic for glide-slope capture and hold, using the signal $d(t)$, is similar to logic for altitude capture and hold using an altitude signal $h(t)$, which was discussed in Section 12.5. However, the response must be more rapid, and the ILS signal, $d(t)$, is much noisier than the altimeter signal, so it must be filtered.

Capturing and holding the glide-slope beam can be done automatically as follows. The airplane is lined up with the runway using VOR/DME and the ILS localizer beam and flown at an altitude of about 1500 ft. When the glide-slope signal is acquired, the control logic is automatically switched from the altitude-hold mode to the glide-slope-hold mode when $d = -d_{sw}$, where d_{sw} is proportional to $\dot{d}$ (see Fig. 12.26). At the switch point, the throttle is then automatically cut back and the elevator is automatically used to pitch the nose down to capture and hold (that is, track) the glide-slope beam.

Glide-Slope Capture and Hold for the Navion

If the Navion approaches along the glide slope with velocity 176 ft/sec, this corresponds to a climb rate of

$$\dot{h}_G = -176(2.5\pi/180) = -7.68 \text{ ft/sec.}$$

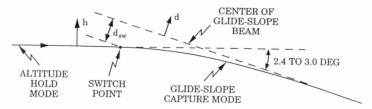

Figure 12.26. Glide-slope capture nomenclature.

Using steady flight along the glide slope at 176 ft/sec as the reference equilibrium condition, the control and state variables in steady level flight at 176 ft/sec are given by (12.6):

$$\delta_e = 0,$$
$$\delta_t = .183(-\dot{h}_G),$$
$$u = w = q = 0,$$
$$\theta = .568(-\dot{h}_G). \tag{12.25}$$

Using (25) as initial conditions with $d(0) = -d_{sw}$ and the altitude-hold logic of Section 12.6, the problem is *identical to the altitude capture* problem except for a difference in sensors (noisy glide-slope beam signal d instead of quiet altimeter signal h).

To filter the noisy glide-slope signal $z_d(t)$, we could use a *complementary filter* such as

$$\dot{\hat{d}} = -w + V\theta + \frac{1}{\tau_d}(z_d - \hat{d}), \tag{12.26}$$

where w/V = perturbation in angle of attack, θ = perturbation in pitch angle (from a vertical gyro), τ_d = filter time constant (to be selected), and V = airplane velocity. If angle of attack is not measured, it can be omitted from (26) with some small loss in accuracy during transients. In Ref. MC, τ_d is selected to be 0.5 sec (cf. DC-8 glide-slope example, pp. 638–660). This is such a short time compared to the time constant of the closed-loop energy mode (1/.267= 3.75 sec) that the effect of the filter may be ignored in simulations. Thus the glide-slope capture path would be very similar to the altitude capture path of Fig. 12.23. The capture is essentially completed in 10 sec, which corresponds to a drop in altitude of about 60 ft.

Glide-Slope Capture and Hold for the 747

To synthesize glide-slope capture logic for the 747 we must first synthesize altitude hold logic for the sea level, low-airspeed flight condition of landing approach. The perturbation equations of motion at sea level at a weight of 564,000 lb and a velocity of 221 ft/sec were given in Problem 12.2.1. However, for approach the engine thrust lag behind the throttle commands is significant, so we use the same first-order dynamic model for δ_t that we used in Sections 12.4 and 12.5. We use a similar performance index:

$$J = \int_0^\infty \left(u^2 + d^2 + 10\delta_t^2 + 10\delta_e^2 + \frac{10}{3}\delta_{tc}^2 \right) dt, \qquad (12.27)$$

where u is in ft/sec, d in ft, (δ_t, δ_{tc}) in ft sec^{-2}, δ_e in crad. With these data, the computed control law is

$$\begin{bmatrix} \delta_e \\ \delta_{tc} \end{bmatrix} = -\begin{bmatrix} -.194 & .620 & -1.370 & -2.429 & -.250 & -.824 \\ .753 & -.765 & 1.784 & 2.952 & .335 & 3.003 \end{bmatrix} \begin{bmatrix} u \\ w \\ q \\ \theta \\ d \\ \delta_t \end{bmatrix},$$

$$(12.28)$$

and the corresponding closed-loop eigenvalues are

$$s = -.406 \pm .597j \,, \ -.436 \pm .458j \,, \ -.554 \,, \ -.203 \ \text{sec}^{-1}. \qquad (12.29)$$

The corresponding open-loop eigenvalues are

$$s = -.480 \pm .608j \,, \ -.250 \,, \ -.001 \pm .152j \,, \ 0 \ \text{sec}^{-1}. \qquad (12.30)$$

Thus the short-period eigenvalues are changed slightly, while the engine-lag, phugoid, and energy modes are all stabilized significantly.

Using steady descent on the glide slope as our reference equilibrium condition, the control and state variables in steady level flight may be obtained from the partial answer to Problem 12.2.1:

$$\delta_e(0-) = .222(-\dot{h}_G),$$
$$\delta_{tc}(0-) = .1426(-\dot{h}_G),$$

$$u(0) = q(0) = 0,$$
$$w(0) = -.0386(-\dot{h}_G),$$
$$\theta(0) = .435(-\dot{h}_G), \tag{12.31}$$

where $\dot{h}_G$ is in ft/sec. For glide-slope inclination = 2.5 deg,

$$-\dot{h}_G = 221(2.5\pi/180) = 9.64 \text{ ft/sec.}$$

A *complementary filter* for estimating d from the noisy glide-slope beam signal, $z_d \equiv d+$ noise, is

$$\hat{h} = h_{IRS} + 2.0(z_d - \hat{d}), \tag{12.32}$$

where h_{IRS} = rate of climb signal for the IRS. This filter is turned on several seconds before estimated glide-slope intercept, so the mean value of the estimate error, $\hat{h} - d$, should be negligible at (and after) the switch time. In use, $\hat{h}$ is fed back, and it is slightly noisy but much less so than z_d. In simulations when we are not interested in the effects of the noise in the z_d signal we can feed back d.

The specific thrust δ_t can be estimated open-loop:

$$\dot{\hat{\delta}}_t = -\frac{1}{4}(\hat{\delta}_t - \delta_{tc}), \tag{12.33}$$

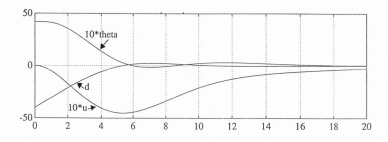

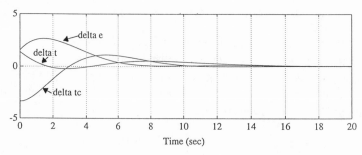

Figure 12.27. 747 glide-slope capture using $d_{sw} = 40$ ft. Units ft, sec, crad.

and the pitch rate is estimated using a fast observer:

$$\dot{\hat{q}} = 10(\dot{\theta} - \hat{q}),$$
$$\text{or } \hat{q} = q^* + 10\theta,$$
$$\text{where } \dot{q}^* = -10\hat{q}.$$

Fig. 12.27 shows a simulated glide-slope capture for the 747, neglecting beam noise and using $d_{sw} = 40$ ft.

12.7 Flare

Introduction

Flare logic may be developed from the glide-slope hold logic of the previous section. At some altitude, h_{sw}, above the runway, the radar altimeter signal, $z_h \triangleq h + v_h$, is substituted for the glide-slope signal, $z_d = d + v_d$, and commands of near-zero altitude ($h_c = -h_f$) and a lower airspeed are executed (see Fig. 12.28). $h_c = -h_f \neq 0$ is used because this causes the aircraft to touch down with a small sink rate (1 to 2 ft/sec), which avoids a long "float" above the runway before touchdown. A lower airspeed is commanded in order to shorten the distance required to brake to a stop on the runway.

We choose the switch altitude h_{sw} and h_f by running several simulations of the flare, varying the parameter $h_{sw} + h_f$. In these simulations we pick the point where $\dot{h} = -1.5$ ft sec^{-1} as the touchdown point, which then determines h_f and h_{sw}.

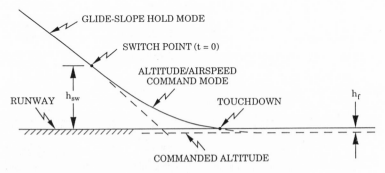

GLIDE-SLOPE HOLD MODE

SWITCH POINT (t = 0)

ALTITUDE/AIRSPEED
COMMAND MODE

RUNWAY h_{sw}

TOUCHDOWN

h_f

COMMANDED ALTITUDE

Figure 12.28. Flare nomenclature (glide-slope angle exaggerated).

Flare Logic for the Navion

Fig. 12.29 shows a simulated flare with zero headwind for the Navion using the h/u command logic of Section 12.5, where $h_{sw} = 24.4$ ft, $h_f = 5.6$ ft, and the commanded airspeed change is -10 ft/sec. The reference equilibrium point for Fig. 12.29 is horizontal flight with airspeed $= 176 - 10 = 166$ ft/sec. Assuming an airspeed of 176 ft/sec on the glide slope and a glide-slope inclination of 2.5 deg, the sink rate is $176(2.5\pi/180) = 7.68$ ft/sec at the begin-flare point. From (12.6) then, the initial conditions for the simulated flare of Fig. 12.29 are

$$
\begin{aligned}
u(0) &= 10 \ \text{ft/sec,} \\
w(0) &= -1.95 \ \text{ft/sec,} \\
q(0) &= 0, \\
\theta(0) &= -5.47 \ \text{crad,} \\
h'(0) &= h_{sw} + h_f,
\end{aligned}
$$

and the elevator and throttle settings just before beginning the flare are

$$
\begin{aligned}
\delta_e(0-) &= .88 \ \text{crad,} \\
\delta_t(0-) &= -1.24 \ \text{ft/sec}^2.
\end{aligned}
$$

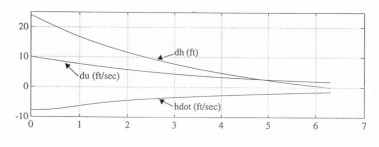

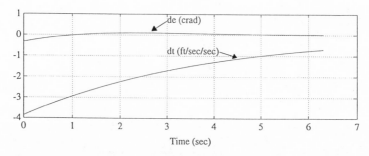

Figure 12.29. Navion landing flare using glide-slope altitude/airspeed command logic (no gusts or sensor noise).

For the simulation we assumed no gusts and no sensor noise, so estimated state is identical to simulated state and the response is the same as the ideal response where perfect estimates of all the states are fed back to elevator and throttle.

To allow for off-nominal descent velocity, we could use

$$h_{sw} = 24.4 \frac{\dot{h}}{7.68} \text{ft},$$

$$h_f = 5.6 \frac{\dot{h}}{7.68} \text{ft},$$

where $\dot{h}$ = actual descent velocity in ft/sec.

Note that thrust is cut sharply at the begin-flare point and then slowly put back. The elevator is changed by about .6 deg to pitch up the nose and is held there. The descent rate changes smoothly from the 7.7 ft/sec of the glide slope to 1.5 ft/sec at $t = 6.3$ sec, the touchdown point ($h = 0$).

Flare Logic Using Exponential Model Following

Flare logic for the 747 can be synthesized in the same way as for the Navion. However, some improvement can be made in the smoothness of the transition to the flare maneuver if we use exponential model following (cf. Appendix C).

Consider a dynamic system

$$\begin{bmatrix} \dot{x} \\ y \end{bmatrix} = \begin{bmatrix} F & G \\ H & L \end{bmatrix} \begin{bmatrix} x \\ u \end{bmatrix}, \tag{12.34}$$

which is initially in an equilibrium condition ($\dot{x} = 0$) such that

$$\begin{bmatrix} 0 \\ y_o \end{bmatrix} = \begin{bmatrix} F & G \\ H & L \end{bmatrix} \begin{bmatrix} x(0) \\ u(0) \end{bmatrix}. \tag{12.35}$$

We wish to command the output y to zero exponentially, that is,

$$y_c = y_o e^{-at}. \tag{12.36}$$

Let $x = x_o e^{-at}$, $u = u_o e^{-at}$, and substitute these and (36) into (34). This gives

$$\begin{bmatrix} aI + F & G \\ \\ H & L \end{bmatrix} \begin{bmatrix} x_o \\ \\ u_o \end{bmatrix} = \begin{bmatrix} 0 \\ \\ y_o \end{bmatrix}, \tag{12.37}$$

which may be solved to yield

$$x_o = M y_o, \quad u_o = N y_o. \tag{12.38}$$

If we use control logic

$$u = u_o e^{-at} - C \left(x - x_o e^{-at} \right), \tag{12.39}$$

where C is a stable gain matrix, then

$$u \ \rightarrow \ u_o e^{-at}, \tag{12.40}$$

$$x \ \rightarrow \ x_o e^{-at}. \tag{12.41}$$

Flare Logic for the 747

Figs. 12.30 and 12.31 show a simulated flare for the 747 using exponential model following with $1/a = 6.5$ seconds. The reference condition is horizontal

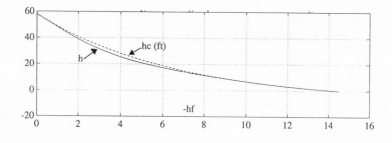

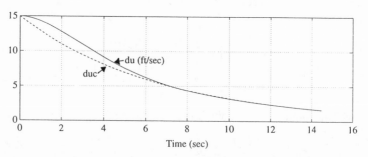

Figure 12.30. 747 landing flare using exponential model-following logic.

flight at touchdown velocity, with initial conditions corresponding to landing approach on a 2.5 deg glide slope with V = 221 ft/sec (corresponding to a sink rate of 9.64 ft/sec). We command a reduction in airspeed to V = 206 ft/sec, and a reduction in sink rate to 1 ft/sec. We found h_{sw} = 58 ft (wheel height above the runway) and h_f = 7 ft to give good performance. The dashed lines are exponential commands or "commanded exponential states," $x = x_o e^{-at}$. The feedback gain matrix C was synthesized using

$$J = \int_0^\infty \left[u^2 + .03h^2 + (\delta t)^2 + (\delta e)^2 + (\delta t_c)^2/3 \right] dt, \qquad (12.42)$$

with units of ft, sec, crad, $(\delta t, \delta t_c)$ in ft/sec^2. Table 12.7 is a Matlab.m file for synthesizing the control logic and simulating the flare.

Note thrust command is cut sharply at the begin-flare point and then slowly put back. The elevator is also changed sharply to pitch up the nose and is then partly returned. The descent rate changes smoothly from 9.6 ft/sec on the glide slope to 1 ft/sec at t = 14.6 sec, the touchdown point (h = 0).

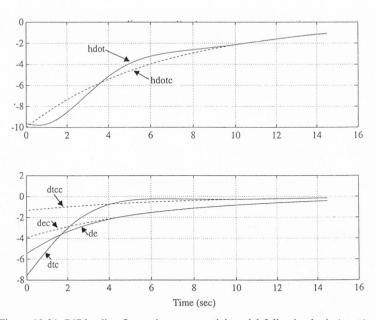

Figure 12.31. 747 landing flare using exponential model-following logic (cont.).

Table 12.7

```
% Landing flare for 747 using exponential model-following.
% x = [u,w,q,theta,h,dt]'; y = [u,h]'; u = [de,dtc]';
% Units ft, sec, crad; xdot = Fx + Gu; y = Hx + Lu;
F=[-.021 .122 0 -.322 0 1;-.209 -.530 2.21 0 0 -.044;...
    .017 -.164 -.412 0 0 .544;0 0 1 0 0 0;0 -1 0 2.21 0 0;...
    0 0 0 0 0 -.25];
G=[.010 -.064 -.378 0 0 0;0 0 0 0 0 .25]';
H3=[1 0 0 0 0 0;0 0 0 0 1 0;0 0 0 0 0 1];L3=zeros(3,2);
A=diag([1 .03 1]);B=diag([1 1/3]);
k=lqr(F,G,H3'*A*H3,B);
a=1/6.5;hsw=58;hf=7;hdot0=-221*2.5*pi/180;du=15;
H=H3([1 2],:);L=zeros(2);
S=[F+a*eye(6) G;H L];T=[zeros(6,2);eye(2)];
MN=S\T;M=MN([1:6],:);N=MN([7 8],:);
% Calculate initial & final equilibrium states/controls:
F4=F([1:4],[1:4]);G4=[G([1:4],1) F([1:4],6)];
H4=[1 0 0 0;0 -1 0 2.21];L4=zeros(2);
xu0=[F4 G4;H4 L4]\[0 0 0 0 0 hdot0]';
xuf=[F4 G4;H4 L4]\[0 0 0 0 -du 0]';
dxu=xu0-xuf;
tf=log((hsw+hf)/hf)/a;
t=tf*[0:.01:1]';
% Feedforward/feedback control law:
% u = N*yc(t) - k*[x - M*yc(t)];
Fcl=F-G*k;Gcl=G*(N+k*M);            % xdot = Fcl*x + Gcl*yc
H1=[H;[0 -1 0 2.21 0 0];-k];L1=[L;[0 0];N+k*M];
yc=exp(-a*t)*[du hsw+hf];
x0=[dxu([1:4]);hsw+hf;dxu(6)];
y=lsim(Fcl,Gcl,H1,L1,yc,t,x0);
xc=yc*M';uc=yc*N';
y(:,2)=y(:,2)-hf*ones(101,1);
yc(:,2)=yc(:,2)-hf*ones(101,1);
axis([0 16 -20 60]);
clg;subplot(211),plot(t,[y(:,2) yc(:,2)]);grid;
text(2,20,'h');text(5,30,'hc (ft)');text(7,-18,'-hf');
xlabel('Time (sec)');
title('747 Landing Flare Using Exponential Model Following')
axis([ 0 16 0 15]);
plot(t,[y(:,1) yc(:,1)]);grid
xlabel('Time (sec)');
text(5,8,'du (ft/sec)');text(3,6,'duc');
pause
axis([0 16 -10 0])
clg;subplot(211),plot(t,[y(:,3) xc*H1(3,:)']);grid
```

```
xlabel('Time (sec)');
title('747 Landing Flare Using Exponential Model Following')
text(3,-3,'hdot');text(6,-6,'hdotc');
axis([ 0 16 -8 2]);
plot(t,[y(:,4) uc(:,1)]);grid;hold on
plot(t,[y(:,5) uc(:,2)]);hold off
xlabel('Time (sec)');
text(1,-7,'dtc');text(3,-4,'de');
text(1,.2,'dtcc');text(.5,-3,'dec');
```

u, w, θ are available from the inertial reference system, q can be estimated from θ using a fast observer, δ_t can be estimated open-loop, and h can be estimated using a complementary filter of the form

$$\dot{\hat{h}} = h_{IRS} + \frac{1}{\tau_h}(z_h - \hat{h}), \tag{12.43}$$

where h_{IRS} is climb rate from the inertial reference system, z_h is altitude from the radar altimeter, and τ_h is selected based on the noise in the radar altimeter measurement (on the order of 1 sec).

The flare can be *wind-proofed*—made nearly independent of wind velocity— by using an estimate of position along the runway x and entering the feedforward commands as a function of x instead of time t, where $\dot{x} = \widehat{V}$ is the estimated ground speed (from the INS), or where $x = \hat{x}$ = estimated distance down the runway from DME or MLS measurements (suggested by A. A. Lambregts of the Boeing Company).

13

Control of Lateral Motions of Aircraft

13.1 Introduction

The lateral motions of an aircraft are controlled by aileron δa and rudder δr. The outputs of interest are roll (or bank) angle ϕ, lateral specific force a_y, sideslip velocity v, yaw rate r, heading angle ψ, and deviation from a desired track δy.

Deviation from a desired track includes deviation from a VOR beam (or on-route computed track) and deviation from the center of the localizer beam on approach to landing (see Fig. 12.25).

When lateral specific force, a_y, is zero in a banked turn, the turn is called "coordinated." Lateral specific force is produced by lateral aerodynamic forces due to sideslip velocity and rudder deflection.

Before discussing stability and stability augmentation, we shall discuss steady turning and sideslipping flight.

13.2 Steady Bank Angle and Lateral Specific Force

Conventional aircraft bank in order to turn, that is, they tilt the wing lift vector in the desired direction of turn. For passenger comfort this is usually done with almost zero sideforce (a "coordinated" turn). Sideforce is aerodynamic, produced by sideslip and rudder deflection.

Fig. 13.1 shows the dynamic equilibrium for a steady right turn at constant altitude. The resultant R of the aerodynamic and thrust forces is equal and opposite to the vector sum of gravitational force mg and the horizontal centrifugal force $mV\dot\psi$, where $\dot\psi$ is the steady turn rate and V is the constant aircraft velocity. R must be increased relative to straight horizontal flight (by increasing angle of attack) so that $R\cos\phi = mg$.

The angle $\bar\phi \cong \phi$ if $|\theta| << 1$ where

$$\tan\bar\phi = V\dot\psi/g. \tag{13.1}$$

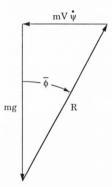

Figure 13.1. Dynamic equilibrium for a steady right turn at constant altitude.

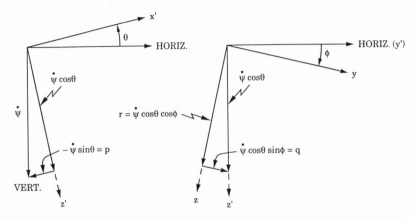

Figure 13.2. Resolution of vertical turn rate $\dot{\psi}$ into body axis components.

Fig. 13.2 shows the resolution of the aircraft angular velocity $\dot{\psi}$, which is vertical, into the body axes, giving

$$
\begin{bmatrix} p \\ q \\ r \end{bmatrix} = \dot{\psi} \begin{bmatrix} -\sin\theta \\ \cos\theta \sin\phi \\ \cos\theta \cos\phi \end{bmatrix} . \tag{13.2}
$$

Fig. 13.3 shows the resolution of the aircraft velocity V, which is horizontal,

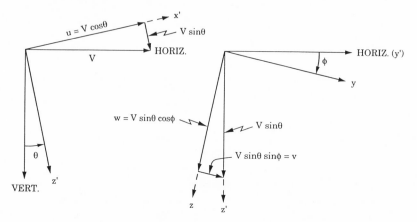

Figure 13.3. Resolution of horizontal velocity V into body axis components.

into the body axes, giving

$$
\begin{bmatrix} u \\ v \\ w \end{bmatrix} = V \begin{bmatrix} \cos \theta \\ \sin \theta \sin \phi \\ \sin \theta \cos \phi \end{bmatrix}. \tag{13.3}
$$

The steady longitudinal state must be changed to make the turn (significantly increased angle of attack, slight change in pitch angle, and small pitch rate).

The steady force balance, with $Y = 0$ for a coordinated turn, from (10.6), is

$$
m(qw - rv) = X - mg \sin \theta + T \cos \epsilon, \tag{13.4}
$$

$$
mru = mg \cos \theta \sin \phi, \tag{13.5}
$$

$$
-mqu = Z + mg \cos \theta \cos \phi - T \sin \epsilon. \tag{13.6}
$$

The quadratic terms on the left side are the components of $mV\dot{\psi}$ in body axes.

From (2), (3), and (5), we have the precise relation for the bank angle (cf. the approximate relation (1)):

$$
\tan \phi = V\dot{\psi} \cos \theta / g. \tag{13.7}
$$

The steady moment balance, from (10.7), is

$$
(I_z - I_y)qr = L, \tag{13.8}
$$

$$
I_{xz}r^2 = M, \tag{13.9}
$$

$$
-I_{xz}qr = N. \tag{13.10}
$$

The small quadratic terms on the left arise because the angular momentum vector is not quite in the same direction as the turn rate vector.

Given V and $\dot{\psi}$, Eqns. (2), (3), (4)–(6), and (8)–(10) are 12 (rather complicated) equations for the 12 quantities $(u, v, w, p, q, r, \theta, \phi, \delta e, \delta t, \delta a, \delta r)$.

These steady turn relations are greatly simplified if we assume that $|\theta| \ll 1$ and $|\phi| \ll 1$, since then the lateral and longitudinal equations are uncoupled; we can use (1) to determine ϕ, and

$$p \cong 0,$$
$$q \cong 0,$$
$$r \cong \dot{\psi}.$$

Also, we have

$$u \cong V,$$
$$v \cong 0,$$
$$w \cong V\theta,$$

and (8)–(10) simplify to

$$L \cong M \cong N \cong 0.$$

Steady Turning of the Navion

From Teper (Ref. TE), the lateral perturbation equations for deviations from rectilinear flight are

$$
\begin{bmatrix} \dot{v} \\ \dot{r} \\ \dot{p} \\ \dot{\phi} \end{bmatrix}
=
\begin{bmatrix}
-.254 & -1.76 & 0 & .322 \\
2.55 & -.76 & -.35 & 0 \\
-9.08 & 2.19 & -8.40 & 0 \\
0 & 0 & 1.0 & 0
\end{bmatrix}
\begin{bmatrix} v \\ r \\ p \\ \phi \end{bmatrix}
+
\begin{bmatrix}
0 & .1246 \\
-.222 & -4.60 \\
29.0 & 2.55 \\
0 & 0
\end{bmatrix}
\begin{bmatrix} \delta a \\ \delta r \end{bmatrix}
$$

$$+ \begin{bmatrix} .254 & 0 \\ -2.55 & .35 \\ 9.08 & 8.40 \\ 0 & 0 \end{bmatrix} \begin{bmatrix} v_w \\ p_w \end{bmatrix}, \qquad (13.11)$$

where (v, v_w) are in ft/sec, (r, p, p_w) in crad/sec, and $(\phi, \delta a, \delta r)$ in crad. v_w is lateral wind velocity and p_w is roll gust velocity, which is produced by variations in vertical wind w_w in the lateral direction.

The lateral specific force at the center of mass, a_y, is given by

$$a_y = -.254(v - v_w) + .1246\delta r, \qquad (13.12)$$

where a_y is in ft/sec^2. Lateral specific force is the sum of the lateral specific force due to sideslip velocity with respect to the air mass plus the lateral specific force due to rudder. If the (x, y, z) body-axis coordinates of the lateral specific force sensor with respect to the center of mass are $(x_s, 0, z_s)$, then its output will be (12) plus $(x_s \dot{r} - z_s \dot{p})$.

The steady-state aileron and rudder deflections required to produce small steady-state changes in bank angle, ϕ, and lateral specific force, a_y, may be determined from (11) and (12) by putting

$$\dot{v} = \dot{r} = \dot{p} = \dot{\phi} = 0. \qquad (13.13)$$

Clearly $p = 0$, and by eliminating $v - v_w$ using (12), we have (for $p_w = 0$)

$$\begin{bmatrix} 1.76 & 0 & 0 \\ .760 & .222 & 3.349 \\ 2.19 & 29.0 & -1.904 \end{bmatrix} \begin{bmatrix} r \\ \delta a \\ \delta r \end{bmatrix} = \begin{bmatrix} .322 & 1.0 \\ 0 & -10.04 \\ 0 & -35.75 \end{bmatrix} \begin{bmatrix} \phi \\ a_y \end{bmatrix}. \qquad (13.14)$$

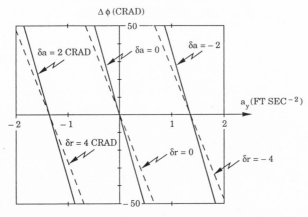

Figure 13.4. Steady-state aileron and rudder settings as functions of bank angle and lateral specific force; Navion at sea level and $V = 176$ ft/sec.

Inverting (14), and using (12) for $v - v_w$, we have

$$
\begin{bmatrix} \delta a \\ \delta r \\ r \\ v - v_w \end{bmatrix} = \begin{bmatrix} -.0165 & -1.475 \\ -.0404 & -3.03 \\ .1830 & .568 \\ -.0198 & -5.42 \end{bmatrix} \begin{bmatrix} \phi \\ a_y \end{bmatrix}. \tag{13.15}
$$

Fig. 13.4 shows steady-state $(\delta a, \delta r)$ as functions of steady-state (ϕ, a_y). The steady-state control deflections are quite small even for large bank angles, and aileron deflection is approximately twice as effective as rudder deflection. If turn coordination were not important (as in an unmanned vehicle), either δa or δr could be used for controlling bank angle.

Thus, in order to hold a steady bank angle of (say) 30 deg, with $a_y = 0$, a *coordinated turn*, we require only small steady deflections of the controls, $\delta a = -.49$ deg, $\delta r = -1.21$ deg. The corresponding yaw angular velocity, r, which is approximately the aircraft turn rate, is $r = 5.49$ deg/sec, and the sideslip velocity is $v - v_w = -1.04$ ft/sec.

Steady Turning of the 747 at 40 kft

From Ref. HE-1, the perturbation equations of lateral motion for the 747 airplane cruising at an altitude of 40 kft and an airspeed of 774 ft/sec (Mach number $M = 0.80$) are

$$
\begin{bmatrix} \dot{v} \\ \dot{r} \\ \dot{p} \\ \dot{\phi} \end{bmatrix} = \begin{bmatrix} -.0558 & -7.74 & 0 & .322 \\ .0773 & -.115 & -.0318 & 0 \\ -.394 & .388 & -.465 & 0 \\ 0 & 0 & 1.0 & 0 \end{bmatrix} \begin{bmatrix} v \\ r \\ p \\ \phi \end{bmatrix} + \begin{bmatrix} 0 & .0564 \\ .00775 & -.475 \\ .143 & .153 \\ 0 & 0 \end{bmatrix} \begin{bmatrix} \delta a \\ \delta r \end{bmatrix}
$$

$$
+ \begin{bmatrix} .0558 & 0 \\ -.0773 & .0318 \\ .394 & .465 \\ 0 & 0 \end{bmatrix} \begin{bmatrix} v_w \\ p_w \end{bmatrix}, \tag{13.16}
$$

where (v, v_w) are in ft/sec, (r, p, p_w) in crad/sec, and $(\phi, \delta a, \delta r)$ in crad.

The lateral specific force, a_y, in ft/sec^2, is given by

$$
a_y = -.0558(v - v_w) + .0564\delta r. \tag{13.17}
$$

As in the Navion example, we can determine the steady $(\delta a, \delta r)$ required to maintain steady (ϕ, a_y) and the corresponding $(r, v - v_w)$:

$$
\begin{bmatrix} \delta a \\ \delta r \\ r \\ v - v_w \end{bmatrix} = \begin{bmatrix} -.1384 & -57.8 \\ -.01484 & -4.69 \\ .0416 & .1292 \\ -.0150 & -22.7 \end{bmatrix} \begin{bmatrix} \phi \\ a_y \end{bmatrix}. \tag{13.18}
$$

Thus, in order to hold a steady bank angle of (say) 30 deg with $a_y = 0$, we require $\delta a = -4.15$ deg, $\delta r = -.44$ deg. The corresponding turn rate is $r = 1.25$ deg/sec, and the sideslip velocity is $v - v_w = -.78$ ft/sec.

Problem 13.2.1 – *Nonlinear Analysis of Steady Turning*

Using the nonlinear kinematic equations of motion (10.3)–(10.4), *show that* a constant altitude turn at constant velocity V and constant turn rate $\dot{\psi}$ requires that

(a) The components of angular velocity in body axes are

$$
\begin{bmatrix} p \\ q \\ r \end{bmatrix} = \begin{bmatrix} -\sin\theta \\ \cos\theta\sin\phi \\ \cos\theta\cos\phi \end{bmatrix} \dot{\psi}.
$$

(b) The components of velocity in body axes are

$$
\begin{bmatrix} u \\ v \\ w \end{bmatrix} = \begin{bmatrix} \cos\theta \\ \sin\theta\sin\phi \\ \sin\theta\cos\phi \end{bmatrix} V.
$$

13.3 Steady Sideslip and Yaw Rate

It is sometimes desirable to *sideslip* an aircraft with respect to the air with zero yaw rate, as when landing in a crosswind. This requires dipping the upwind wing so that the lift vector has an upwind component to counter the sideforce due to sideslipping. The steady-state aileron and rudder deflections required to produce small changes in sideslip velocity with respect to the air, $v - v_w$, and yaw rate, r, may be determined from (15) by rearranging dependent and independent variables:

$$
\begin{bmatrix} \delta a \\ \delta r \\ \phi \\ a_y \end{bmatrix} = \begin{bmatrix} .265 & -.0613 \\ .542 & -.1621 \\ .580 & 5.53 \\ -.1866 & -.0202 \end{bmatrix} \begin{bmatrix} v - v_w \\ r \end{bmatrix}. \tag{13.19}
$$

Thus, for a sideslip velocity of 10 ft/sec and $r = 0$, we have

$$\delta a = 2.65 \text{ crad} \equiv 1.52 \text{ deg},$$
$$\delta r = 5.42 \text{ crad} \equiv 3.11 \text{ deg},$$
$$\phi = 5.80 \text{ crad} \equiv 3.32 \text{ deg},$$
$$a_y = -1.87 \text{ ft/sec}^2.$$

Note the aircraft is sideslipping to the right, and the right wing tip is down ($\phi > 0$).

An aircraft can also make a steady *skidding turn* with zero bank angle by using aerodynamic sideforce due to sideslip; this is an uncomfortable maneuver for the pilot and passengers and produces a much slower turn rate than bank-to-turn.

To produce a turn rate of $r = .65$ deg/sec (considerably less than the coordinated turn in Section 13.2) with $\phi = 0$ for the Navion requires $a_y = 2.0$ ft/sec^2, sideslip velocity is $v - v_w = -10.84$ ft/sec, $\delta a = -1.69$ deg, and $\delta r = -3.47$ deg.

The "De-Crab" Maneuver

If there is a *crosswind* on landing approach, an aircraft keeping its wings level ($\phi = 0$) is "crabbed" relative to the runway, since normally there is no sideslip with respect to the air so that

$$v = v_w. \tag{13.20}$$

Thus the aircraft has a yaw angle with respect to the runway centerline, of

$$\psi_{ss} = -v_w/V, \tag{13.21}$$

where v_w is the crosswind velocity and V is the aircraft velocity.

Just before touchdown most (but not all) aircraft are *de-crabbed* to $\psi = 0$, so that the wheels are lined up with the runway. To maintain this as an equilibrium condition, the upwind wing must be lowered ($\phi \neq 0$) so that the component of gravitational force $g\phi$ is equal and opposite to the sideforce due to sideslip with respect to the air and rudder deflection ($v = 0, v_w \neq 0, r = 0$).

Finally, just before touchdown, the aircraft is usually leveled so that all the main landing gear wheels touch simultaneously.

Problem 13.3.1 – *Steady-State Lateral Controls for 747 in Landing Configuration*

The following equations of motion are for the 747 airplane at sea level in the landing configuration (flaps and landing gear down) at a weight of 564,000 lb with $V = 221$ ft/sec:

$$
\begin{bmatrix} \dot{v} \\ \dot{r} \\ \dot{p} \\ \dot{\phi} \end{bmatrix}
=
\begin{bmatrix}
-.0890 & -2.19 & 0 & .319 \\
.0760 & -.217 & -.166 & 0 \\
-.602 & .327 & -.975 & 0 \\
0 & .15 & 1.0 & 0
\end{bmatrix}
\begin{bmatrix} v \\ r \\ p \\ \phi \end{bmatrix}
+
\begin{bmatrix}
0 & .0327 \\
.0264 & -.151 \\
.227 & .0636 \\
0 & 0
\end{bmatrix}
\begin{bmatrix} \delta a \\ \delta r \end{bmatrix}
$$

$$
+
\begin{bmatrix}
.0890 & 0 \\
-.0760 & .166 \\
.602 & .975 \\
0 & 0
\end{bmatrix}
\begin{bmatrix} v_w \\ p_w \end{bmatrix} ,
\tag{13.22}
$$

where (v, v_w) are in ft/sec, (r, p, p_w) in crad/sec, and $(\phi, \delta a, \delta r)$ in crad. The lateral specific force, a_y, in ft/ sec^2, is given by

$$
a_y = -.0890(v - v_w) + .0327\delta r.
$$

(a) Show that the steady values of $\delta a, \delta r, r, v - v_w$ are determined by ϕ and a_y as follows (for $p_w = 0$):

$$
\begin{bmatrix} \delta a \\ \delta r \\ r \\ v - v_w \end{bmatrix}
=
\begin{bmatrix}
-.542 & -42.4 \\
-.344 & -16.74 \\
.1457 & .457 \\
-.1263 & -17.39
\end{bmatrix}
\begin{bmatrix} \phi \\ a_y \end{bmatrix} .
$$

(b) For sideslip velocity, $v - v_w$, and yaw rate, r, as the commanded variables, show that

$$
\begin{bmatrix} \delta a \\ \delta r \\ \phi \\ a_y \end{bmatrix} = \begin{bmatrix} 2.394 & -1.648 \\ .922 & -1.560 \\ .1845 & 7.03 \\ -.0589 & -.0510 \end{bmatrix} \begin{bmatrix} v - v_w \\ r \end{bmatrix}.
$$

(c) From (b), show that the steady de-crabbed state ($v = r = 0$) of the 747 with a constant crosswind $v_w = 10$ ft/sec corresponds to $\delta a = 23.94$ crad, $\delta r = 9.22$ crad, $\phi = -1.845$ crad, $a_y = .589$ ft/sec^2. The ailerons are close to saturation! Presumably, this is why *the 747 is not de-crabbed.* Instead the landing gear was designed to take the side loads associated with landing in the crabbed condition (wings level but aircraft yawed with respect to the runway).

Problem 13.3.2 – *Steady-State Lateral Controls for Space Shuttle at 60 kft*

The lateral equations of motion of the space shuttle at $h = 60$ kft, velocity 1452 ft/sec (Mach 1.5), and $\alpha = 6.6$ deg are (Ref. ST)

$$
\begin{bmatrix} \dot\beta \\ \dot p \\ \dot r \\ \dot\phi \end{bmatrix} = \begin{bmatrix} -.090 & 0 & -1 & .023 \\ -5.00 & -.490 & .203 & 0 \\ .587 & .053 & -.117 & 0 \\ 0 & 1 & 0 & 0 \end{bmatrix} \begin{bmatrix} \beta \\ p \\ r \\ \phi \end{bmatrix} + \begin{bmatrix} -.002 & .008 \\ 4.91 & 1.555 \\ -.397 & -.594 \\ 0 & 0 \end{bmatrix} \begin{bmatrix} \delta a \\ \delta r \end{bmatrix},
$$

$$
a_y = V\dot\beta + r = -.1307\beta - .0029\delta a + .0116\delta r,
$$

where units are ft, sec, and mrad.

(a) Find the steady-state values of $\delta a, \delta r, r, v - v_w$ as functions of ϕ and a_y (for $p_w = 0$).

(b) Find the steady-state values of $\delta a, \delta r, \phi, a_y$ as functions of $v - v_w$ and r (for $p_w = 0$).

13.4 Observability and Controllability of Lateral Modes

Navion at Sea Level

The modal form of the Navion lateral EOM is

$$
\begin{bmatrix} \dot{\xi}_{D1} \\ \dot{\xi}_{D2} \\ \dot{\xi}_R \\ \dot{\xi}_S \end{bmatrix} =
\begin{bmatrix} -.487 & 2.33 & 0 & 0 \\ -2.33 & -.487 & 0 & 0 \\ 0 & 0 & -8.43 & 0 \\ 0 & 0 & 0 & -.00876 \end{bmatrix}
\begin{bmatrix} \xi_{D1} \\ \xi_{D2} \\ \xi_R \\ \xi_S \end{bmatrix}
$$

$$
+ \begin{bmatrix} -1.88 & -4.48 \\ -.44 & .73 \\ -28.47 & 2.77 \\ 2.76 & -1.34 \end{bmatrix}
\begin{bmatrix} \delta a \\ \delta r \end{bmatrix}
+ \begin{bmatrix} -2.93 & -.154 \\ .512 & -.182 \\ 9.26 & 8.23 \\ 0 & .942 \end{bmatrix}
\begin{bmatrix} v_w \\ p_w \end{bmatrix}, \quad (13.23)
$$

where (ξ_{D1}, ξ_{D2}) are the dutch roll modal coordinates, ξ_R is the roll mode coordinate, and ξ_S is the spiral mode coordinate. These coordinates are related to the state variables by

$$
\begin{bmatrix} v \\ r \\ p \\ \phi \end{bmatrix} =
\begin{bmatrix} .095 & .794 & .013 & .051 \\ 1.000 & 0 & .041 & .176 \\ -.092 & -.884 & 1.000 & -.009 \\ -.355 & -.114 & -.119 & 1.000 \end{bmatrix}
\begin{bmatrix} \xi_{D1} \\ \xi_{D2} \\ \xi_R \\ \xi_S \end{bmatrix}, \quad (13.24)
$$

and the inverse transformation is

$$
\begin{bmatrix} \xi_{D1} \\ \xi_{D2} \\ \xi_R \\ \xi_S \end{bmatrix} = \begin{bmatrix} -.041 & .940 & -.058 & -.164 \\ 1.248 & -.134 & -.016 & -.040 \\ 1.098 & .-028 & .981 & .042 \\ -.026 & .346 & .098 & .941 \end{bmatrix} \begin{bmatrix} v \\ r \\ p \\ \phi \end{bmatrix}.
\tag{13.25}
$$

From (23) we see that

(a) The dutch roll mode is controlled better by rudder than by aileron since

$$
\left[(4.48)^2 + (.73)^2\right]^{\frac{1}{2}} \div \left[(1.88)^2 + (.44)^2\right]^{\frac{1}{2}} = 2.35.
$$

(b) The roll mode is controlled much better by aileron than by rudder since $28.47/2.77 = 10.3$.

(c) The spiral mode is controlled better by aileron than by rudder since $2.76/1.34 = 2.05$.

(d) The dutch roll mode is disturbed more by lateral gusts than by roll gusts since

$$
\left[(2.93)^2 + (.512)^2\right]^{\frac{1}{2}} \div \left[(.154)^2 + (.182)^2\right]^{\frac{1}{2}} = 12.5.
$$

(e) The roll mode is about equally disturbed by lateral gusts and roll gusts since $9.26/8.23 = 1.12$.

(f) The spiral mode is undisturbed by lateral gusts but is disturbed by roll gusts.

From (24) we may deduce that

(a) The dutch roll mode is strongly observed in yaw rate r, roll rate p, and sideslip velocity v.

(b) The roll mode is most strongly observed in roll rate p.

(c) The spiral mode is most strongly observed in roll angle ϕ.

(d) Lateral specific force, $a_y = -.254(v - v_w) + .1246\delta r$, observes mainly dutch roll.

We conclude that measurements of ϕ and a_y would observe all three modes adequately and that feedback of a_y or r to δr and ϕ to δa should give good stabilization.

747 at 40 kft; Fifth-Order Model Including Heading Angle

In order to synthesize heading-hold logic we must include heading angle, ψ, as a fifth state variable where

$$\dot{\psi} = r. \tag{13.26}$$

(26) is coupled one-way to the other four EOM, so it introduces a fifth eigenvalue at $s = 0$. We shall call the corresponding mode the *heading mode*. The modal form of the later EOM is

$$
\begin{bmatrix} \dot{\xi}_{D1} \\ \dot{\xi}_{D2} \\ \dot{\xi}_R \\ \dot{\xi}_S \\ \dot{\xi}_H \end{bmatrix} =
\begin{bmatrix} .029 & .854 & 0 & 0 & 0 \\ -.854 & .029 & 0 & 0 & 0 \\ 0 & 0 & -.684 & & \\ 0 & 0 & 0 & -.00989 & \\ 0 & 0 & 0 & 0 & 0 \end{bmatrix}
\begin{bmatrix} \xi_{D1} \\ \xi_{D2} \\ \xi_R \\ \xi_S \\ \xi_H \end{bmatrix}
$$

$$
+ \begin{bmatrix} -.033 & 1.367 \\ .019 & 1.228 \\ .230 & -1.705 \\ .960 & -11.92 \\ .921 & -11.45 \end{bmatrix}
\begin{bmatrix} \delta a \\ \delta r \end{bmatrix}
+ \begin{bmatrix} 2.09 & -.128 \\ 2.88 & .045 \\ 2.53 & .776 \\ -.010 & 3.30 \\ 0 & 3.16 \end{bmatrix}
\begin{bmatrix} v_w \\ p_w \end{bmatrix}, \tag{13.27}
$$

where ξ_H is the heading mode coordinate. These coordinates are related to the state variables by the eigenvector matrix

$$
\begin{bmatrix} v \\ r \\ p \\ \phi \\ \psi \end{bmatrix} = \begin{bmatrix} .170 & -.130 & .030 & .001 & 0 \\ -.121 & -.173 & -.010 & .010 & 0 \\ .029 & .854 & .564 & -.002 & 0 \\ 1.000 & 0 & -.825 & .233 & 0 \\ -.207 & .134 & .014 & -.972 & 1.000 \end{bmatrix} \begin{bmatrix} \xi_{D1} \\ \xi_{D2} \\ \xi_R \\ \xi_s \\ \xi_H \end{bmatrix} , \qquad (13.28)
$$

and the inverse transformation is

$$
\begin{bmatrix} \xi_{D1} \\ \xi_{D2} \\ \xi_R \\ \xi_S \\ \xi_H \end{bmatrix} = \begin{bmatrix} 3.28 & -2.87 & -.079 & .100 & 0 \\ -2.56 & -2.53 & 2.70 & .120 & 0 \\ 3.70 & 4.08 & 1.390 & -.174 & 0 \\ -.997 & 26.8 & 5.26 & 3.25 & 0 \\ 0 & 25.7 & 5.05 & 3.16 & 1.000 \end{bmatrix} \begin{bmatrix} v \\ r \\ p \\ \phi \\ \psi \end{bmatrix} . \qquad (13.29)
$$

From (27) we note that

(a) The dutch roll mode is slightly unstable and the spiral mode is nearly neutrally stable.

(b) The dutch roll mode is much better controlled by rudder than by aileron, since

$$
[(1.367)^2 + (1.228)^2]^{\frac{1}{2}} \div [(.033)^2 + (.019)^2]^{\frac{1}{2}} = 48.3.
$$

(c) The roll mode is better controlled by rudder than by aileron since $1.705/.230 = 7.4$.

(d) The spiral and heading modes are better controlled by rudder than by aileron since $11.92/.960 = 12.4$ and $11.45/.921 = 12.4$.

(e) The dutch roll mode is disturbed much more by lateral gusts than by roll gusts since

$$
[(2.09)^2 + (2.88)^2]^{\frac{1}{2}} \div [(.128)^2 + (.045)^2]^{\frac{1}{2}} = 26.2.
$$

(f) The roll mode is disturbed more by lateral gusts than by roll gusts, since $2.53/.776 = 3.26$.

(g) The spiral and heading modes are almost undisturbed by lateral gusts but they are disturbed by roll gusts.

From (28) we may deduce that:

(a) The dutch roll mode is most strongly observed in roll rate p and roll angle ϕ.

(b) The roll mode is most strongly observed in roll angle ϕ and roll rate p.

(c) The spiral mode is observed in heading angle ψ and roll angle ϕ.

(d) The heading mode is observed in heading angle only.

(e) Lateral specific force (hence sideslip velocity) observes mainly the dutch roll mode.

We conclude that measurements of ϕ and ψ observe all of the modes and that rudder δr controls all of the modes.

Problem 13.4.1 – *Modal Form for 747 in Landing Configuration*

(a) Using the lateral equations of motion given in Problem 13.2.1, find the modal form and discuss controllability, observability, and disturbability of the modes.

(b) Add the yaw angle, ψ, as a state, where

$$\dot{\psi} = r,$$

and repeat (a). Note the new mode has eigenvalue zero (the heading mode) and show that it is undisturbable by v_w.

Problem 13.4.2 – *Modal Form for Space Shuttle at 60 kft*

Using the lateral equations for the space shuttle at 60 kft given in Problem 13.3.2, find the modal form and discuss controllability, observability, and disturbability of the modes.

13.5 Control of Bank Angle and Sideforce

Introduction

The spiral and dutch roll modes usually need stabilization as indicated in the examples of Section 13.3. The spiral mode is observable with a roll angle sensor such as a vertical gyro, and the dutch roll mode is observable with a yaw rate gyro. The spiral mode is controllable with aileron, and the dutch roll mode is controllable with rudder. Also there is a *wide spectral separation* between the spiral mode and the dutch roll mode. Hence the spiral mode can be stabilized by feeding back roll angle, ϕ, to aileron, δa, and the dutch roll mode can be stabilized by feeding back yaw rate, r, to rudder, δr (see Problem 13.4.1). The resulting SAS stabilizes yaw rate but not yaw (heading) angle. Thus, when zero yaw rate is commanded, small errors in implementation of the command and in the mathematical model of the airplane produce a small hang-off error in the yaw rate, and heading angle slowly drifts. To correct this, some feedback of deviation from the desired heading angle must be added; this is done in Section 13.6. By the same reasoning, a heading-hold SAS will have a slow drift off a desired track, so deviation from the desired track must be added as a feedback; this is done in Section 13.7.

A Bank Angle/Sideforce SAS for the Navion

It is desirable to recover from a disturbance that banks the aircraft using zero sideforce. Thus a good "wings-level" SAS can be designed using a performance index of the form

$$ J = \int_{o}^{\infty} (A_\phi \phi^2 + A_a a_y^2 + \delta a^2 + \delta r^2)dt. \tag{13.30} $$

For the Navion (units ft, sec, crad), this indicates that we are willing to use 1 crad of aileron δa and 1 crad of rudder δr if the bank angle is $1/\sqrt{A_\phi}$ crad or the lateral specific force is $1/\sqrt{A_a}$ ft/sec^2.

Fig. 13.5 shows the closed-loop root locus vs. $A = A_\phi = A_a$. Note that there is a reflected transmission zero at $s = -3.12$ rad/sec, but it has no significant influence on the root locus until $A \gg 1$.

We chose $A = .25$ since this puts the modified spiral mode at $s = -1.55$ rad/sec, which gives a settling time of about $3/1.55 \cong 2$ sec.

Fig. 13.6 shows the response to the worst initial condition vector of unit length, namely $[\ v(0)\ \ r(0)\ \ p(0)\ \ \phi(0)\ ] = [\ -.18\ \ .21\ \ .09\ \ .96\ ]$, which is mostly bank angle with yaw rate in the same direction and sideslip in the opposite direction. The settling time is about 3 sec, because it excites the modified dutch roll mode as well as the modified spiral mode.

Fig. 13.7 shows the response to a bank angle command of 10 crad with a lateral specific force command of zero. The rise time is about 2 sec, and the settling time is about 3 sec as predicted. The maximum magnitude of lateral

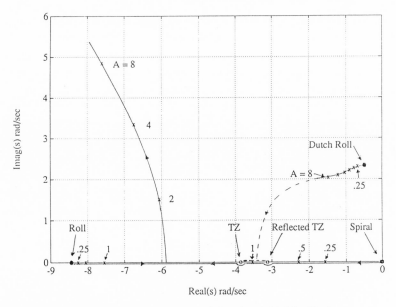

Figure 13.5. Locus of closed-loop poles vs. A for (ϕ, a_y) SAS for Navion.

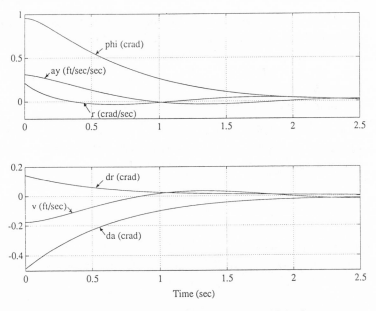

Figure 13.6. Response of Navion to worst initial conditions with bank angle/sideforce SAS.

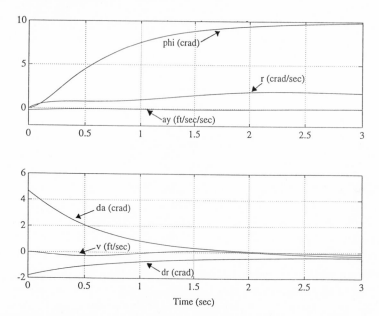

Figure 13.7. Response of Navion to commands for bank angle = 10 crad and sideforce = 0.

specific force is less than .5 ft/sec². The maximum (aileron, rudder) deflections are about (5, 2) crad respectively. The equilibrium turn rate is about 2 crad/sec, and the sideslip velocity magnitude is less that .5 ft/sec.

A Bank Angle/Sideforce SAS for the 747 at 40 kft

Assuming an inertial reference system (IRS), good estimates are available of v, ϕ, and ψ. Using fast observers (lead networks), reasonably good estimates of p and r can be obtained from ϕ and ψ, such as

$$\dot{\hat{p}} = 20(\dot{z}_\phi - \hat{p}),$$
$$\dot{\hat{r}} = 20(\dot{z}_\psi - \hat{r}).$$

Thus, full-state feedback is feasible.

The outputs we wish to control are ϕ and a_y (roll angle and lateral specific force), so we consider the quadratic performance index

$$J = \int_0^\infty [A(\phi^2 + 64a_y^2) + \delta a^2 + \delta r^2]dt.$$

Fig. 13.8 shows a closed-loop root locus vs. A. The open-loop dutch roll mode is unstable; however, for $A = 0+$, it is reflected across the imaginary axis. We chose $A = 1$, for which the LQR gains and closed-loop eigenvalues are

$$K = \begin{bmatrix} -.137 & .631 & .428 & .342 \\ .612 & -5.242 & -1.121 & -.654 \end{bmatrix},$$

$$ev = -.651 \pm .919j \, , \, -.877 \pm .476j.$$

Note that the slowest closed-loop eigenvalues have real parts $s = -.651$, which means a settling time from an initial disturbance of about $3/.65 \cong 5$ sec.

Fig. 13.9 shows the 747 response to the worst initial condition vector of unit magnitude, namely $x(0) = [-.151 \ .920 \ .294 \ .209]$, which is principally yaw rate but has some sideslip, roll rate, and bank angle. The settling time is about 7 sec, a little longer than our estimate from the eigenvalues.

Fig. 13.10 shows the 747 response to commands for bank angle = 10 crad and sideforce = 0. The rise time is about 5 sec, while the maximum magnitude of a_y is about 1 ft/sec². The maximum rudder angle is about 9 crad (about 5

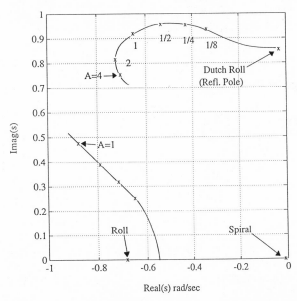

Figure 13.8. Locus of closed-loop poles vs. A for (ϕ, a_y) SAS for 747 at 40 kft.

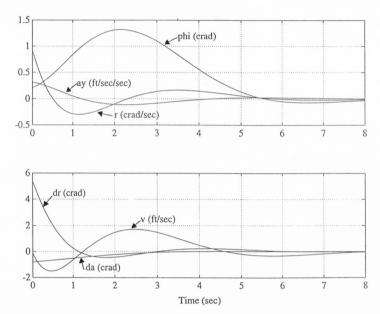

Figure 13.9. Response of 747 to worst initial condition vector of unit magnitude.

deg), and there is considerable sideslip (max 8 ft/sec). The yaw rate starts in the right direction, then reverses, then comes back, a sort of "snaking" behavior.

Problem 13.5.1 – *Control of Bank Angle and Sideforce for 747 in Landing Configuration*

(a) Synthesize (ϕ/a_y) hold logic for the 747 in landing configuration using both δa and δr, and the data of Problem 13.3.1. Assume all states measured or estimated accurately.

(b) Plot the response to the commands $\phi_c = 10$ crad, $a_y = 0$.

Problem 13.5.2 – *Control of Bank Angle and Sideforce for the Space Shuttle at 60 kft*

(a) Synthesize (ϕ/a_y) hold logic for the space shuttle at 60 kft using both δa and δr, and the data of Problem 13.3.2. Assume all states measured or estimated accurately.

(b) Plot the response to the commands $\phi_c = 10$ crad, $a_{yc} = 0$.

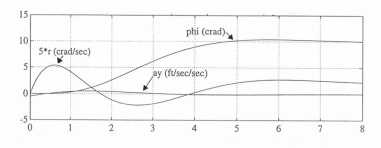

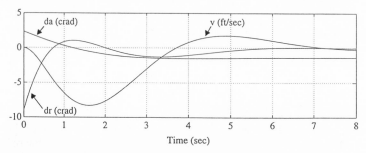

Time (sec)

Figure 13.10. Response of 747 at 40 kft with LQ regulator to commands for
bank angle = 10 crad and sideforce = 0.

Problem 13.5.3 – *Stabilizing the Navion with Proportional Feedback*

The spiral mode has a wide spectral separation from the dutch roll/roll modes, so that
they can be stabilized separately as two SISO designs.

(a) Show that the open-loop transfer function from aileron δa to roll angle ϕ is

$$\frac{\phi(s)}{\delta a(s)} = \frac{29.0[(s + .499)^2 + (2.08)^2]}{(s + .00885)[(s + .486)^2 + (2.33)^2](s + 8.43)},$$

$$\cong \frac{3.44}{s} \text{ For } |s| < .3.$$

(b) Plot a root locus versus k_ϕ using the proportional feedback law

$$\delta a = -k_\phi \phi.$$

(c) Show that the transfer function from rudder angle δr to yaw rate r with the aileron
loop closed using $k_\phi = .4$ is

$$\frac{r(s)}{\delta r(s)} = \frac{-4.60[(s + .196)^2 + (.550)^2](s + 8.39)}{(s + .213)[(s + .511)^2 + (2.35)^2](s + 8.18)}.$$

(d) Plot a root locus versus k_r with the aileron loop closed using the proportional
feedback law

$$\delta r = k_r r.$$

Note that this feedback adds damping to the dutch roll mode, increases the damping in the spiral mode, and changes the roll mode only slightly.

(e) Show that the closed-loop eigenvalues using $k_\phi = .4$ and $k_r = 1$ are

$$s = -1.679 \pm 1.728j, -8.09, -.360 \sec^{-1}.$$

(f) Plot the responses to commands for $\phi = 10$ crad and $a_y = 0$ using the SAS of (e) and compare them to the responses using the LQR-designed SAS in Fig. 13.7.

(g) Plot the response of the aircraft with the SAS of (d) to a sharp-edged side gust pulse of initial magnitude 5 ft/sec, which then decays exponentially with a time constant of 1 sec.

(h) Plot the response of the aircraft with the SAS of (d) to a sharp-edged roll gust pulse of initial magnitude 1 crad/sec, which then decays exponentially with a time constant of 1 sec.

Problem 13.5.4 – *Control of Bank Angle for the 747 Using Differential Engine Thrust*

In an emergency when the aerodynamic controls malfunction, many multi-engine aircraft can still be controlled by using differential engine thrust on the right and left engines (e.g. the United Airlines DC-10 in the 1980's when hydraulic lines were severed by turbine blades in a failure of the center engine). See problem 12.4.7 for longitudinal control using thrust only.

(a) Synthesize bank angle control logic for the 747 at sea level in its landing configuration using differential thrust on the right and left engines. Assume the inboard engines are 77 ft apart (l_1), the outboard engines are 133 ft apart (l_2), the yaw moment of inertia $I_z = 43.1 \times 10^6$ slug ft^2, and the weight $mg = 564000$ lb. Let $\Delta T = $ (increase, decrease) in (right, left) engine specific force in ft/sec/sec, so that $N_{\Delta T} = 100 * m(l_1 + l_2)/I_z$ in crad/sec/sec.

(b) Plot the response to a command for a 10 crad bank to the right, not exceeding $m\Delta T = 5000$ lb. Note lateral specific force can not be controlled, so the turn is not coordinated.

13.6 Control of Heading and Sideforce

Heading-hold logic requires feedback of deviation from desired heading, ψ, a fifth state variable, where

$$\dot\psi \cong r. \tag{13.31}$$

This introduces a fifth open-loop eigenvalue at $s = 0$, and we shall call the corresponding mode the *heading mode*. This mode is observable only with a measurement of ψ, and is controllable by either rudder or aileron.

Heading/Sideforce-Hold Logic for the Navion

Fig. 13.11 shows a closed-loop root locus versus A for the performance index

$$J = \int_0^\infty \left[A(\psi^2 + 3a_y^2) + \delta_a^2 + \delta_r^2 \right] dt, \tag{13.32}$$

where a_y = lateral specific force. The weighting factor on a_y^2 was chosen so that one degree of ψ is weighted almost the same as one ft/sec^{-2} of a_y. The spiral and heading modes are stabilized, and the damping of the dutch roll mode is increased. The roll mode is virtually unchanged and is not shown in the figure.

If we choose $A = 1$, the state feedback gains, K, and the closed-loop eigenvalues are

$$K = \begin{bmatrix} .234 & -.017 & .028 & .244 & .350 \\ -.255 & -.375 & .003 & .139 & .907 \end{bmatrix}, \tag{13.33}$$

$$ev = \begin{bmatrix} -8.434 & -1.303 \pm 2.481j & -.450 \pm .479j \end{bmatrix}. \tag{13.34}$$

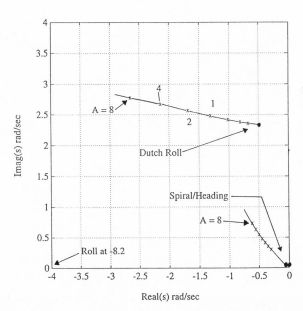

Figure 13.11. Locus of closed-loop roots vs. LQ weight A—heading/sideforce SAS for the Navion.

Note that the slowest closed-loop eigenvalues have real parts = $-.450$ rad/sec, which should give a settling time of about $3/.45 \cong 7$ sec.

The worst initial conditions with unit magnitude for the closed-loop system are:

$$x(0) = [.370 \quad .142 \quad .010 \quad .139 \quad .907], \qquad (13.35)$$

which correspond to large heading angle with sideslip, yaw rate, and roll rate all in the same direction. Fig. 13.12 shows the response of the closed-loop system with the worst unit initial conditions; a large bank is necessary to recover zero heading angle.

Fig. 13.13 shows the response to the commands $\psi_c = 10$ crad, $a_{yc} = 0$. The aircraft is banked to the right (maximum $\phi \cong 16$ crad) and is sideslipping slightly to the right, while the maximum magnitude of a_y is about 2 ft/sec^2.

Heading/Sideforce-Hold Logic for the 747 at 40 kft

We assume an inertial reference system (IRS) that gives relatively noise-free estimates of v, ϕ, and ψ. The angular velocities can be estimated by fast observers,

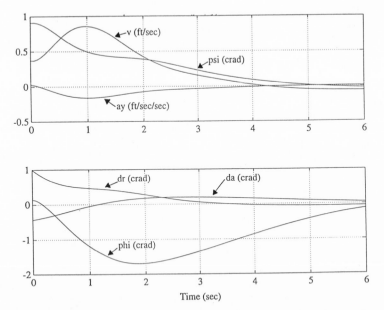

Figure 13.12. Response of the Navion with LQR heading/sideforce SAS to the worst unit initial conditions.

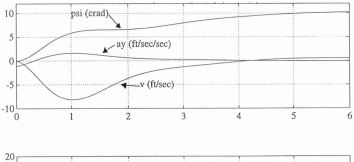

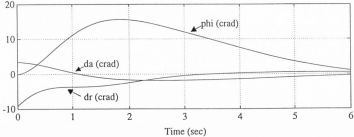

Time (sec)

Figure 13.13. Response of the Navion with LQR heading/sideforce SAS to commands for a heading change of 10 crad and zero sideforce.

say

$$\dot{\hat{p}} = 20(\dot{z}_\phi - \hat{p}), \tag{13.36}$$

$$\dot{\hat{r}} = 20(\dot{z}_\psi - \hat{r}), \tag{13.37}$$

where z_ϕ and z_ψ are the values of ϕ and ψ from the IRS and $(\hat{p}, \hat{r})$ are estimates of roll rate and yaw rate. Thus we have relatively accurate estimates of all of the state variables.

Fig. 13.14 shows a closed-loop root locus versus A for the performance index (32). The spiral and heading modes are stabilized, and the damping of the dutch roll mode is increased. The roll mode is virtually unchanged.

If we choose $A = 1$, the state feedback gains, K, and the closed-loop eigenvalues are

$$K = \begin{bmatrix} -.002 & .110 & .032 & .021 & .081 \\ .018 & -2.152 & -.211 & -.129 & -.992 \end{bmatrix}, \tag{13.38}$$

$$ev = [-.307 \pm .864j \quad -.166 \pm .265j \quad -.686]. \tag{13.39}$$

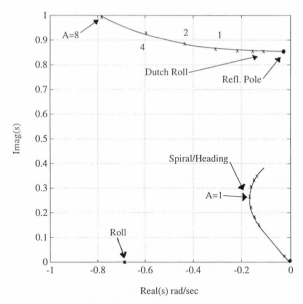

Figure 13.14. Locus of closed-loop roots vs. LQ weight A—heading/sideforce SAS for the 747 at 40 kft.

Note that the slowest closed-loop eigenvalues have real parts = $-.166$ rad/sec, which should give a settling time of about $3/.166 = 18$ sec.

The worst initial conditions with unit magnitude for the closed-loop system are

$$x(0) = [.037 \quad .648 \quad .091 \quad .079 \quad .751]. \tag{13.40}$$

They correspond to large heading angle with sideslip, yaw rate, and roll rate all in the same direction. Fig. 13.15 shows the response of the closed-loop system with the worst unit initial conditions, which differs significantly from the Navion in having larger sideslip and bank angle excursions.

Fig. 13.16 shows the response to the commands $\psi_c = 10$ crad, $a_{yc} = 0$. The aircraft is banked to the right (maximum $\phi \cong 44$ crad) and sideslipped to the right, while the maximum magnitude of a_y is about 1 ft/sec^2. 10 crad of rudder and less than 1 crad of aileron are used.

Problem 13.6.1 – *Control of Heading and Sideforce for the 747 in Landing Configuration*

Synthesize (ψ, a_y) command logic for the 747 in the landing configuration using the data from Problem 13.3.1, and plot the response to $\psi_c = 10$ crad, $a_{yc} = 0$.

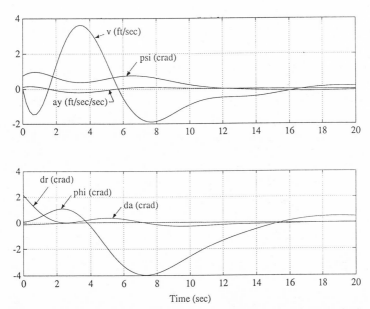

Figure 13.15. Response of the 747 at 40 kft with LQR heading/sideforce SAS to the worst unit initial conditions

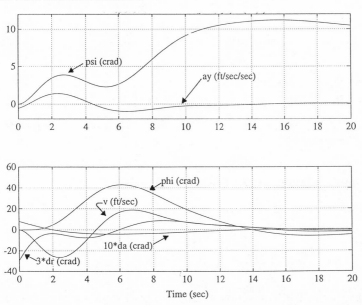

Figure 13.16. Response of the 747 at 40 kft with LQR heading/sideforce SAS to commands for a heading change of 10 crad and zero sideforce.

Problem 13.6.2 – *Control of Heading and Sideforce for the Space Shuttle at 60 kft*

Synthesize (ψ, a_y) command logic for the space shuttle at 60 kft using the data from Problem 13.3.2, and plot the response to $\psi_c = 10$ crad, $a_{yc} = 0$.

Problem 13.6.3 – *Digital Control of Heading and Sideforce for the Space Shuttle at 60 kft*

Repeat problem 13.6.2 synthesizing *discrete-step* (ψ, a_y) command logic with sample time $T_s = 1$ sec, and plot the response to $\psi_c = 10$ crad, $A_{yc} = 0$. Discretize the performance index using 'cvrt', and use 'dlqr' with the cross-coupling term ND (see sections 5 and 6 of Appendix D); note the response is almost identical to the response in Problem 13.6.2, despite the rather long sample time.

13.7 Control of Track and Sideforce

Track/sideforce-hold logic requires feedback of lateral deviation from a desired track, y, where

$$\dot{y} = v + V\psi. \tag{13.41}$$

This introduces a sixth state variable y and a second open-loop eigenvalue at $s = 0$ (the track mode). This mode is observable only with a measurement of y but is controllable by either rudder or aileron.

Estimates of y can be made by

(a) Dead reckoning using airspeed/compass, doppler radar, or an inertial reference system.

(b) Computed deviation from a desired track, using radio navigation aids such as ILS localizer/DME, VOR/DME or TACAN, LORAN, OMEGA, and GPS.

Control of Track and Sideforce for the Navion

We shall consider two outputs, track deviation y and lateral specific force a_y. Our performance index is

$$J = \int_0^\infty (A_y y^2 + A_a a_y^2 + \delta_a^2 + \delta_r^2)dt. \tag{13.42}$$

If y is 5 ft or $a_y = .5$ ft sec^{-2}, we would be willing to use one crad of δ_a and δ_r, so we take $A_y = 1/(5)^2 = .04$, $A_a = 1/(.5)^2 = 4.0$. With these as nominal values, we use the Navion lateral equations of motion and calculate the closed-loop poles vs. an overall output-weighting factor A having values $1/8, 1/4, 1/2, 1,$ 2, 4, and 8. The results are shown in the symmetric root locus of Fig. 13.17. The spiral, heading, and track modes are stabilized, moving away from $s = 0$ in a "star" pattern having six points; the damping of the dutch roll mode is increased, while the roll mode (at $s = -8.24$ rad/sec) is virtually unchanged (hence it is not shown).

We chose $A = 1$ for which the LQR feedback gains K and the closed-loop eigenvalues ev are

$$
K = \begin{bmatrix} .540 & .061 & .035 & .332 & 1.042 & .164 \\ -.282 & -.288 & -.006 & -.147 & -.685 & -.111 \end{bmatrix}, \quad (13.43)
$$

$$
ev = [\,-8.43 \quad -1.049 \pm 2.310j \quad -.291 \pm .497j \quad -.576\,]. \quad (13.44)
$$

Note that the slowest closed-loop eigenvalues have real parts $= -.291$ rad/sec, which should give a settling time of about $3/.291 \cong 10$ sec.

The worst initial conditions with unit magnitude for this closed-loop system are

$$
x(0) = [\,.437 \quad .080 \quad .016 \quad .172 \quad .859 \quad .186\,]. \quad (13.45)
$$

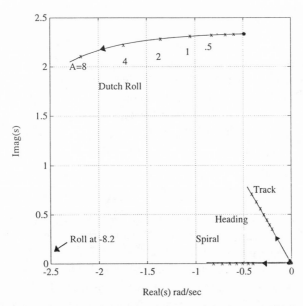

Figure 13.17. Locus of closed-loop roots vs. LQ weight A—track/sideforce SAS for the Navion.

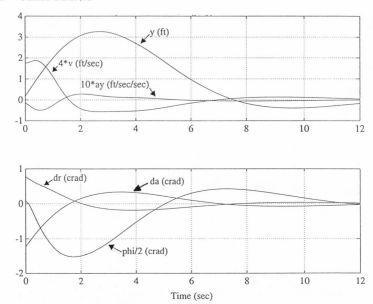

Figure 13.18. Response of the Navion with LQR track/sideforce SAS to the worst unit initial conditions.

Note that they correspond to large lateral velocity $(v + V\psi)$ and lateral acceleration of the center of mass. Fig. 13.18 shows the response of the closed-loop system with the worst unit initial conditions, which involves a large excursion in y due to the initial $\dot{y}$ and $\ddot{y}$. The bank angle ϕ brings the aircraft back to the track and then has to reverse to bring the flight path direction parallel to the track; this is why the settling time is longer than predicted (almost two transient periods).

Track-command logic may be generated from track-hold logic by simply replacing y by $y - y_c$ where y_c = commanded track deviation. Fig. 13.19 shows the response to a command for a 10 ft change in track (sidestep to the right of 10 ft) while holding sideforce = 0. The simulation is done simply by putting $y(0) = -10$ with the other initial states all zero. The response is an "S-turn" or "sidestep" maneuver, which is primarily controlled by aileron δa; the aircraft is first banked to the right to create positive $\dot{y}$, then banked to the left to create negative $\dot{y}$. The heading angle ψ is increased and then decreased to zero. The specific sideforce a_y is held to about .1 ft/sec^2.

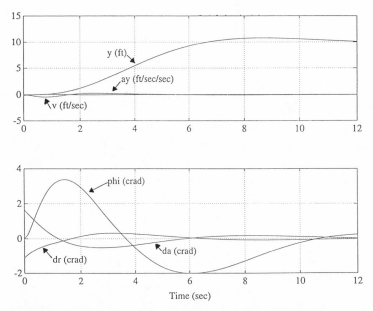

Figure 13.19. Response of the Navion with LQR track/sideforce SAS to commands for 10 ft displacement to the right and zero sideforce.

De-Crab Maneuver for the Navion

The *de-crab maneuver is a sideslip* maneuver, which is made just before touchdown when landing in a crosswind. On approach the aircraft is crabbed into the wind, that is, zero sideslip with respect to the air and zero bank angle. If the aircraft touches down in this crabbed condition, the wheels are not aligned with the runway and a lateral skid occurs that abruptly aligns the aircraft with the runway and takes a lot of rubber off the tires. Hence some aircraft are de-crabbed—commanded to have $\psi = 0$ with respect to the runway just before touchdown. To do this, the upwind wing is lowered slightly and the aircraft is sideslipped into the wind.

We should really like to have a (y, v) SAS to command sideslip v and maintain alignment along the runway $(y = 0)$. However, we can also do it with the (y, ay) SAS by commanding the ay that corresponds to the desired v. This can be determined by solving the equations $(\dot{v}, \dot{r}, \dot{p})$ all equal to zero with $(r = p = 0)$ and v specified. This yields the steady-state values of $\phi, \delta a, \delta r$ in the de-crabbed state. The steady-state value of ay is then found as $-g * \phi$. Note that the values of (v, ψ) in the ground frame are different from the values in the frame moving with the air mass.

Fig. 13.20 shows a de-crab maneuver for the Navion on approach in a cross-wind of 10 ft/sec. The crab angle is initially $\psi(0) = -10/176$ rad (-5.68 crad). The final bank angle is $\phi = -5.79$ crad, and the final sideslip velocity with respect to the air must be 10 ft/sec so that there is no sideslip with respect to the runway. The steady lateral specific force is 1.872 ft/sec². The aircraft drifts off the runway centerline less than 1 ft during the maneuver.

Some aircraft landing gear and tires are designed to withstand lateral skidding so that de-crab is not necessary (the 747 is an example). Also some aircraft have been designed with "castering" wheels, which line up with the runway before the whole aircraft is lined up (the B-52 is an example); alignment is performed using aerodynamic forces after touchdown.

Control of Track and Sideforce for the 747 at 40 kft

We assume an IRS that gives relatively noise-free estimates of (v, ϕ, ψ) and fast observers that estimate (p, r) from (ϕ, ψ) as in (36)–(37). We add a *complementary filter* to estimate y, which we call $\hat{y}$, using a measurement of y that

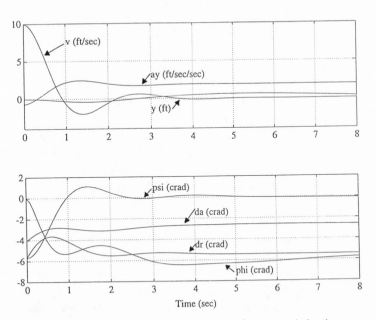

Figure 13.20. De-crab maneuver for Navion in 10 ft/sec crosswind, using $y_c = 0, a_{cy} = 1.87$ ft/sec.

we call z_y, and the estimates of v and ψ as follows:

$$\dot{\hat{y}} = \hat{v} + V\hat{\psi} + k(z_y - \hat{y}), \qquad (13.46)$$

where k depends on how noisy the z_y signal is. For the ILS localizer/DME estimate, k should be about 2 rad/sec, as with the glide-slope complementary filter of (12.26).

We use the performance index (42) but with $A_y = 1/10^2$ and $A_a = 1/.1^2$, which indicates we are willing to use 1 crad of δa and δr if there is a track error of 10 ft or a lateral specific force of .1 ft/sec^2. Fig. 13.21 is a locus of closed-loop roots vs. the overall weighting factor A where A takes on the values 0, 1/8, 1/4, 1/2, 1, 2, 4, and 8. The spiral, heading, and track modes are stabilized, moving away from s = 0 in a "star" pattern having six points; the damping of the dutch roll mode is increased, while the damping of the roll mode (initially at s = -.684 rad/sec) is decreased slightly.

We chose $A = 1$ for which the LQR feedback gains K and the closed-loop

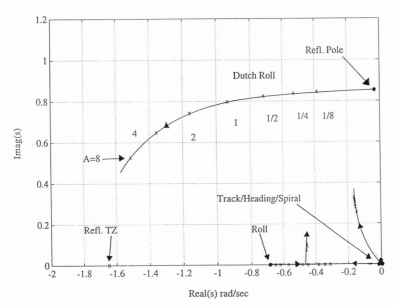

Figure 13.21. Locus of closed-loop roots vs. LQ weight A—track/sideforce SAS for the 747 at 40 kft.

eigenvalues ev are

$$K = \begin{bmatrix} .109 & .590 & .332 & .259 & 1.821 & .032 \\ -.046 & -5.392 & -.891 & -.614 & -5.027 & -.083 \end{bmatrix}, \quad (13.47)$$

$$ev = [\,-.933 \pm .797j \quad -.156 \pm .308j \quad -.483 \quad -.448\,]. \quad (13.48)$$

Note that the slowest closed-loop eigenvalues have real parts $= -.152$ rad/sec, which should be given a settling time of about $3/.15 \cong 22$ sec.

The worst initial conditions with unit magnitude for this closed-loop system are

$$x(0) = [\,.082 \quad .202 \quad .115 \quad .106 \quad .963 \quad .022\,]^T. \quad (13.49)$$

Note that they correspond to large lateral velocity $(v + V\psi)$ and lateral acceleration of the center of mass. Fig. 13.22 shows the response of the closed-loop system with the worst unit initial conditions, which involves a large excursion in y due to the initial $\dot{y}$ and $\ddot{y}$. The bank angle ϕ brings the aircraft back to the track and then reverses to bring the flight path direction parallel to the track.

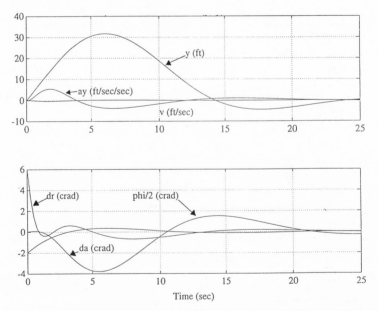

Figure 13.22. Response of the 747 at 40 kft with LQR track/sideforce SAS to the worst unit initial conditions.

Fig. 13.23 shows the response to a command for a 100-ft change in track (sidestep to the right of 10 ft) while holding sideforce = 0. The simulation is done by putting $y(0) = -100$ with the other initial states all zero. The response is an "S-turn" or "sidestep" maneuver, which, in this case, is primarily controlled by rudder δr; the aircraft is first sideslipped to the left, which creates a rolling moment that banks the aircraft to the right to create positive $\dot{y}$, then this procedure is repeated to the left to create negative $\dot{y}$. The heading angle ψ is increased and then decreased to zero. The specific sideforce a_y is held to about 1 ft/sec^2.

Localizer Capture and Hold

It is desirable to cross the localizer beam at a large angle so that capture occurs some distance from the airport. The localizer beam can be detected approximately ± 2 deg about its centerline and nearly 7 miles from the runway threshold. Fig. 13.24 shows a sketch of a localizer capture path that starts with a 30-deg crossover angle. For a smooth capture a circular arc path is commanded that comes in tangent to the localizer centerline; then the control mode is switched

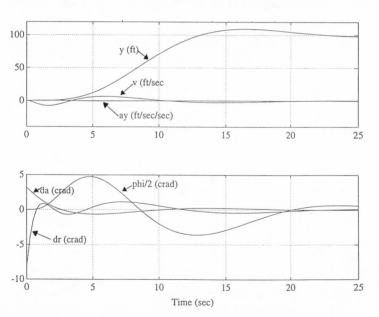

Figure 13.23. Response of the 747 at 40 kft with LQR track/sideforce SAS to commands for 100-ft displacement to the right and zero sideforce.

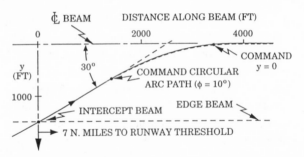

Figure 13.24. Localizer capture path for Navion, crossing at 30 deg, turning with a bank angle of 10 deg.

to localizer hold when y reaches a threshold value. At the beam-intercept point, a complementary filter is started that uses heading and sideslip estimates with the localizer signal to estimate y; when $\hat{y}$ reaches a value such that a nominally banked turn path would end up tangent to the localizer centerline, a *circular-arc track command* is begun.

For the Navion, we chose a circular-arc path corresponding to a moderate bank angle of $\phi_s = 10$ deg. This gives a steady turn rate of .0319 rad/sec = 1.83 deg/sec, which means a turn radius, R, of $176/.0319 \equiv 5510$ ft. Thus the maneuver is carried by replacing y and ψ in the track/sideforce control hold law by $y - y_c$ and $\psi - \psi_c$ where

$$y_c = 5510(1 - \cos \psi_c), \tag{13.50}$$

$$\psi_c = -\pi/6 + .0319t. \tag{13.51}$$

Although our track command system was designed for step commands, it gives satisfactory results for a slowly changing command like this one. In this case, we are commanding a steady bank angle (see Section 13.5), *but* we have "double-integral-error" control since $\psi - \psi_c$ and $y - y_c$ are being fed back.

Simulation of the localizer capture is shown in Figure 13.25. The results look quite satisfactory.

Problem 13.7.1 – *Control of Track and Sideforce for the 747 in Approach Configuration*

(a) Synthesize (y, a_y) hold logic for the 747 in approach configuration using δ_a and δ_r and the data of Problem 13.3.1. Assume all states measured or estimated accurately.

(b) Calculate and plot the response to $(y_c, a_{yc} = (10, 0)$.

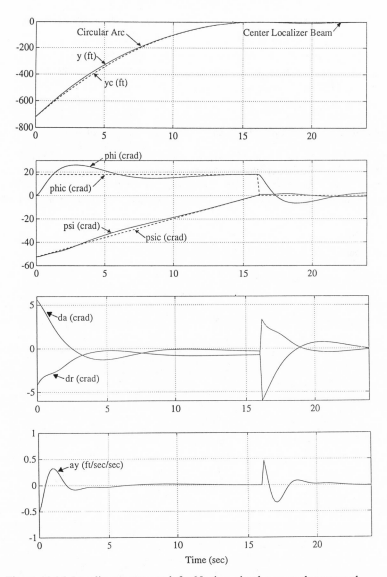

Figure 13.25. Localizer capture path for Navion, circular-arc path command, then $y = 0$ command.

Problem 13.7.2 – *Control of Track and Sideforce for the Space Shuttle at 60 kft*

(a) Synthesize (y, a_y) hold logic for the space shuttle at 60 kft using δ_a and δ_r and the data of Problem 13.3.2. Assume all states measured or estimated accurately.

(b) Calculate and plot the response to $(y_c, a_{yc} = (10, 0)$.

Problem 13.7.3 – *De-Crab Maneuver for 747 in Approach Configuration*

Using Problem 13.3.1(c) and Problem 13.7.1 synthesize a de-crab command controller. Then calculate and plot the response to a de-crab command for a crosswind of 10 ft/sec. In addition to the large steady aileron deflection required, note there is a large initial rudder kick that would saturate the rudder—still another reason why the 747 is not de-crabbed.

14

Control of Helicopters
near Hover

14.1 Introduction

In *forward flight*, the dynamics and control of rotary-wing aircraft are similar to those of fixed-wing aircraft. The rotor is, in effect, an augmented wing with a circular planform. However, forward speed is limited by the stalling of the retreating blades and compressibility effects on the advancing blades. The rotor induces severe vibration on the fuselage in fast forward flight, which is very fatiguing for the crew and passengers. Dynamic modeling is complicated by these effects, the impingement of blade vortex-wakes on the other blades, and by the flexibility of the rotor blades (cf. BRA and GEM).

Near hover, the dynamics and control of rotary-wing aircraft are significantly different from those of fixed-wing aircraft. Hence we shall discuss control only near hover in this chapter. The differences are caused by the gyroscopic and torquing effects associated with the rotor, which introduce significant coupling of the lateral and longitudinal motions.

We shall use a dynamic model that approximates the fuselage as a rigid body and the rotor as a set of rigid blades of negligible inertia. Thus the rotor tip path plane (TPP) can be tilted "instantaneously" by cyclic pitch changes, and the rotor thrust can be changed instantaneously by collective pitch changes. Tilt of the TTP with respect to the fuselage provides pitching and rolling moments on the fuselage, since then the thrust axis does not pass through the center of mass. Tilt of the TPP with respect to inertial space provides components of thrust for maneuvering in the horizontal direction, while increasing or decreasing thrust moves the vehicle vertically.

This instantaneous tilt model is sufficiently accurate for synthesis of low-bandwidth logic for stability augmentation. Higher bandwidth models treat the rotor as a second rigid body (cf. HA-3), or each blade as a rigid body, or each blade as a flexible beam, in order of increasing accuracy (and complexity).

The *natural motions* near hover differ from those of fixed-wing aircraft as follows:

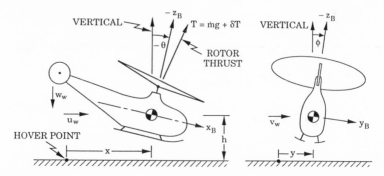

Figure 14.1. Nomenclature for motions of a helicopter near hover.

- The short period mode is replaced by two real modes, the pitch and heave modes.
- The phugoid mode is usually slightly unstable.
- The dutch roll mode is replaced by a lateral phugoid mode, which is usually slightly unstable.
- The spiral mode is replaced by a yaw mode.
- There is significant lateral-longitudinal coupling.

Neglecting lateral-longitudinal coupling is useful for developing qualitative ideas, but *synthesis of satisfactory control logic requires use of the coupled model.*

14.2 Small Deviations from Steady Hover

Near hover, the rotor acts like a vertical propellor with the thrust approximately equal to the vehicle weight (see Fig. 14.1 for nomenclature).

Large helicopters have governors that control the engine throttle to maintain nearly constant rotor angular velocity. In small helicopters this is done by the pilot using a twisting motion on a sleeve mounted on the collective stick (like the throttle on a motorcycle).

The collective pitch of the rotor blades is controlled by the *collective stick* δc, which is located to the left of the pilot (like the hand brake on a car). Pulling up on the stick increases the pitch of all the rotor blades, which increases the rotor thrust; it also increases rotor torque, so the governor (or the pilot) must increase throttle to maintain rotor speed. Thus, in a helicopter with a governor, the collective stick behaves like a throttle.

The cyclic pitch of the rotor blades is controlled by the *cyclic stick*, which is directly in front of the pilot. Forward motion of the stick, positive longitudinal cyclic stick, δe, tilts the swash plate laterally so that the pitch of the advancing blades (on the right) is decreased and the pitch of the retreating blades (on the left) is increased. This puts a forward torque on the rotor, which tilts the rotor TPP forward, which tilts the rotor thrust forward, which puts a pitching moment on the fuselage. Thus fore and aft stick motion produces the same effect in rotary wing aircraft as in fixed-wing aircraft where the stick moves the elevators to produce a pitching moment; thus we use the same symbol, δe, for it.

Motion of the stick to the right, positive lateral cyclic stick, δa, tilts the swash plate longitudinally so the the pitch of the rearward-pointing blades is decreased and the pitch of the forward-pointing blades is increased. This puts a rightward torque on the rotor, which tilts the rotor TPP to the right, which tilts the rotor thrust to the right, which puts a positive roll moment on the fuselage. The result is the same as deflecting ailerons in a fixed-wing aircraft, so we use the same symbol, δa.

The pitch of the tail rotor is controlled by the *pedals*. Pushing the right pedal causes a positive (nose-right) yawing moment, the same result obtained by moving the pedals to deflect the rudder in a fixed-wing aircraft, so we use the same symbol, δr.

The perturbation equations of motion are similar to those derived earlier for fixed-wing aircraft except that the controls are collective stick, longitudinal cyclic stick, lateral cyclic stick, and pedals, and there is significant cross-coupling between longitudinal and lateral motions. If we use the state vector

$$x = [\,u \quad w \quad q \quad \theta \quad v \quad r \quad p \quad \phi\,]^T \tag{14.1}$$

and the control vector

$$\delta = [\,\delta e \quad \delta c \quad \delta a \quad \delta r\,]^T, \tag{14.2}$$

then $\dot{x} = Fx + G\delta$ has the F matrix

$$F = \begin{bmatrix} X_u & X_w & X_q & -g & X_v & X_r & X_p & 0 \\ Z_u & Z_w & Z_q & -g\theta_o & Z_v & Z_r & Z_p & -g\phi_o \\ M_u & M_w & M_q & 0 & M_v & M_r & M_p & 0 \\ 0 & 0 & 1 & 0 & 0 & -\phi_o & 0 & 0 \\ Y_u & Y_w & Y_q & 0 & Y_v & Y_r & Y_p & g \\ N_u & N_w & N_q & 0 & N_v & N_r & N_p & 0 \\ L_u & L_w & L_q & 0 & L_v & L_r & L_p & 0 \\ 0 & 0 & 0 & 0 & 0 & \theta_o & 1 & 0 \end{bmatrix} \tag{14.3}$$

and G is an 8×4 matrix with the fourth and eighth rows equal to zero.

The equations giving position and heading angle are

$$\dot{x} = u, \tag{14.4}$$

$$\dot{h} = -w, \tag{14.5}$$

$$\dot{\psi} = r + \phi_o q, \tag{14.6}$$

$$\dot{y} = v, \tag{14.7}$$

where θ_o and ϕ_o are the trim pitch and roll angles of the body axes with respect to horizontal and (u_o, v_o, w_o) are the nominal velocity components (= 0 in hover). The stability derivatives $L_{(\,)}$ and $N_{(\,)}$ are really $\overline{L}_{(\,)}$ and $\overline{N}_{(\,)}$, as defined in Section 10.1.

Example–OH-6A near Hover

For a small utility helicopter, the OH-6A at 2,550 lb gross weight, Heffley et al (Ref. HE-2) give the equations of motion shown in Table 14.1.

The inherently large gyroscopic cross-coupling terms are $M_p = .3763$ and $L_q = -1.136$. The terms $N_u = -2.62$ and $N_w = 3.1$ are also large coupling terms for the OH-6A but are not necessarily large for other helicopters.

TABLE 14.1

EQUATIONS OF MOTION FOR OH-6A NEAR HOVER
x = [u,w,q,theta;v,r,p,phi]'; u = [de,dc;da,dr]';
w = [uw,ww;vw,pw]'; units ft, sec, crad; d's in deci-in,
xdot = Fx + Gu + Ga*w

```
F =
-0.0257   0.0113   0.0130  -.3216   0.0004  -0.0006  -0.0081      0
-0.0422  -0.3404   0.0001  -.0093  -0.0440   0.0147   0.0005  .0171
 1.2600  -0.6000  -1.7645      0   -0.2600   0.0719   0.3763      0
      0        0   0.9986      0        0    0.0532      0        0
 0.0158  -0.0194  -0.0084      0   -0.0435   0.0034  -0.0134  .3216
-2.6200   3.1000  -0.1724      0   -0.1700  -0.8645  -1.0750      0
 0.0300  -0.1900  -1.1360      0   -4.6200  -0.2873  -4.9200      0
      0        0  -0.0015      0        0    0.0289   1.0000      0


G =
    0.0860    0.0216   -0.0028   -0.003
   -0.0016   -0.7343    0.0011   -0.003
   -7.4080   -0.7850    0.3500   -0.096
```

```
        0            0            0            0
    0.0038      -0.0057       0.0514       0.153
    0.4930       9.5070       1.9820     -25.680
    1.8740       1.2060      12.7900      -0.781
        0            0            0            0
```

Ga =

```
    0.0257      -0.0113      -0.0004       0.0081
    0.0422       0.3404       0.0440      -0.0005
   -1.2600       0.6000       0.2600      -0.3763
        0            0            0            0
   -0.0158       0.0194       0.0435       0.0134
    2.6200      -3.1000       0.1700       1.0750
   -0.0300       0.1900       4.6200       4.9200
        0            0            0            0
```

The natural modes and modal controllability and disturbability are given in Table 14.2.

TABLE 14.2

NATURAL MODES OF OH-6A NEAR HOVER and
MODAL CONTROLLABILITY and DISTURBABILITY
x = T*m; mdot = Fm*m + Gm*u + GAm*w
Fm = diag(ev); T = evec; Gm = T\G; Gam = T\Ga

```
        ROLL        PITCH        YAW      LATERAL PHUGOID
ev =
      -4.9262     -2.0103     -0.8209     0.0143 +/- 0.5123i
evec =
       0.0035     -0.0870     -0.0406    -0.2156 -/+ 0.1449i
       0.0000     -0.0042     -0.0389     0.0071 -/+ 0.0100i
      -0.1251      1.0000      0.0297    -0.2418 -/+ 0.1016i
       0.0225     -0.4877     -0.1010    -0.2082 +/- 0.3616i
       0.0158     -0.0236     -0.0187     0.2705 +/- 0.2408i
       0.2622     -0.3407      1.0000     1.0000
       1.0000     -0.3199     -0.0557    -0.2755 -/+ 0.1739i
      -0.2046      0.1648      0.0327    -0.3523 +/- 0.4708i
```

```
   LONGITUDINAL PHUGOID    HEAVE
ev =
    -0.0001 +/- 0.4082i   -0.2294
evec =
    -0.2582 -/+ 0.1802i   -0.0360
```

```
 0.0340 -/+ 0.0325i     0.1593
-0.1915 -/+ 0.0881i    -0.0486
-0.2153 +/- 0.3382i    -0.0203
 0.0338 +/- 0.0576i    -0.0247
 1.0000                 1.0000
-0.0512 -/+ 0.0293i    -0.0318
-0.0715 +/- 0.0540i     0.0125
```

	de	dc	da	dr
Gm =				
R	-0.82	1.48	13.11	-0.23
P	-6.84	-1.23	1.71	0.07
Y	-4.24	11.58	0.61	-2.29
La	-1.60-/+1.42i	0.20-/+0.14i	-1.55-/+1.66i	-0.11-/+0.02i
Lo	5.49+/-4.70i	-0.96-/+0.11i	1.00+/-1.22i	0.40+/-0.14i
H	-5.17	-1.36	-0.38	-0.78

	uw	ww	vw	pw
Gam =				
R	-0.34	0.25	4.78	4.87
P	-1.28	0.84	0.79	0.16
Y	1.38	-4.15	-0.41	0.22
La	-0.10-/+0.17i	-0.06-/+0.01i	-0.53-/+0.61i	-0.68-/+0.71i
Lo	0.48+/-0.64i	0.13-/+0.06i	0.33+/-0.43i	0.70+/-0.76i
H	0.13	1.12	0.00	-0.41

- The *roll mode* ($s = -4.9262$ rad/sec) is the fastest mode. It is primarily controlled by lateral stick δa and is disturbed primarily by sidewind v_w and roll wind p_w.

- The *pitch mode* ($s = -2.0103$ rad/sec) is the next fastest mode. It has significant yaw rate and roll rate in it. It is primarily controlled by longitudinal cyclic stick δe and is disturbed primarily by horizontal and vertical wind u_w, w_w but is also significantly disturbed by sidewind.

- Surprisingly, the *yaw mode* ($s = -.8209$ rad/sec) is more controllable with collective stick δc and longitudinal cyclic stick δe than it is with pedal δr and lateral cyclic stick δa. Also it is disturbed more by vertical and horizontal wind than it is by lateral and roll wind.

- The *lateral phugoid mode* ($s = .0143 \pm .5123j$ rad/sec) is slightly unstable. It contains significant longitudinal components, which is still another indication of the strong lateral-longitudinal coupling. It is just slightly more controllable with lateral cyclic stick than it is with longitudinal cyclic stick; however, it is primarily disturbed by sidewind and roll wind.

- The *longitudinal phugoid mode* ($s = -.0001 \pm .4082j$ rad/sec) is almost neutrally stable and involves significant yaw rate. It is primarily controlled by longitudinal stick δe and is disturbed by u_w, v_w, and p_w.

- The *heave mode* ($s = -.2294$ rad/sec) is predominantly vertical velocity and yaw rate. It is primarily controlled by collective pitch and is primarily disturbed by vertical wind.

Yaw rate r is the largest element in the eigenvectors of all modes except the pitch and roll modes; clearly it is much easier to yaw than to roll or pitch due to the "gyroscopic stiffness" of the rotor.

The modal frequencies are fairly close to those found in an approximate uncoupled analysis, but the eigenvectors are significantly different.

14.3 Steady-State Control near Hover

Using the coupled model of Table 14.1, we find

$$
\begin{bmatrix} \delta e_s \\ \delta c_s \\ \delta a_s \\ \delta r_s \end{bmatrix}
=
\begin{bmatrix}
.1763 & -.0287 & -.0116 & .0191 \\
-.0567 & -.4617 & -.0575 & .0200 \\
-.0303 & .0598 & .3683 & .0004 \\
-1.220 & -.4615 & .0030 & -.2430
\end{bmatrix}
\begin{bmatrix} u_c \\ w_c \\ v_c \\ r_c \end{bmatrix},
\qquad (14.8)
$$

$$
\begin{bmatrix} \theta_s \\ \phi_s \end{bmatrix}
=
\begin{bmatrix}
-.0352 & -.0036 & -.0089 & .0034 \\
.0107 & .0649 & .0754 & -.0016
\end{bmatrix}
\begin{bmatrix} u_c \\ w_c \\ v_c \\ r_c \end{bmatrix}.
\qquad (14.9)
$$

The coupling between longitudinal and lateral motions is again shown here; large steady-state pedal δr_s, a lateral control, is required to hold longitudinal velocity commands u_c, w_c.

14.4 Control of Velocity Vector and Yaw Rate

14.4.1 *Hold Logic*

To hold velocity u, w, v and yaw rate r near zero, using all four controls $\delta c, \delta e, \delta a$, and δr, we use the quadratic performance index

$$J = \int_0^\infty (u^2 + w^2 + v^2 + r^2 + \delta e^2 + \delta c^2 + \delta a^2 + \delta r^2)dt. \qquad (14.10)$$

Using an LQR computer code and the coupled model of Table 14.1, we find the closed-loop eigenvalues

Mode	Open-Loop	Closed-Loop
Yaw	$-.821 \text{ sec}^{-1}$	-27.496
Roll	-4.926	-4.813
Pitch	-2.010	-2.092
Phugoid	$.000 \pm .408j$	$-.651 \pm .902j$
Lateral Phugoid	$.014 \pm .512j$	$-.520 \pm .762j$
Heave	$-.229$	$-.769$

and the associated feedback law

$$\begin{bmatrix} \delta e \\ \delta c \\ \delta a \\ \delta r \end{bmatrix} = - \begin{bmatrix} .791 & .107 & -.187 & -.479 & .275 & .015 & .004 & .111 \\ .074 & -.573 & -.010 & -.025 & .110 & .327 & -.007 & .024 \\ -.248 & .023 & .017 & .146 & .625 & .053 & .081 & .446 \\ .151 & -.321 & -.006 & -.023 & .086 & -.911 & .036 & .036 \end{bmatrix} \times \begin{bmatrix} u \\ w \\ q \\ \theta \\ v \\ r \\ p \\ \phi \end{bmatrix}. \qquad (14.11)$$

Examining these gains again shows the longitudinal-lateral coupling; there are significant gains from v to δe, from v and r to δc, from u and θ to δa, and from u and w to δr.

14.4.2 *Response to Worst-Case Initial Conditions*

With the loop closed, the worst initial condition vector of unit magnitude is

$$x(0) = [\,.902 - .053 - .052 - .230 - .339 - .004 - .020 - .109\,]^T. \quad (14.12)$$

This is mostly forward velocity with a leftward velocity about one-third as big; there is also a bank angle to the left, and the nose is slightly pitched down, giving an acceleration to the left and forward.

Fig. 14.2 shows the closed-loop response to the worst initial condition vector of unit magnitude. The largest control amplitudes are on δe and δa, which tilt the helicopter to push the nose up and right so that the rotor thrust brings the vehicle back to zero speed.

14.4.3 *Response to Commands*

Fig. 14.3 shows the closed-loop response to a command to go from hover to forward velocity $u_c = 10$ ft/sec, while holding the other two components of velocity and the yaw rate zero ($w_c = v_c = r_c = 0$). The autopilot used in the simulation involves feedforward of the steady-state results of (8) and (9), with the hold logic (11), that is, a change of setpoint.

The primary control is longitudinal cyclic stick δe which starts with an amplitude of 10 deci-in. However, maximum $|\delta a|$ is about 3 deci-in, maximum $|\delta c|$ is about 1 deci-in, and maximum $|\delta r|$ is about 2 deci-in. Maximum v and w are less than 1 ft/sec, and maximum r is less than 1 crad/sec. The maneuver is accomplished by tilting the nose down almost 15 crad to accelerate the vehicle forward, then returning the pitch angle almost to zero. A slight transient roll angle to the right occurs.

Fig. 14.4 shows the closed-loop response to a command to go from hover to vertical velocity $w_c = -10$ ft/sec while holding the other two components of velocity and the yaw rate zero ($u_c = v_c = r_c = 0$).

The primary control is collective stick, δc, which starts with an amplitude of 10.3 deci-in. Due to cross-coupling, the pedal transient is significant, with a maximum amplitude of 3.6 deci-in. There are also small transients in both longitudinal and lateral cyclic stick ($\delta e, \delta a$). Maximum u is .6 ft/sec, and maximum v is .2 ft/sec. Transient pitch and roll angles (θ, ϕ) are less than 1.2 crad.

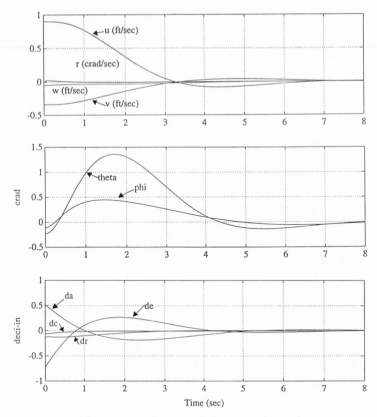

Figure 14.2. Response of the OH-6A helicopter near hover to the worst initial condition vector of unit magnitude with LQR SAS.

Problem 14.4.1 – *Response of the OH-6A to a Lateral Velocity and Yaw Rate Commands*

(a) Calculate and plot the response to a sideslip command, $v_c = 10$ ft/sec, while holding forward velocity, vertical velocity, and yaw rate equal to zero.

(b) Calculate and plot the response to a yaw rate command, $r_c = 1$ crad/sec, while holding forward velocity, vertical velocity, and lateral velocity equal to zero.

Problem 14.4.2 – *Decoupled Approximation to the Longitudinal Motions of the OH-6A*

The decoupled approximation splits the system into the two subsystems used in Chapters 12 and 13, namely u, w, q, θ controlled by $\delta e, \delta c$ and v, r, p, ϕ controlled by $\delta a, \delta r$.

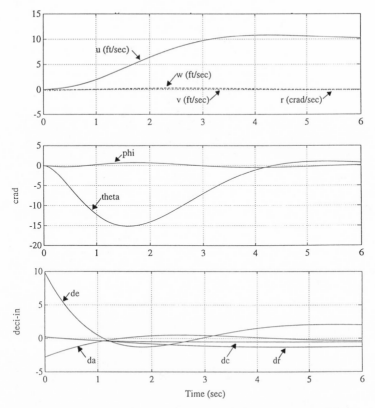

Figure 14.3. Response of the OH-6A helicopter in hover to command for forward velocity, with no change in vertical velocity, lateral velocity, or yaw rate; with LQR SAS.

(a) Find the modal form of the longitudinal approximation and compare the eigensystem with the appropriate modes of the coupled modal form in the text.

(b) Compare the controllability and disturbability of the longitudinal approximation with the coupled version in the text.

(c) Compare the steady-state control predicted by the longitudinal approximation with the coupled version in the text.

(d) Using the appropriate truncated form of the performance index (14.10), determine the feedback gains from u, w, q, θ to $\delta e, \delta c$ and compare with the coupled version.

(e) Determine the worst initial conditions for the closed-loop longitudinal system and plot the response to these conditions.

(f) Determine and plot the response to commands $u_c = 10, w_c = 0$ ft/sec.

(g) Determine and plot the response to commands $u_c = 0, w_c = -10$ ft/sec.

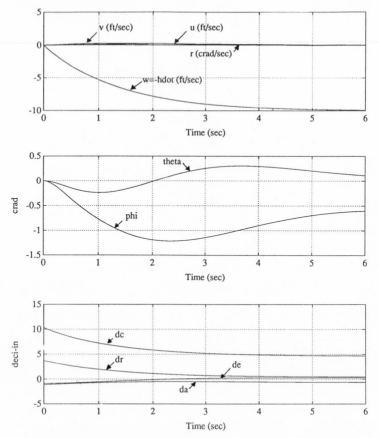

Figure 14.4. Response of the OH-6A helicopter in hover to command for vertical velocity with no change in forward velocity, lateral velocity, or yaw rate; with LQR SAS.

(h) Evaluate a further approximation that splits the longitudinal system into two sub-subsystems, vertical velocity w controlled by collective stick δc, and u, q, θ motions controlled by longitudinal cyclic stick δe.

Problem 14.4.3 – *Decoupled Approximation to the Lateral Motions of the OH-6A*

(a) Find the modal form of the lateral approximation and compare the eigensystem with the appropriate modes of the coupled modal form in the text.

(b) Compare the controllability and disturbability of the lateral approximation with the coupled version in the text.

(c) Compare the steady-state control predicted by the lateral approximation with the coupled version in the text.

(d) Using the appropriate truncated form of the performance index (14.10), determine the feedback gains from v, r, p, ϕ to $\delta a, \delta r$ and compare with the coupled version.

(e) Determine the worst initial conditions for the closed-loop lateral system and plot the response to these conditions.

(f) Determine and plot the response to commands $v_c = 10$ ft/sec and $r_c = 0$.

(g) Determine and plot the response to commands $v_c = 0$ and $r_c = 1$ crad/sec.

(h) Evaluate a further approximation that splits the lateral system into two subsystems, yaw rate r controlled by pedal δr, and v, p, ϕ motions controlled by lateral cyclic stick δa.

Problem 14.4.4 – *OH-6A Decoupled SAS Applied to the Coupled Model*

Apply the longitudinal and lateral stability augmentation systems designed in Problems 14.4.2 and 14.4.3 to the coupled model of Table 14.1.

(a) Determine the closed-loop eigenvalues and compare with those obtained in (14.10).

(b) Determine the worst initial conditions for the closed-loop system and plot the response to these conditions; compare to Fig. 14.2.

(c) Determine and plot the response to commands $u_c = 10$ ft/sec, $w_c = 0$, and compare to Fig. 14.3.

(d) Determine and plot the response to commands $u_c = 0, w_c = -10$ ft/sec, $v_w = 0, r_c = 0$, and compare to Fig. 14.4. Use the coupled model for the feedforward commands; integral output error would have to be used in practice.

15

Aeroelastic Systems

15.1 Introduction

15.1.1 *When Flexibility Must Be Considered in Control Design*

Flexibility or fuel slosh must be considered in designing the control system of an aircraft, launch vehicle, or missile if the frequency of the lowest vibration or slosh mode is less than about ten times the desired control bandwidth (see Chapter 9 where examples are given of controlling spacecraft with flexibility and fuel slosh).

Large aircraft and launch vehicles have low frequency deformation modes, and the bandwidth of fighter control systems is large, so flexibility is often of concern to the control system designer. Most aircraft have baffles in the fuel tanks so that fuel slosh frequencies tend to be higher than deformation frequencies.

Ride quality and metal fatigue of large flexible aircraft are affected by excitation of deformation modes of the fuselage and the wings. Active control has been used on the B-70, B-52, C-141, 747, and the B-1 aircraft to reduce the amplitudes of such modes.

15.1.2 *Aeroelastic Modes*

Aerodynamic forces may be thought of as feedback forces dependent on vehicle shape, air density, and velocity. Fig. 15.1 shows a block diagram of a flexible vehicle with aerodynamic forces and a control system. The aerodynamic forces depend on the state of the vehicle (x), and the state of the vehicle depends on aerodynamic forces (f) and actuator forces or control-surface deflections (u).

The aerodynamic feedback loop modifies the vaccuum rigid-body and vibration modes. For an aircraft the *modified rigid-body* modes are the short-period, phugoid, dutch roll, roll, and spiral modes. Only *quasi-steady aerodynamic forces* are needed to describe these modes, that is, the aerodynamic feedback block in Fig. 15.1 can be approximated as a *pure gain matrix*.

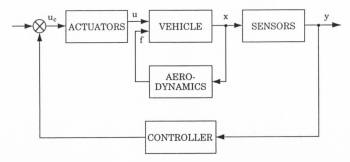

Figure 15.1. Block diagram of a flexible aircraft with aerodynamic force and control feedback loops.

The *modified vibration modes* are called "aeroelastic" modes, and *unsteady aerodynamic forces* are needed to describe them, that is, the aerodynamic feedback block in Fig. 15.1 is a *dynamic transfer function matrix*. In addition to the quasi-steady aerodynamic forces, there are significant dynamic lags. Section 15.2 gives an example of aeroelastic mode determination.

At subsonic speeds the aerodynamic lags are largely due to downwash velocities induced on the lifting surfaces by vorticity in the wake, which was shed off the trailing edges of the lifting surfaces a short time before. The aerodynamic forces must be modeled as a dynamic system if the characteristic times of the structural response are comparable to the aerodynamic time lags. This is the case when the lifting surfaces are thin and have vacuum vibration periods that are less than the time to travel about thirty chord lengths.

Vibration modes can be omitted from the control design model if they have frequencies that are very high compared to the closed-loop control bandwidth. We inherently assumed this in Chapters 10 to 14. However, this truncation of the more accurate aeroelastic model must be done with care since either the control or the aerodynamic feedback can destabilize one or more of the vacuum modes.

15.1.3 *Active Ride-Quality Control and Flutter Suppression*

The small damping of aeroelastic modes can cause significant ride quality and fluctuating load problems. A U.S. Air Force program in the seventies called "Load Alleviation and Mode Stabilization (LAMS)" investigated these problems on a "control-configured-vehicle" version of the B-52 aircraft (Ref. CCV). Load alleviation was an important goal for the C-141 Starlifter aircraft after flight tests showed metal fatigue on the wing structure after only a few hundred

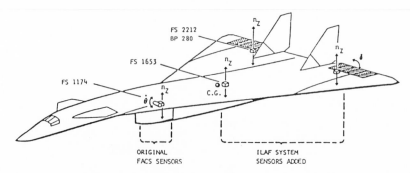

Figure 15.2. Wykes's spillover suppression system for the B-70 (from Ref. WY).

hours of flight. In both cases, the damping of lightly damped aeroelastic modes was increased by adding high bandwidth control systems.

An unstable aeroelastic mode is called a "flutter mode," and the velocity at which it occurs is called the flutter speed. If a stable aeroelastic mode is destabilized by the control system this is called "spillover."

The first ride-quality/spillover-suppression system used in flight was reported by Wykes in 1968 (Ref. WY) on the experimental B-70 aircraft (see Fig. 15.2 taken from his paper). Using a conventional SAS (the "original FACS sensor" signals fed back to the elevators), he predicted that the first fuselage bending mode of the B-70 would be destabilized. To suppress this spillover, he added "Identically Located Accelerometers and Forcers (ILAF)" by feeding back vertical velocity (from integrating accelerometers on the hinge lines) to the elevators (see Fig. 15.2). In his paper Wykes clearly described the guaranteed damping feature of what we now call "co-located sensors and actuators." This was a predecessor of the "low-authority control, high-authority control" concept given by Aubrun for flexible spacecraft (Ref. AU) in 1980. Without any SAS the fuselage bending mode is strongly excited when the aircraft flies through turbulent air at low altitudes, producing an almost debilitating vibration in the cockpit.

The first flutter suppression system demonstrated in flight was reported by Rogers in 1975 (Ref. RO) on a modified B-52 aircraft (see Fig. 15.3, taken from his paper). An artificial flutter mode was created by adding large wing-tip tanks, which had very forward mass centers. This increase in the wing torsion moment of inertia caused a flutter mode well within the flight envelope at a frequency of 2.4 Hz. This mode was relatively benign, in that the amplitude

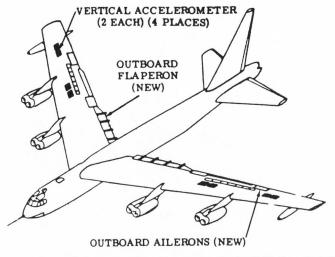

Figure 15.3. Rogers's flutter suppression system for the B-52 (from Ref. RO).

grew so slowly that the pilot could detect it and slow the airplane down to a velocity where the mode was stable before any damage was done. The active flutter control system, shown in Fig. 15.3, used high bandwidth flaperons and wing-mounted accelerometers and allowed the aircraft to be flown 12 knots above the open-loop flutter speed.

15.2 A Simple System with a Flutter Mode

The synthesis of an aeroelastic model of an aircraft, launch vehicle, or missile is a complicated task, involving finite-element codes to predict the vacuum vibration modes and unsteady-aerodynamics codes to then compute the dynamic model of the aerodynamic feedback forces.

To give a flavor of this task, we consider here a simple aeroelastic system that exhibits a flutter mode, namely the "typical section" (see Fig. 15.4 from Ref. ED-1).

The typical-section model may be thought of as a two-dimensional wing mounted in a wind tunnel with vertical and angular compliance (k_h and k_θ), or it may be thought of as an approximate model of a three-dimensional half wing mounted as a cantilever beam from one sidewall of the wind tunnel; the typical section is then the section of the wing at the three-quarter semispan (see Fig. 15.5), and the vertical compliance arises from beam bending and the angular compliance arises from beam torsion.

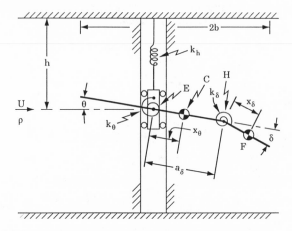

Figure 15.4. Nomenclature for "typical-section" aeroelastic model.

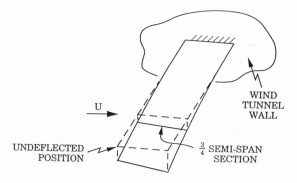

Figure 15.5. Cantilever wing interpretation of "typical-section" model.

Using the 2-D wing interpretation, free-body diagrams of the wing plus flap and the flap alone are shown in Fig. 15.6. Point E is the support point for the angular pivot (the "elastic axis" in the halfwing interpretation). Point C is the center of mass of the wing plus flap when $\delta = 0$; H is the flap hingeline; F is the center of mass of the flap alone.

Assuming both θ and δ to be small compared to 1 radian, the linearized equations of motion are

$$0 = m(\ddot{h} + x_\theta \ddot{\theta}) + m_f x_\delta \ddot{\delta} + L + k_h h, \tag{15.1}$$

$$0 = \bar{J}_\theta \ddot{\theta} + x_\theta m(\ddot{h} + x_\theta \ddot{\theta}) + (a_\delta + x_\delta) m_f x_\delta \ddot{\delta} + \bar{J}_\delta \ddot{\delta} - M + k_\theta \theta, \tag{15.2}$$

$$0 = \bar{J}_\delta(\ddot{\theta} + \ddot{\delta}) + m_f x_\delta [\ddot{h} + a_\delta \ddot{\theta} + x_\delta(\ddot{\theta} + \ddot{\delta})] - M_f + k_\delta \delta + Q, \tag{15.3}$$

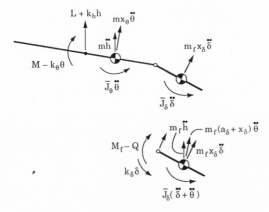

Figure 15.6. Free-body diagrams of wing plus flap and flap alone.

where

h = vertical displacement of E,

θ = pitch angle of the wing,

δ = flap deflection angle,

m = mass per unit span of wing plus flap,

m_f = mass per unit span of flap,

$\bar{J}_\theta$ = moment of inertia per unit span of wing plus flap about center of mass,

$\bar{J}_\delta$ = moment of inertia per unit span of flap, about flap center of mass,

x_θ = distance from elastic axis aft to center of mass of wing plus flap,

x_δ = distance from hinge line to center of mass of the flap,

a_δ = distance from elastic axis to hinge line,

(k_θ, k_δ) = spring constants for torsional springs connecting wing to wall, flap to wing,

k_h = spring constant of linear spring connecting wing to wall,

L = aerodynamic lift per unit span of wing plus flap,

(M, M_f) = aerodynamic moment per unit span on (wing plus flap, flap alone) about the (elastic axis, hinge line),

Q = control moment per unit span applied to flap from wing.

We define a vector Λ as

$$\Lambda = [\; -bL \quad M \quad M_f \;], \tag{15.4}$$

where

$$b = \text{the semi-chord of the wing plus flap.} \tag{15.5}$$

Using unsteady incompressible flow theory, Theodorsen and Garrick gave expressions for Λ in the early 1930s (Ref. THE and JO) for oscillatory (h, θ, δ). The theory was extended to arbitrary (h, θ, δ) histories by Edwards in 1984 (Ref. ED).

Λ is usually divided into a noncirculatory part, Λ_{nc}, and a circulatory part, Λ_c:

$$\Lambda = \Lambda_{nc} + \Lambda_c. \tag{15.6}$$

The *noncirculatory part* contains "apparent additional mass" as well as some damping and stiffness terms:

$$\Lambda_{nc} = -M_A \ddot{x} - B_A \dot{x} - K_A x, \tag{15.7}$$

where

$$x = [\; h/b \quad \theta \quad \delta \;]^T, \tag{15.8}$$

$$M_A = \rho b^4 \begin{bmatrix} \pi & -\pi a & -T_1 \\ -\pi & \pi(a^2 + 1/8) & 2T_{13} \\ -T_1 & 2T_{13} & -T_3/\pi \end{bmatrix}, \tag{15.9}$$

$$B_A = \rho U b^3 \begin{bmatrix} 0 & \pi & T_4 \\ 0 & \pi(1/2 - a) & T_{16} \\ 0 & T_{17} & T_{19}/\pi \end{bmatrix}, \tag{15.10}$$

$$K_A = \rho U^2 b^2 \begin{bmatrix} 0 & 0 & 0 \\ 0 & 0 & T_{15} \\ 0 & 0 & T_{18}/\pi \end{bmatrix}, \tag{15.11}$$

and

$$\begin{aligned}
\rho &= \text{air density,} \\
T_1 &= -(2+c^2)\sqrt{1-c^2}/3 + c\cos^{-1}c, \\
T_3 &= -(1-c^2)(5c^2+4)/8 + c(7+2c^2)\sqrt{1-c^2}\cos^{-1}c/4, \\
T_4 &= c\sqrt{1-c^2} - \cos^{-1}c, \\
T_5 &= -(1-c^2) - (\cos^{-1}c)^2 + 2c\sqrt{1-c^2}\cos^{-1}c, \\
T_7 &= -(1/8+c^2)\cos^{-1}c + c\sqrt{1-c^2}(7+2c^2)/8, \\
T_8 &= -(1+2c^2)\sqrt{1-c^2}/3 + c\cos^{-1}c, \\
T_9 &= (1-c^2)^{\frac{3}{2}}/6 + aT_4/2, \\
T_{10} &= \sqrt{1-c^2} + \cos^{-1}c, \\
T_{11} &= (2-c)\sqrt{1-c^2} + (1-2c)\cos^{-1}c, \\
T_{13} &= -T_7/2 - (c-a)T_1/2, \\
T_{15} &= T_4 + T_{10}, \\
T_{16} &= T_1 - (c-a)T_4 - T_8 + T_{11}/2, \\
T_{17} &= -T_1 - T4/2 - (1-c^2)^{\frac{3}{2}}/3, \\
T_{18} &= T_5 - T_4T_{10}, \\
T_{19} &= -T_4T_{11}/2,
\end{aligned}$$

and

$c = $ distance from the center of the wing plus flap to the hinge line in units of b,

$a = $ distance from the elastic axis aft to the center of the wing plus flap in units of b.

The Laplace transform of the *circulatory part* is

$$\Lambda_c(s) = [sB_c(s) + K_c(s)]x(s), \qquad (15.12)$$

where

$$B_c(s) = \rho U b^3 C(\bar{s}) R S_1, \qquad (15.13)$$

$$K_c(s) = \rho U^2 b^2 C(\bar{s}) R s_2, \qquad (15.14)$$

$$R = \left[2\pi \quad -2\pi(a+1/2) \quad T_{12} \right]^T, \qquad (15.15)$$

$$S_1 = \begin{bmatrix} 1 & 1/2 - a & T_{11}/2\pi \end{bmatrix},$$ (15.16)

$$S_2 = \begin{bmatrix} 0 & 1 & T_{10}/\pi \end{bmatrix},$$ (15.17)

$$C(\bar{s}) = \frac{K_1(\bar{s})}{[K_o(\bar{s}) + K_1(\bar{s})]},$$ (15.18)

and

$$\bar{s} \triangleq sb/U,$$ (15.19)

$$T_{12} = (2+c)\sqrt{1-c^2} - (1+2c)\cos^{-1} c.$$ (15.20)

The complex Theodorsen function, $C(\bar{s})$, involves the Bessel functions K_0 and K_1. Edwards and Breakwell (loc. cit.) pointed out that this function has a distributed singularity along the negative real axis in the complex $\bar{s}$ plane with a branch point at the origin. Rock showed that this becomes an infinite distribution of poles and zeros for the case where the wing is in a wind-tunnel (Ref. ROC).

Edwards computed the real and imaginary parts of $C(\bar{s}) = F(\bar{s}) + jG(\bar{s})$ using the infinite-series representation of the Bessel functions throughout the entire complex $\bar{s}$ plane (excluding the negative real axis), where F = real part, G = imaginary part. This is shown in Fig. 15.7 in the solid lines.

A second-order approximation to $C(\bar{s})$, first given by R. T. Jones (Ref. JO), is

$$C(\bar{s}) \cong \frac{1}{2} + \frac{1}{2} \frac{1 + \bar{s}/.0629}{(1 + \bar{s}/.0455)(1 + \bar{s}/.300)}.$$ (15.21)

This is shown in Fig. 15.7 in the dotted lines. Clearly in the range $|\bar{s}| < 1$—reduced frequency < 1—the Jones approximation is reasonably accurate. Since most aeroelastic system bandwidths are well within $|\bar{s}| < 1$, the Jones approximation is frequently used.

Using the Jones approximation, we find that the transfer functions from $x(s)$ to $\Lambda_c(s)$ are second-order. Thus, in the time domain, two aerodynamic states are introduced, making the aeroelastic system of order eight. This aerodynamic lag "compensator" has six inputs (x and $\dot{x}$) and three outputs (Λ).

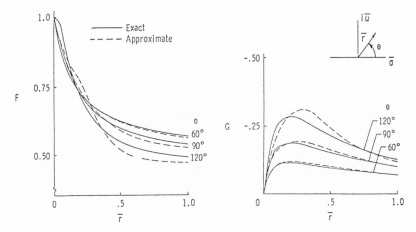

Figure 15.7. Comparison of the Theodorsen function and Jones's second-order approximation.

Example 15.2.1

Edwards (Ref. ED-1) wrote the Laplace transform of the equations of motion in the following normalized matrix form

$$\left[(M_s + \eta M_a)s^2 + (B_s + \eta \bar{U} B_a)s + (K_s + \eta \bar{U}^2 K_a)\right] x(s) = L_c(s) + GQ(s),$$

$$(15.22)$$

where

$$
\begin{aligned}
M_s &\quad \text{is in units of } mb^2, \\
K_s &\quad \text{is in units of } k_h b^2, \\
M_a &\quad \text{is in units of } \rho b^4, \\
B_a &\quad \text{is in units of } \rho U b^2, \\
K_a &\quad \text{is in units of } \rho U^2 b^2, \\
L_c &\quad \text{is in units of } k_h b^2, \\
s &\quad \text{is in units of } \omega_h,
\end{aligned}
$$

and

$$
\begin{aligned}
\omega_h &= \sqrt{k_h/m}, \\
\eta &= \rho b^2/m, \\
\bar{U} &= U/(b\omega_h),
\end{aligned}
$$

$$G = [\, 0 \ \ 0 \ \ 1\,]^T.$$

Thus

$$L_c(s) = \eta RC(s/\bar{U})(\bar{U}S_1 s + \bar{U}^2 S_2)x(s). \tag{15.23}$$

He computed the following example:

$$
\begin{aligned}
k_\theta &= k_h b^2, \\
k_\delta &= .225 k_h b^2, \\
m_f &= .2m, \\
x_\theta &= .2b, \\
x_\delta &= .0125b, \\
a &= -.4b, \\
c &= .6b, \\
J_\theta &= \bar{J}_\theta + m x_\theta^2, \\
&= m r_\theta^2, \\
r_\theta^2 &= .25b^2, \\
J_\delta &= \bar{J}_\delta + m x_\delta^2, \\
&= m_f r_\delta^2, \\
r_\delta^2 &= .05b^2, \\
\rho b^2/m &= 1/(40\pi).
\end{aligned}
$$

Fig. 15.8 shows a locus of the aeroelastic poles vs. $\bar{U}$. At $\bar{U} = 0$, the inertial coupling of the plunge (h), torsion (θ), and flap (δ) motions produce three undamped vibration modes. Apparent additional mass makes these frequencies slightly lower than they would be in a vacuum.

As $\bar{U}$ increases, the damping of the torsion and flap modes increases, but the damping of the plunge mode first increases then decreases, going unstable at $\bar{U} \cong 6.0$, indicating a *flutter* at that velocity (and higher).

A MATLAB .m file for calculating this root locus is given in Table 15.1.

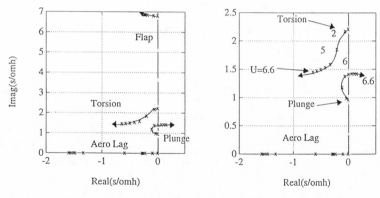

Figure 15.8. Locus of poles of a three degree-of-freedom section vs. $U/(b\omega_h)$.

TABLE 15.1

```
% MATLAB .m file for calculating equations of motion for typical
% section and root locus vs. U/(b*omh); omh^2 = kh/m, t in 1/omh,
% mf in m. (h,xth,xde,rth,rde,c,a) in b = semi-chord;
% eta = rho*b^2/m. (kth,kde,b*L,M,Mf,Q) in kh*b^2.

% Enter data:

mf=.125;xth=.2;xde=.1;rth=.5;rde=sqrt(.05);c=.6;a=-.4;
kth=1;kde=.225;eta=1/(40*pi);

% Compute normalized structural mass and stiffness matrices:

Ms=[1 xth mf*xde;xth rth^2 mf*(rde^2+(c+a)*xde)];
Ms=[Ms;mf*xde mf*(rde^2+(c+a)*xde) mf*rde^2];
Ks=diag([1 kth kde]);

% Compute non-circulatory aerodynamic forces & moment matrices:

ep=acos(c);d=sqrt(1-c^2);
T1=c*ep-d*(2+c^2)/3;T3=-d^2*(5*c^2+4)/8+c*(7+2*c^2)*d*ep/4;
T4=c*d-ep;T5=-d^2-ep^2+2*c*d*ep;
T7=-(c^2+1/8)*ep+c*d*(7+2*c^2)/8;T8=c*ep-(1+2*c^2)*d/3;
T9=d^3/6+a*T4/2;T10=d+ep;T11=(1-2*c)*ep+(2-c)*d;
T12=d*(2+c)-ep*(1+2*c);T13=-T7/2-(c-a)*T1/2;T15=T4+T10;
T16=T1-(c-a)*T4-T8+T11/2;T17=-T1-T4/2-d^3/3;
T18=T5-T4*T10;T19=-T4*T11/2;
Ma=eta*[pi -pi*a -T1;-pi*a pi*(a^2+1/8) 2*T13;-T1 2*T13 -T3/pi];
M=Ms+Ma;
```

```
% Compute circulatory aerodynamic lift and moment matrices:

R=[-2*pi 2*pi*(a+.5) -T12]';
S1=[0 1 T10/pi];
S2=[1 .5-a T11/(2*pi)];
p1=.0455;p2=.300;z=.0629;
n1=(1-p1/z)/(1-p1/p2);n2=(1-p2/z)/(1-p2/p1);

% Non-circulatory System with Circulatory Aerodynamic Feedback:
% x=[h/b,theta,delta]'; xdot=v; M*vdot=-B*v-K*x+.5*eta*R*u;
% y=Ub^2*S1*x+Ub*S2*v; xa=[xa1,xa2]'; (xa)dot=Fa*xa+Ga*y;
%   u=Ha*xa+y.

U=[0 2 5 5.8 6 6.2 6.4 6.6]';
for i=1:8,Ub=U(i)
  Ba=eta*Ub*[0 pi T4;0 pi*(.5-a) T16;0 T17 T19/pi];
  Ka=eta*Ub^2*[0 0 0;0 0 T15;0 0 T18/pi];
  B=Ba;K=Ks+Ka;
  F=[zeros(3) eye(3);-M\K -M\B];G=[0;0;0;M\R*eta];
    H=[Ub^2*S1,Ub*S2];
  Fa=Ub*diag([-p1 -p2]);Ga=Ub*[p1*n1;p2*n2];Ha=[.5 .5];La=.5;
  F8=[F+G*La*H,G*Ha;Ga*H,Fa];
eva(:,i)=eig(F8);
end;eva

% Plot root locus:

clg;axis([-2 .5 0 7]);
plot(real(eva),imag(eva),'x');grid
xlabel('Real(s/omh)');ylabel('Imag(s/omh)')
title('Root Locus vs. U/(b*omh)')
axis([-2 .5 0 2.5]);axis('square')
plot(real(eva),imag(eva),'x');grid
xlabel('Real(s/omh)');ylabel('Imag(s/omh)')
```

15.3 Flutter Suppression for a Simple System

Using the computer code of Table 15.1, we generated a state variable model $(\dot{x} = Fx + GQ)$ of Edward's typical section for $\bar{U} = 6.6$, which is 10% above the flutter speed.

The state vector is $x = [\, h/b \quad \theta \quad \delta \quad \dot{h}/b \quad \dot{\theta} \quad \dot{\delta} \quad x_1 \quad x_2 \,]$. Jones' second-order approximation to the Theodorsen function is used in this model (x_1 and

x_2 are the aerodynamic "lag" states). Time is in units of $1/\omega_h$.

$$
F = \begin{bmatrix}
0 & 0 & 0 & 1 & 0 & 0 & 0 & 0 \\
0 & 0 & 0 & 0 & 1 & 0 & 0 & 0 \\
0 & 0 & 0 & 0 & 0 & 1 & 0 & 0 \\
-1.180 & -.460 & -.026 & -.208 & -.237 & .037 & -.0316 & -.0316 \\
.916 & -3.291 & .227 & .238 & -.315 & -.154 & .0361 & .0361 \\
1.331 & 5.61 & -47.3 & -.098 & -.294 & -.517 & -.0149 & -.0149 \\
0 & 4.266 & 2.345 & .646 & .582 & .096 & -.300 & 0 \\
0 & 58.12 & 31.96 & 8.81 & 7.93 & 1.310 & 0 & -1.980
\end{bmatrix},
$$

$$
G/100 = [\,0 \quad 0 \quad 0 \quad -.0133 \quad -.0626 \quad 2.048 \quad 0 \quad 0\,]^T.
$$

** In the transfer function from Q to h the aeroelastic plunge and torsion modes have the largest residues, followed by the aerodynamic lag modes; the smallest residue is from the flap mode.

Fig. 15.9 is a locus of closed-loop poles vs. A using an LQ regulator with performance index

$$
J = \int_0^\infty [Ah^2 + (100Q)^2]dt. \tag{15.24}
$$

The slower aerodynamic lag pole is "trapped" by a nearby zero (this shows that we could have used a first-order model for the aerodynamic lag, omitting this mode). The unstable plunge poles move toward a pair of complex *reflected zeros*. Since these are not zeros of the transfer function, this mode dominates the response of the closed-loop system, and very little change in the response of the closed-loop system is achieved by increasing A even by several orders of magnitude. For $A = 1$, the LQR state feedback gains are ($100Q = -kx$)

$$
k = [\, -57.2 \quad 9.72 \quad -5.70 \quad -4.93 \quad -44.4 \quad -1.093 \quad -1.746 \quad -.956 \,],
$$
$$
\tag{15.25}
$$

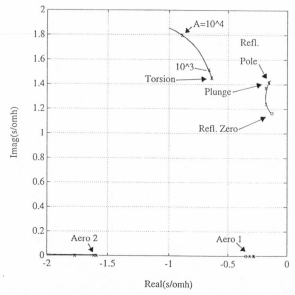

Figure 15.9. Locus of closed-loop poles vs. A for $J = \int_0^\infty (Ah^2 + (100Q)^2)dt$.

indicating a strong feedback of h and $\dot{\theta}$. The aerodynamic states would have to be estimated, since they cannot be measured directly; they are observable with measurements of h or $\dot{\theta}$.

The closed-loop eigenvalues (in rad/centi-sec) for $A = 1$ are

$$ev = [\ -.242 \pm 6.878i \quad -.151 \pm 1.417i \quad -.637 \pm 1.449i \quad -1.586 \quad -.278 \ ].$$

$$(15.26)$$

Figs. 15.10 and 15.11 show the open- and closed-loop responses to an initial value of $h = 1$, with all other states $= 0$. Without control, the system is unstable, that is, there is a flutter mode.

An unusual aspect of this control problem is that *the unstable complex zeros of the transfer function from Q to h place an upper limit on the closed-loop bandwidth.* Increasing the flap authority cannot attenuate the response significantly faster than shown in Fig. 15.11. This is explained by the fact that flap motion itself cannot stabilize the motion; it can only create pitch and plunge motions, which in turn provide the damping to stabilize the flutter mode.

Rock (Ref. ROC) and Stoltz (Ref. STO) designed, built, and tested control systems for a wind-tunnel model of a typical section quite similar to the example given here, and both demonstrated successful active flutter suppression.

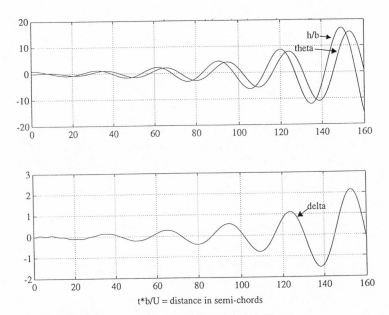

Figure 15.10. Flutter response of typical section to $h(0) = 1$.

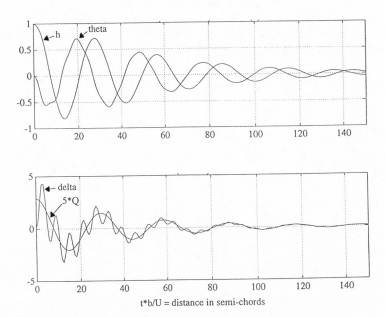

Figure 15.11. Flutter suppression response of typical section to $h(0) = 1$.

The design of an active flutter control system for an aircraft is a big project. In addition to the expert aerodynamicists, structural dynamics engineers, and control engineers needed for the control logic design, expert designers are also needed for the high bandwidth actuators and sensors.

Problem 15.3.1 – *Control of Typical Section with an External Vertical Force*

The typical section model will exhibit flutter with only the plunge h and torsion θ degrees of freedom (DOF). Furthermore the flutter mode is usually controllable with a vertical force through the elastic axis, which could be supplied from outside the wind tunnel through the supporting structure (as done in Ref. ROC).

Eliminate the flap DOF from Edwards's model used in the example (omit the third and sixth rows and columns of the F matrix).

(a) Determine the eigenvalues and eigenvectors of this 2 DOF system, and compare them with the eigensystem of the 3 DOF example.

(b) Using a vertical force per unit span f_z through the elastic axis as the control (in units of bk_h), the G matrix for $\bar{U} = 6.6$ is

$$G = [\, 0 \ \ 0 \ \ 1.1813 \ \ -.9162 \ \ 0 \ \ 0 \,].$$

Calculate the zeros of the transfer function from f_z to h and compare with the zeros of the transfer function from Q to h in the 3 DOF example. Note there is no bandwidth limitation for this type of control (no unstable zeros).

(c) Plot a locus of closed-loop poles vs. A where the performance index is

$$J = \int_0^\infty (Ah^2 + f_z^2)dt.$$

(d) Choose a value for A and plot the response to $h(0) = 1$ with the other three initial conditions zero. Compare with the open-loop response of the truncated model (with $\delta = 0$) to the same initial conditions.

Appendix A:
Linear System Representations

A.1 Representations of Linear Dynamic Systems

Six representations of time-invariant linear dynamic systems are used in this book:

(1) State space:

$$\dot{x} = Ax + Bu, \tag{A.1}$$

$$y = Cx + Du. \tag{A.2}$$

This form is obviously nonunique since the choice of state variables is arbitrary.

(2) Complex modal state space: Same form as state space representation except that the elements are complex numbers and A_m is diagonal for distinct eigenvalues; complex eigenvalues occur in complex conjugate pairs. The B_m and C_m matrices are not unique since only their column-row outer products (the residue matrices) are unique. This is identical to the factored-residue/pole representation of (6).

If T represents the (complex) eigenvector matrix of A, then

$$A_m = T^{-1}AT, \quad B_m = T^{-1}B, \quad C_m = CT. \tag{A.3}$$

(3) Real modal state space: Same form as state space representation except A_{mr} is block-diagonal for distinct eigenvalues (2 by 2 real blocks for complex eigenvalues, single real blocks for real eigenvalues). The B_{mr} and C_{mr} matrices are not unique, but they are real and do not contain repeated columns and rows as in the complex modal state representation. For more details, see the next section.

(4) Transfer function:

$$y(s) = \left[\frac{b(0)s^n + b(1)s^{n-1} + \cdots + b(n)}{s^n + a(1)s^{n-1} + \cdots + a(n)} \right] \cdot u(s). \tag{A.4}$$

The $b(k)$ are real matrices with n_o rows and n_c columns where n_o = number of outputs and n_c = number of controls.

(5) Zero/pole:

$$y_i(s) = \left[\frac{K_{ij}(s - z_{ij}(1)) \cdots (s - z_{ij}(m_{ij}))}{(s - p(1)) \cdots (s - p(n))}\right] \cdot u_j(s). \qquad (A.5)$$

This is a "factored" form of the transfer function representation.

(6) Residue/pole:

$$y(s) = \left[D + \sum_{k-1}^{n} \frac{R(k)}{s - p(k)}\right] \cdot u(s). \qquad (A.6)$$

The D and $R(k)$ are complex matrices. The R(k) matrices are rank one, so each one can be factored into the outer product of two vectors; these vectors form the columns and rows of the B_m and C_m matrices in the factored-residue/pole representation.

The control engineer should be familiar with all of these representations. Current professional software is capable of transforming from one representation to another quickly and accurately.

A.2 Complex Modal State and Residue/Pole

The complex modal state representation is a"factored form" of the residue/pole representation (Ref. BE), and is especially useful for MIMO systems. For distinct poles it may be written as

$$A_m = -\text{diag}[p(1) \cdots p(n)],$$

$$B_m = \begin{bmatrix} B_m(1,1) \cdots & B_m(1,nc) \\ \vdots & \vdots \\ B_m(n,1) \cdots & B_m(n,nc) \end{bmatrix},$$

$$C_m = \begin{bmatrix} C_m(1,1) \cdots & C_m(1,n) \\ \vdots & \vdots \\ C_m(no,1) \cdots & C_m(no,n) \end{bmatrix}, \qquad (A.7)$$

where n = number of poles, nc = number of controls, no = number of outputs.

The residue/pole form is obtained from the complex modal state form by calculating outer products of the columns of C_m with corresponding rows of B_m; for the kth mode the residue matrix is:

$$R(k) = \begin{bmatrix} C_m(1, k) \\ \vdots \\ C_m(no, k) \end{bmatrix} [B_m(k, 1) \cdots B_m(k, nc)]. \tag{A.8}$$

Thus the complex modal state realization is a "factored" form of the residue/pole realization. It follows that *for distinct poles p_k, each residue matrix R(k) has rank one (Ref. BE)*, that is, it can be factored into the outer product of two vectors. This factorization can be done by inspection for small residue matrices or by using the *singular value decomposition* algorithm (Ref. GO). For repeated poles, the story is more complicated (Ref. WA).

A.3 Real Modal State

A complex modal state representation is inefficient for storage since all complex numbers are stored twice (original plus complex conjugate). Complex number arithmetic on a computer may take twice as long as necessary since some algorithms simply treat a complex number $z = a + jb$ as a two-by-two matrix:

$$z = \begin{bmatrix} a & b \\ -b & a \end{bmatrix}.$$

Thus $z_1 \cdot z_2$ is handled as a product of two 2-by-2 matrices:

$$z_1 \cdot z_2 = \begin{bmatrix} a_1 & b_1 \\ -b_1 & a_1 \end{bmatrix} \begin{bmatrix} a_2 & b_2 \\ -b_2 & a_2 \end{bmatrix} = \begin{bmatrix} a_1 a_2 - b_1 b_2 & a_1 b_2 + a_2 b_1 \\ -a_1 b_2 - a_2 b_1 & a_1 a_2 - b_1 b_2 \end{bmatrix}.$$

This is usually not much of a problem with the speed of modern computers, but the real modal state representation is useful when computing time is important, as in real-time controllers or in analysis and synthesis of very large systems. Also some software algorithms only work for real representations.

For distinct eigenvalues, a real modal state vector ξ may be obtained from the complex modal state vector ξ_c by a block-diagonal transformation T_c,

$$\xi_c = T_c \cdot \xi, \tag{A.9}$$

where T_c has unity blocks for real modes and the following 2-by-2 block for complex modes:

$$\frac{1}{2} \begin{bmatrix} 1 & -j \\ 1 & j \end{bmatrix}. \tag{A.10}$$

This transformation results in a real modal dynamics matrix that has 2-by-2 real blocks for complex modes (Ref. MA-1, pp. 404–406):

$$\begin{bmatrix} \sigma + j\omega & 0 \\ 0 & \sigma - j\omega \end{bmatrix} \longrightarrow \begin{bmatrix} \sigma & \omega \\ -\omega & \sigma \end{bmatrix}. \tag{A.11}$$

In the two corresponding columns of the eigenvector matrix the first column is the real part of the eigenvector and the second column is the imaginary part, so the eigenvector matrix no longer has repeated columns. Also, the input and output matrices no longer have repeated columns and rows.

Minimum-Realization Real Modal Form

Another version of the real modal form, which we shall call the minimum-realization real modal form, uses 2-by-2 blocks for complex modes in the block-diagonal dynamics matrix of the form

$$\begin{bmatrix} 0 & 1 \\ a_1 & a_2 \end{bmatrix}, \tag{A.12}$$

where

$$a_1 = -\sigma^2 - \omega^2,$$
$$a_2 = 2\sigma.$$

This eliminates two multiplications for each complex mode in implementing the compensator.

Many more multiplications can be eliminated by the following normalization of the input and output matrices:

- For real modes, the largest element in the column of C_m corresponding to that mode is used to divide that column and multiply the corresponding row in K_m yielding a 1 in the C_m column.
- For complex modes, the largest "complex" element in the two columns of C_m corresponding to that mode is used to divide that "complex column" and multiply the corresponding "complex row" in K_m yielding a (0,1) row in the two columns of C_m.

A.4 Factored Residue/Pole

Since only the products of corresponding columns of C_m and rows of B_m are important in the input/output relations, we can normalize these vectors to have equal length.

If the vectors are normalized to *unit* length, call them $\hat{c}(k)$ and $\hat{b}(k)$, we have an elegant and informative representation of a MIMO system:

$$y(s) = \left[D + \sum_{k=1}^{n} \frac{r(k) \cdot \hat{c}(k) \cdot \hat{b}(k)}{s - p(k)} \right] \cdot u(s), \qquad (A.13)$$

where the scalar $r(k)$ is the product of the magnitude of the two vectors (hence it is also equal to the square root of the sum of the squares of the elements of the residue matrix).

If the output and control-input variables have been scaled so that one unit of each is of comparable importance to the designer (which has been done in all of the examples in this book), then

- The scalar $r(k)$ is a measure of the importance of the kth mode in the input/output relations.
- The elements of the unit vector $\hat{b}(k)$ indicate the relative amounts of each control input to mode k (a measure of modal controllability).
- The elements of the unit vector $\hat{c}(k)$ indicate the relative amounts of mode k that appear in each output (a measure of modal observability).

A.5 Model Order Reduction

Frequency Separation

If there is a large frequency separation in the modes of the system, the fast (heavily damped) modes may be approximated as quasi-steady. To approximate the ith mode as quasi-steady, place $\dot{m}_i = 0$, solve for m_i in terms of the controls, and substitute this expression for m_i in the output. Clearly this is equivalent to omitting the "s" term in the transfer functions associated with fast poles.

Modal Truncation

Using the canonical complex modal state realization, let r_{max} = the maximum over $k = 1, \cdots, n$ of $|r(k)|$. If $|r(k1)|$ is much smaller than r_{max}, and if the $k1^{th}$ mode is reasonably well-damped, then it is relatively unimportant in all of the input/output relations and may be omitted (truncated out) to obtain a simpler approximate representation of the system.

Factor Truncation

Factor truncation is based on omitting "principal components" from the balanced-realization form (Ref. MO). A linear transformation of the compensator states is determined such that the observability- controllability measures of the new states (the factors or principal components) are uncoupled. The new states with the lowest measures can then be omitted without affecting the observability-controllability measures (hence the stability) of the remaining states.

An efficient algorithm for determining the balanced realization was given by Laub (Ref. LA) as follows:

- Given (A, B, C) where

$$\dot{x} = Ax + Bu, \tag{A.14}$$

$$y = Cx, \tag{A.15}$$

calculate the matrices (P, Q) from the steady-state Lyapunov equations

$$AP + PA^T + BB^T = 0, \tag{A.16}$$

$$A^T Q + QA + C^T C = 0. \tag{A.17}$$

- Calculate R where

$$R^T R = P. \tag{A.18}$$

- Calculate $[U, \Sigma^2]$ = the eigensystem of RQR^T, where Σ^2 is a diagonal matrix whose elements are the eigenvalues, and the columns of U are the corresponding eigenvectors. The eigenvalues are the measures of observability/controllability.

- The new states (principal components) x_B are given by the transformation $x = Tx_B$, where

$$T = R^T U \Sigma^{-1/2}, \tag{A.19}$$

that is,

$$\dot{x}_B = (T^{-1}AT)x_B + (T^{-1}B)u, \tag{A.20}$$

$$y = (CT)x_B. \tag{A.21}$$

This algorithm is implemented in the currently available professional software codes.

A.6 Example of Different Representations for a MIMO System

Listed below are the six representations of the longitudinal model of the Navion (a small aircraft) with $x = [u, w, q, \theta]^T$, $u = [\delta e, \delta t]^T$, $y = [u, h]^T$. The units are ft, sec, and crad.

State Space Representation

$$\begin{bmatrix} \dot{x} \\ y \end{bmatrix} = S \begin{bmatrix} x \\ u \end{bmatrix}, \quad S = \begin{bmatrix} A & B \\ C & D \end{bmatrix}$$

$$S = \begin{bmatrix} -.045 & .036 & 0 & -.322 & 0 & 1.000 \\ -.370 & -2.020 & 1.760 & 0 & -.282 & 0 \\ .191 & -3.960 & -2.980 & 0 & -11.000 & 0 \\ 0 & 0 & 1.000 & 0 & 0 & 0 \\ 1.000 & 0 & 0 & 0 & 0 & 0 \\ 0 & -1.000 & 0 & 1.760 & 0 & 0 \end{bmatrix}$$

Complex Modal State Space Representation

$$A_m = \text{diag}\,[-2.505 + 2.595i - 2.505 - 2.595i - 0.017 + 0.213i - 0.017 - 0.213i]$$

$$B_m = \begin{bmatrix} -5.491 - 1.216i & -0.072 + 0.032i \\ -5.491 + 1.216i & -0.072 - 0.032i \\ 0.008 - 1.228i & 0.499 + 0.040i \\ 0.008 + 1.228i & 0.499 - 0.040i \end{bmatrix}$$

$$C_m = \begin{bmatrix} -0.003 - 0.020i & -0.003 + 0.020i & 1.000 & 1.000 \\ -0.219 + 0.305i & -0.219 - 0.305i & -0.105 - 1.165i & -0.105 + 1.165i \end{bmatrix}$$

Real Modal State Space Representation

$$A_{mr} = \begin{bmatrix} -2.505 & 2.595 & 0 & 0 \\ -2.595 & -2.505 & 0 & 0 \\ 0 & 0 & -.017 & .213 \\ 0 & 0 & -.213 & -.017 \end{bmatrix}$$

$$B_{mr} = \begin{bmatrix} .0160 & .9983 \\ -2.4568 & .0798 \\ -2.1962 & -.0287 \\ .4863 & -.0127 \end{bmatrix}$$

$$C_{mr} = \begin{bmatrix} 1.0000 & .0000 & -.0143 & -.0976 \\ -.1051 & 1.1651 & -1.0947 & 1.5260 \end{bmatrix}$$

Transfer Function Representation

$$y(s) = \frac{N(s)}{d(s)} \cdot u(s)$$

$$N(s) = \begin{bmatrix} -.010s^2 + 2.815s + 6.795 & s^3 + 5.000s^2 + 12.99s \\ .282s^3 + .853s^2 - 37.10s - 0.605 & .370s^2 + 1.103s + 3.258 \end{bmatrix}$$

$$d(s) = s^4 + 5.045s^3 + 13.228s^2 + .674s + .5960$$

Zero/Pole Representation

$$y(s) = \frac{N(s)}{d(s)} \cdot u(s)$$

$$N(s) = \begin{bmatrix} -.010(s - 297.7)(s + 2.395) & s[(s + 2.500)^2 + (2.596)^2] \\ .282(s + 13.08)(s - 10.07)(s + .0163) & .370[(s + 1.490)^2 + (2.566)^2] \end{bmatrix}$$

$$d(s) = \left[(s + 2.505)^2 + (2.595)^2\right] \left[(s + .017)^2 + (.213)^2\right]$$

Residue/Pole Representation

$$y(s) = \sum_{k-1}^{4} \frac{R(k)}{s - p(k)} \cdot u(s)$$

$$R(1) = \begin{bmatrix} -.008 + .111i & .001 + .001i \\ 1.573 - 1.410i & .006 - .029i \end{bmatrix}$$

$$p(1) = -2.505 + 2.595i$$

$$R(3) = \begin{bmatrix} -.008 - 1.228i & .499 + .040i \\ -1.432 - .120i & -.006 - .586i \end{bmatrix}$$

$$p(3) = -.017 + .213i$$

and the corresponding elements for (2) and (4) are the conjugates of (1) and (3).

Factored-Residue/Pole Representation

$$r(1) = 2.1154$$

$$p(1) = -2.505 + 2.595i$$

$$\hat{c}(1) = \begin{bmatrix} -.0076 + .0519i \\ -.5821 - .8114i \end{bmatrix}$$

$$\hat{b}(1) = [-.9763 + .2161i, -.0128 - .0057i]$$

$$r(3) = 2.0417$$

$$p(3) = -.017 + .213i$$

$$\hat{c}(3) = \begin{bmatrix} .6498 + .0000i \\ -.0683 + .7571i \end{bmatrix}$$

$$\hat{b}(3) = [.0060 + .9260i, .3763 - .0301i]$$

and the corresponding elements for (2) and (4) are the conjugates of (1) and (3).

Appendix B:
Steady-State Control

B.1 Steady-State Control

A linear dynamic system of the form

$$\dot{x} = Fx + Gu, \tag{B.22}$$

$$y = Hx + Lu, \tag{B.23}$$

is in equilibrium when $\dot{x} = 0$. Clearly one equilibrium solution is $u = 0$, which implies $x = y = 0$. Other equilibrium solutions may exist when u, and hence x and y, are nonzero.

B.1.1 *Number of Controls Equals Number of Outputs*

If the number of controls equals the number of outputs, then specification of the desired steady-state outputs $y = y_c$ may determine the required steady-state controls u from

$$0 = Fx + Gu, \tag{B.24}$$

$$y_c = Hx + Lu. \tag{B.25}$$

If this system is nonsingular, it is "steady-state controllable." Since the solutions are proportional to y_c, it is convenient to write them in the form

$$x = N_{xy}y_c, \tag{B.26}$$

$$u = N_{uy}y_c, \tag{B.27}$$

where

$$\begin{bmatrix} N_{xy} \\ N_{uy} \end{bmatrix} = \begin{bmatrix} F & G \\ H & L \end{bmatrix}^{-1} \begin{bmatrix} 0 \\ I \end{bmatrix}.$$

Examples of steady-state control are given in Sections 12.2 and 13.1.

B.1.2 *Number of Controls Exceeds Number of Outputs*

If the number of controls is greater than the number of outputs, there may be an infinite number of control vectors u that produce a specified output vector y_c. In that case, we may wish to determine the "smallest" control vector that produces the desired output, that is, to find u such that Equations (3) and (4) are satisfied and

$$J = u^T Bu/2 \qquad (B.7)$$

is minimized. This optimization problem is easily solved by adjoining (3) and (4) to (7) with Lagrange multiplier vectors λ and μ (see, for example, Ref. BR-1, Chapter 1):

$$J' = J + \lambda^T (Fx + Gu) + \mu(Hx + Lu - y_c). \qquad (B.8)$$

Setting the gradients of (8) with respect to x and u equal to zero, along with (3) and (4), yields the system of equations

$$
\begin{bmatrix}
F & G & 0 & 0 \\
H & L & 0 & 0 \\
0 & 0 & F^T & H^T \\
0 & B & G^T & L^T
\end{bmatrix}
\begin{bmatrix}
x \\
u \\
\lambda \\
\mu
\end{bmatrix}
=
\begin{bmatrix}
0 \\
y_c \\
0 \\
0
\end{bmatrix}. \qquad (B.9)
$$

If this system is nonsingular, it is "steady-state controllable." The solutions of (9) for x and u are of the form (5) and (6).

B.1.3 *Number of Outputs Exceeds Number of Controls*

If the number of outputs is greater than the number of controls, then, in general, no control vector u can exactly produce a specified output vector y_c. However, it may be of interest to determine the u that produces the "smallest" output error in the sense of minimizing

$$J = (y - y_c)^T A(y - y_c)/2 + u^T Bu/2, \qquad (B.10)$$

where y is defined in (2) and (3) is satisfied. This optimization problem is easily solved by adjoining (2) and (3) to (10) with Lagrange multiplier vectors λ and μ (see, for example, Ref. BR-1, Chapter 1):

$$J' = J + \lambda^T (Fx + Gu) + \mu(Hx + Lu - y). \qquad (B.11)$$

Setting the gradients of (11) with respect to x and u equal to zero, along with (3) and (4), yields the system of equations

$$
\begin{bmatrix}
F & G & 0 & 0 \\
H & L & 0 & -A^{-1} \\
0 & 0 & F^T & H^T \\
0 & B & G^T & L^T
\end{bmatrix}
\begin{bmatrix}
x \\
u \\
\lambda \\
\mu
\end{bmatrix}
=
\begin{bmatrix}
0 \\
y_c \\
0 \\
0
\end{bmatrix},
\tag{B.12}
$$

$$
y = y_c + A^{-1}\mu.
\tag{B.13}
$$

B.2 Nonzero Set Points and Output Commands

B.2.1 Systems with State Feedback

Output command logic is obtained from hold logic by simply changing the set point (Ref. KW):

$$
u = -Cx \quad \text{(hold logic)},
\tag{B.14}
$$

$$
u - u_s = -C(x - x_s) \quad \text{(output command logic)},
\tag{B.15}
$$

where

$$
u_s = N_{uy}y_c,
\tag{B.16}
$$

$$
x_s = N_{xy}y_c.
\tag{B.17}
$$

Combining (15) through (17), we have

$$
u = (N_{uy} + CN_{xy})y_c - Cx.
\tag{B.18}
$$

The gain matrix $N_{uy} + CN_{xy}$ is a *feedforward gain matrix* since it feeds the command signal directly "forward" to the controls u. Examples are given in Sections 12.5 through 12.7 and in Sections 13.5 throught 13.7.

The relation (18) is convenient for simulation in preliminary design studies. However, it is *not recommended for implementation on the real system* because it is too dependent on the model (it uses the inverse of the model to predict the steady state). Better results will usually be obtained with the real system by redesigning the controller to use *integral-error feedback*.

B.2.2 *Systems with Estimated State Feedback*

For a system using feedback of estimated state, the relation (18) is modified by replacing x by $\hat{x}$ = the estimate of x, that is,

$$u = (N_{uy} + CN_{xy})y_c - C\hat{x}, \qquad (B.19)$$

where

$$\dot{\hat{x}} = F\hat{x} + Gu + K(y - H\hat{x}), \qquad (B.20)$$

is the estimator.

If the estimator-regulator is put into *compensator form*, then there is no u input, and the correct compensator form of (20) is obtained by using (19) in (20) to eliminate u:

$$\dot{\hat{x}} = (F - KH - GC)\hat{x} + Ky + G(N_{uy} + CN_{xy})y_c. \qquad (B.21)$$

Note the extra term proportional to y_c in (21).

In (21) we have assumed that the controlled output quantities y are also the measured quantities. If the measured quantities are different than the commanded quantities, (21) must be suitably modified (see next subsection).

If the estimator has been in use before the output command is given, so that $\hat{x}$ is closely tracking the state x, then the response using either (20) or (21) should be very close to the response using state feedback as in (15).

B.2.3 *Systems with Compensators*

Given a plant model

$$\dot{x} = Fx + Gu, \qquad (B.22)$$

$$y = Hx + Lu = \text{controlled outputs}, \qquad (B.23)$$

$$z = H_z x + L_z u = \text{sensed outputs}, \qquad (B.24)$$

and a stabilizing hold compensator

$$\dot{x}_c = Ax_c + Bz, \qquad (B.25)$$

$$-u = Cx_c + Dz = \text{control inputs}, \qquad (B.26)$$

we wish to modify the hold compensator to obtain an *output command compensator*. We assume y and u have the same dimensions.

We first determine u_s and x_s for commanded $y = y_c$ from the steady-state equations of motion (3) and (4). From (24), (5), and (6) z_s can be determined:

$$z_s = (H_z N_{xy} + L_z N_{uy})y_c \overset{\Delta}{=} N_{zy}y_c. \tag{B.27}$$

The *output command compensator* is then

$$\dot{x}_c = Ax_c + B(z - z_s),$$
$$-(u - u_s) = Cx_c + D(z - z_s),$$

or

$$\dot{x}_c = Ax_c + Bz - BN_{zy}y_c, \tag{B.28}$$
$$-u = Cx_c + Dz - (N_{uy} + DN_{zy})y_c. \tag{B.29}$$

If we let $x = x_s + \delta x, u = u_s + \delta u, y = y_c + \delta y$, and $z = z_s + \delta z$, then the relations for $(\delta x, \delta y, \delta z, x_c, \delta u)$ are exactly the same as for (x, y, z, x_c, u) in the hold compensator, confirming that all we have done is to change the set point.

Description in Terms of Transfer Functions

Let the open-loop transfer functions be

$$y(s) = Y(s) \cdot u(s), \tag{B.30}$$
$$z(s) = Z(s) \cdot u(s), \tag{B.31}$$

and let the hold compensator transfer functions be

$$u(s) = -U(s) \cdot z(s). \tag{B.32}$$

The values of y and z for a constant input u of magnitude u_s are given by

$$y_s = Y(0) \cdot u_s, \tag{B.33}$$
$$z_s = Z(0) \cdot u_s. \tag{B.34}$$

Hence

$$u_s = [Y(0)]^{-1} \cdot y_c \overset{\Delta}{=} N_{uy} \cdot y_c, \tag{B.35}$$
$$z_s = Z(0) \cdot [Y(0)]^{-1} \cdot y_c \overset{\Delta}{=} N_{zy} \cdot y_c. \tag{B.36}$$

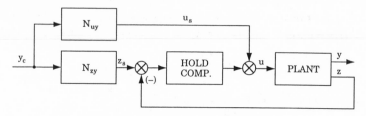

Figure B.1. Block diagram of output command compensator.

An output command compensator is then easily found as

$$u(s) - u_s = -U(s) \cdot [z(s) - z_s]$$

or

$$u(s) = N_{uy} \cdot y_c(s) - U(s) \cdot \left[z(s) - N_{zy} \cdot y_c(s) \right] . \qquad (B.37)$$

A block diagram of the scheme is shown in Figure B.1.

Eliminating $z(s)$, the transfer function from $y_c(s)$ to $u(s)$ is

$$u(s) = [I + U(s) \cdot Z(s)]^{-1} \cdot \left[N_{uy} + U(s) \cdot N_{zy} \right] \cdot y_c(s). \qquad (B.38)$$

Hence, the transfer function from y_c to y is

$$y(s) = Y(s) \cdot [I + U(s) \cdot Z(s)]^{-1} \cdot \left[N_{uy} + U(s) \cdot N_{zy} \right] \cdot y_c(s). \qquad (B.39)$$

Appendix C:
Synthesis of Analog
Control Logic

C.1 Frequency Response and Root Locus

Historically, the *frequency response* (FR) method was the first method used for analyzing control logic (1930–1940; see Ref. JA). It is still the most popular method of analysis and will always be an important method (Ref. MI).

Typically, a closed loop is cut (conceptually) at some point and a sinusoidal signal is assumed at the input side; the steady-state magnitude $|G(j\omega)|$ and phase $\arg[G(j\omega)]$ at the output side are then calculated and plotted versus frequency ω. For stability of the closed-loop system, the phase lag must be less than 180 deg at the frequency where the magnitude ratio goes through unity (called the "cross-over" frequency). For minimum phase systems, only the magnitude plot is needed; for stability, the slope of the $\log_{10}|G(j\omega)|$ vs. $\log_{10}\omega$ plot must be greater than -2 (preferably about -1) at cross-over.

There now exist many computer codes that quickly and accurately calculate and plot $\log_{10}|G(j\omega)|$ and $\arg[G(j\omega)]$ vs. $\log_{10}\omega$ given the poles, zeros, and gain of $G(s)$.

The *phase margin* is the difference between 180 deg and the phase lag at cross-over. A phase margin of 35 deg or more is considered desirable for stability *robustness*, that is, one should allow for the possibility that the plant being controlled has up to 35 deg more phase lag at the cross-over frequency than the plant model predicts.

The *gain margin* is the reciprocal of the magnitude ratio at the frequency where phase lag is 180 deg, and it is another commonly used measure of stability robustness. It must be above unity for stability, and a value of 2 or more is considered desirable.

Synthesis of control logic (compensator synthesis) is based on *successive loop closures*, selecting certain types of feedback in each loop, such as proportional, derivative, integral, lead, lag, notch, etc. The parameters in these feedbacks are adjusted until satisfactory closed-loop response is attained. For multiloop controllers, this requires iteration.

There are many good references on frequency response methods (such as MC), so we shall not describe them further.

An extension of FR analysis to multiple-input, multiple-output systems has been made in Ref. DO using plots of the smallest singular-value of a complex transfer function matrix versus frequency. It is a generalization of the Nyquist plot, where the smallest singular value is the shortest distance from the complex FR locus to the point at $-1 + j(0)$.

The *root locus* method was introduced about 1950 (Ref. EV) and allows one to quickly sketch the locus of closed-loop poles versus some parameter of the system, often a control gain. *Synthesis* is done in the same way as in the FR method, but root loci are used instead of FR plots to ensure stability of the loops. Again, iteration is required for multiloop controllers.

There are also many good references on root locus methods (e.g. BL and MC), so we shall not describe them further. An example is given in Section 12.4.

Many computer codes are now available that quickly and accurately calculate and plot a root locus vs. a parameter.

C.2 Pole Placement

Using *full-state feedback* for single-input, single-output (SISO) systems, one can place the closed-loop poles at any desired location, provided that the system is completely controllable. An algorithm for determining the state-feedback gains given the desired poles was given by Ackermann (Ref. ACK). Many current software programs use this algorithm.

Estimators (or "observers") using a single measurement can also be synthesized by placing the estimate-error poles.

Using *feedback of estimated state*, a SISO compensator can be synthesized by pole placement.

The problem with this method is that it is usually not clear where to place the poles for good response and satisfactory control activity. The method does not generalize easily to MIMO systems; in fact, it is a very bad method to use in MIMO systems since the control eigensystem (or its equivalent) must be specified, which is an almost impossible task without causing the controls to fight each other rather than cooperate.

C.3 Linear-Quadratic-Regulator (LQR) Synthesis

This method was introduced about 1960 (Refs. KAL-1 and KAL-2) and makes use of a quadratic performance index to synthesize state-feedback gains that minimize this index. It is especially useful for MIMO systems, since it produces gains that coordinate the multiple controls. All loops are closed simultaneously instead of successively as in FR and root locus methods.

Estimators with multiple measurement inputs can also be synthesized using these methods (Section C.4). MIMO compensators are obtained by using the "regulator" gains on the estimated-states (see C.5).

Consider a time-invariant linear system described by

$$\dot{x} = Fx + Gu, \tag{C.1}$$

where x is the state vector and u is the control vector. An output vector y is to be controlled, where

$$y = Mx + Lu, \tag{C.2}$$

and we wish to minimize

$$J = \frac{1}{2} \int_o^\infty \left(y^T Ay + u^T Bu \right) dt. \tag{C.3}$$

The solution to this minimization problem is discussed in many places (e. g. BR-3 and HA-3). The MacFarlane-Potter method (Refs. MA-2 and PO) is based on finding the eigensystem of the Euler-Lagrange equations

$$\begin{bmatrix} \dot{x} \\ \dot{\lambda} \end{bmatrix} = \begin{bmatrix} \bar{F} & -G\bar{B}^{-1}G^T \\ -M^T\bar{A}M & -\bar{F}^T \end{bmatrix} \begin{bmatrix} x \\ \lambda \end{bmatrix}, \tag{C.4}$$

where

$$\begin{aligned} \bar{B} &= B + L^T AL, \\ \bar{A} &= A + AL\bar{B}^{-1}L^T A, \\ \bar{F} &= F - G\bar{B}^{-1}L^T AM. \end{aligned}$$

The eigenvalues of this system are symmetric about both the real and imaginary axes in the complex s-plane, so half of the eigenvalues have negative real parts.

Write the modal coordinate transformation as

$$\begin{bmatrix} x \\ \lambda \end{bmatrix} = \begin{bmatrix} X_- & X_+ \\ \Lambda_- & \Lambda_+ \end{bmatrix} \begin{bmatrix} \xi_- \\ \xi_+ \end{bmatrix}. \tag{C.5}$$

where the columns of X_- and Λ_- are the eigenvectors associated with the eigenvalues with negative real parts, and the columns of X_+ and Λ_+ are the eigenvectors associated with the eigenvalues with positive real parts. The solution is then

$$u = -Cx, \tag{C.6}$$

where

$$C = \bar{B}^{-1}(L^T AM + G^T S), \tag{C.7}$$
$$S = \Lambda_-(X_-)^{-1}, \tag{C.8}$$

and the eigenvalues of the closed-loop system (eigenvalues of $F - GC$) are the eigenvalues with negative real parts of the Euler-Lagrange equations. Note that S is the solution to the steady-state Riccati equation

$$0 = -S\bar{F} - S\bar{F}^T - M^T \bar{A}M + SG\bar{B}^{-1}G^T S, \tag{C.9}$$

There are now many software codes that determine C, S, and the closed-loop eigenvalues quickly and accurately.

C.4 Linear-Quadratic-Gaussian (LQG) Estimator Synthesis

Consider a time-invariant linear system described by

$$\dot{x} = Fx + \Gamma w, \tag{C.10}$$
$$z = Hx + Nw + v, \tag{C.11}$$

where

$$x = \text{is the state vector,}$$
$$w = \text{is a random disturbance vector,}$$
$$z = \text{is a vector of measured quantities,}$$
$$v = \text{is a random noise vector.}$$

We assume that $w(t)$ and $v(t)$ are independent white-noise processes (cf. Appendix E.3) with spectral density matrices Q and R, respectively. Then the optimal estimate of the state vector, $\hat{x}$, is given by (Ref. KAL-2)

$$\dot{\hat{x}} = F\hat{x} + K(z - H\hat{x}), \tag{C.12}$$
$$\hat{x}(0) = 0, \tag{C.13}$$

where

$$K = PH^T R^{-1}, \tag{C.14}$$

$$0 = FP + PF^T + \Gamma Q \Gamma^T - PH^T R^{-1} HP. \tag{C.15}$$

P, and hence K, may be determined using the same computer code used to find the optimal regulator gain matrix C, by making the following changes in notation ("duality," discussed in BR-1; compare (15) with (9)):

Regulator	F	G	M	L	A	B	C	S
Estimator	F^T	H^T	Γ^T	N^T	Q	R	K^T	P

The matrix P is the steady-state estimate-error covariance matrix

$$P = E(\hat{x} - x)(\hat{x} - x)^T. \tag{C.16}$$

Again, currently available software finds P, K, and the closed-loop eigensystem of the estimator quickly and accurately.

C.5 Linear-Quadratic Compensator Synthesis

Consider a time-invariant linear system described by

$$\dot{x} = Fx + Gu + \Gamma w,$$

$$y = Hx + Lu, \tag{C.17}$$

$$z = H_z x + L_z u + v. \tag{C.18}$$

For the assumptions made in developing the optimal estimator, the expected value of J in (3) is minimized for the system (17), where only the measurements $z(t)$ in (18) are available, by feeding back the estimated state with the optimal regular gains

$$u = -K_r \hat{x}, \tag{C.19}$$

where

$$\dot{\hat{x}} = F\hat{x} + Gu + K_e(z - H_z \hat{x} - L_z u). \tag{C.20}$$

This is called the "certainty-equivalence principle" (Refs. JOS and GUN).

From (19) and (20), the transfer functions from the measurements z to the controls u are

$$u(s) = -K_r(Is - F + K_e H_z + GK_r - K_e L_z K_r)^{-1} K_e z(s). \tag{C.21}$$

Current software quickly and accurately determines these transfer functions given the state feedback gains K_r and the estimator gains K_e. A MATLAB .m file for synthesizing a continuous MIMO LQG compensator is listed in Table C.1.

TABLE C.1 - MATLAB .m File for Synthesizing a Continuous MIMO LQG Compensator

```
function [Sc,rmsy,rmsu,ev]=mimocomy(S,A,B,Ga,Q,R)
% Synthesizes continuous MIMO LQG compensator and determines RMS
% response of closed-loop system. Inputs are defined by: xdot =
% Fx + Gu + Ga*w; y=Hx + Lu; z=Hz*x + Lz*u + v=measurements,
% S = [F G;H L;Hz Lz]; density (w,v) = (Q,R).
% J = integral(y'Ay + u'Bu)dt;
% (xe)dot = F*xe + Gu + ke*(z-Hz*xe-Lz*u); u = - kr*xe.
% Sc is compensator system matrix defined below.
% (rmsy,rmsu) are rms values of (output,control) vectors.
% ev are the closed-loop eigenvalues.
[ns,nw]=size(Ga);[nz,nz]=size(R);
[ny,ny]=size(A);[nu,nu]=size(B);
F=S(1:ns,1:ns);G=S(1:ns,ns+1:ns+nu);
H=S(ns+1:ns+ny,1:ns);L=S(ns+1:ns+ny,ns+1:ns+nu);
Hz=S(ns+ny+1:ns+ny+nz,1:ns);Lz=S(ns+ny+1:ns+ny+nz,ns+1:ns+nu);
kr=lqr(F,G,H'*A*H,B+L'*A*L,H'*A*L);
[ke,P]=lqe(F,Ga,Hz,Q,R);
evr=eig(F-G*kr);eve=eig(F-ke*Hz);
ev=[evr,eve];
Sc=[F-G*kr-ke*Hz+ke*Lz*kr,ke;kr,zeros(nu,nz)];
Xe=lyap(F-G*kr,ke*R*ke');
rmsu=sqrt(diag(kr*Xe*kr'));
rmsy=sqrt(diag((H-L*kr)*Xe*(H-L*kr)'+H*P*H'));
```

Large MIMO compensators are implemented most efficiently in the *minimum-realization real modal form* (see Section A.3).

Furthermore, from many years of experience with classical compensator synthesis, it is well known that compensators that have lower order than the system model often give good closed-loop performance. Thus, we expect that in most cases we can apply *order-reduction* methods (see Section A.4.1) to LQG compensators and still achieve good closed-loop performance. Three useful order-reduction concepts follow:

(1) *Frequency separation*, where very fast modes in the compensator can be approximated as quasi-steady. The resulting lower-order compensator

has some direct feedthrough of measurements to the controls (these must be measurements where the additive noise is negligible).

(2) *Modal truncation*, where the compensator is put into *modal form* and stable modes with small residues are omitted (see Section A.4.1).

(3) *Factor truncation*, where the compensator is put into *balanced-realization form* and factors with small singular values are then omitted (see Section A.4.2).

C.6 Optimal Lower-Order LQG Compensators

A full-order compensator that has been reduced to order m is *not* the optimal compensator of order m for the expected value of the original quadratic performance index (it is often quite close). The optimal compensator of order m can be determined using *parameter optimization*. A computer code (called "SANDY") for doing this (given in Ref. LY) uses gradient methods for parameter optimization. In order to use this code, the user must

(a) supply an initial guess of the compensator parameters. Good initial guesses can be obtained by frequency separation, modal truncation, or factor truncation.

(b) put the initial guess into a minimum-realization form; otherwise the optimal compensator parameters will not be unique.

An example of this procedure and the use of SANDY is given in Section 12.12.

C.7 Asymptotic Model Following

Exponential model following is used in Section 12.7 to synthesize landing flare logic for the 747 aircraft. Here, we consider the more general case where we wish the plant

$$\dot{x} = Fx + Gu, \tag{C.22}$$

with outputs

$$y = Hx + Ju, \tag{C.23}$$

to follow the outputs of a model plant described by

$$\dot{x}_m = F_m x_m + G_m u_m, \tag{C.24}$$

$$y_m = H_m x_m + J_m u_m, \tag{C.25}$$

that is, we wish to make $y \to y_m$. We can do this by generalizing the feedforward idea expressed in (12.39) as follows; let

$$x = Xx_m + Lu_m + \delta x, \tag{C.26}$$

$$u = Px_m + Uu_m + \delta u. \tag{C.27}$$

Equating y to $y_m + \delta y$ yields

$$\delta y = H\delta x + J\delta u, \tag{C.28}$$

$$HX + JP = H_m, \tag{C.29}$$

$$HL + JU = J_m. \tag{C.30}$$

Substituting (26) and (27) into (22) yields

$$X(F_m x_m + G_m u_m) + L\dot{u}_m + \delta\dot{x} = F(Xx_m + Lu_m + \delta x) + G(Px_m + Uu_m + \delta u).$$

If we choose the coefficients of x_m and u_m to vanish, then

$$\delta\dot{x} = F\delta x + G\delta u - L\dot{u}_m, \tag{C.31}$$

$$XF_m = FX + GP, \tag{C.32}$$

$$XG_m = FL + GU. \tag{C.33}$$

Equations (29), (30), (32), and (33) determine the feedforward gain matrices X, L, P, U, provided the dimensions of y, y_m, u, and u_m are all equal:

$$\left.\begin{array}{rcl} FX - XF_m + GP & = & 0 \\[2mm] HX + JP & = & H_m \end{array}\right\} \Rightarrow X, P, \tag{C.34}$$

$$\begin{bmatrix} F & G \\ H & J \end{bmatrix} \begin{bmatrix} L \\ U \end{bmatrix} = \begin{bmatrix} XG_m \\ J_m \end{bmatrix} \Rightarrow L, U. \tag{C.35}$$

The first equation in (34) is an unsymmetric Lyapunov equation, which makes the solution of (34) nontrivial. After X is determined, L and U may be found quite simply from (35).

Equations (31) and (28) describe the *perturbations* away from exact model following:

$$\delta\dot{x} = F\delta x + G\delta u - L\dot{u}_m, \tag{C.36}$$

$$\delta y = H\delta x + J\delta u. \tag{C.37}$$

Using quadratic synthesis, we can find a stabilizing gain matrix C such that

$$\delta u = -C\delta x \qquad\qquad (C.38)$$

will cause $\delta x \rightarrow 0, \delta u \rightarrow 0 \Rightarrow \delta y \rightarrow 0$ for $\dot{u}_m = 0$. Thus, for $\dot{u}_m = 0$, we have *asymptotic model-following*. Examples are given in Ref. TR. When $\dot{u}_m \neq 0$, $L\dot{u}_m$ acts as a forcing function in (36) that will cause $\delta y \neq 0$.

A PC-MATLAB .m file MODFOL that finds the feedforward gains given the plant and model matrices is listed in Table C.2. It is based on the Kronecker product, which converts the system into a set of linear equations (suggested by Douglas Bernard).

TABLE C.2 - MATLAB .m File for Asymptotic Model-Following

```
function [XX,UX,XU,UU]=modfol(A,B,C,D,Am,Bm,Cm,Dm)
% Computes feedforward gain matrices for asymptotic
% model-following. y --> ym, where u = uf - K*(x - xf) and
% uf = UX*xm + UU*um, xf = XX*xm + XU*um, K = state feedback
% matrix (determined elsewhere). xdot = Ax + Bu , y = Cx + Du ,
% xmdot = Am*xm + Bm*um , ym = Cm*xm + Dm*um .
[n,m]=size(B);
[nm,m]=size(Bm);
inm=eye(nm);
ab=[kron(inm,A)-kron(Am',eye(n)),kron(inm,B);kron(inm,C),
    kron(inm,D)];
c2=Cm(:,1);
for i=2:nm,c2=[c2;Cm(:,i)];end
xb=ab\[zeros(n*nm,1);c2];
for j=1:nm
   for i=1:n,
      XX(i,j)=xb(i+n*(j-1));
   end
end
for j=1:nm
   for i=1:m
      UX(i,j)=xb(n*nm+i+m*(j-1));
   end
end
xa=[A,B;C,D]\[XX*Bm;Dm];
XU=xa([1:n],:);
UU=xa([1+n:n+m],:);
```

C.8 Asymptotic Disturbance Rejection

We wish the output vector y to tend asymptotically to zero in the presence of a disturbance whose state vector is w, where

$$\dot{x} = Fx + Gu + \Gamma w, \tag{C.39}$$

$$\dot{w} = F_w w, \tag{C.40}$$

$$y = Hx + Lu + Jw, \tag{C.41}$$

and the dimensions of y and u are equal. Note the disturbance vector w is uncontrollable by u.

We can do this in a manner similar to model following (Section C.7) by letting

$$x = Xw + \delta x, \tag{C.42}$$

$$u = Uw + \delta u. \tag{C.43}$$

Substituting (42) and (43) into (39) and (41). and using (40) gives

$$XF_w w + \delta \dot{x} = F(Xw + \delta x) + G(Uw + \delta u) + \Gamma w, \tag{C.44}$$

$$y = H(Xw + \delta x) + L(Uw + \delta u) + Jw. \tag{C.45}$$

If we choose X and U so that the coefficients of w vanish in (44) and (45), we have

$$\left. \begin{array}{rrrrr} FX & - & XF_w & + & GU & = & -\Gamma \\ \\ HX & & & + & LU & = & -J \end{array} \right\} \Rightarrow X, U, \tag{C.46}$$

and

$$\delta \dot{x} = F\delta x + G\delta u, \tag{C.47}$$

$$y = H\delta x + L\delta u. \tag{C.48}$$

Thus if we can stabilize δx by choosing C, where

$$\delta u = -C\delta x, \tag{C.49}$$

then $\delta x \to 0$, $\delta u \to 0$, $y \to 0$, giving asymptotic disturbance rejection. The control law may be written as

$$u = Uw - C(x - Xw), \tag{C.50}$$

or

$$u = (U + CX)w - Cx. \tag{C.51}$$

The .m file in Table C.2 is easily modified to determine the gain matrices X and U.

Since the disturbance state vector w is usually not measurable directly, it must be *estimated* using measurements of the plant state vector x. Thus (51) will be implemented as

$$u = (U + CX)\hat{w} - C\hat{x}, \tag{C.52}$$

where $(\hat{x}, \hat{w})$ are outputs of an estimator.

Appendix D:
Synthesis of Digital
Control Logic

D.1 Introduction

The control systems for spacecraft and many aircraft use digital computers instead of analog computers to implement the control logic. Clearly, most of the control systems in the future will use "digital control" of this type. Hence we shall discuss here some aspects of digital control. For a more complete treatment, see one of the textbooks on this subject (e.g. Refs. AST, FPW, KAT, or KU).

Fig. D.1 shows a block diagram of a digital flight control system. The vehicle is, of course, an "analog" system, and most of the sensors in current use are analog sensors, which means they have a continuously changing output as a function of time. In order to use digital logic, these analog signals must be sampled and converted into numbers. Thus an analog-to-digital (A/D) converter is required. Furthermore, the output of the digital computer, a string of numbers, must be converted into an analog signal to be sent to the analog actuators. Currently, the most common D/A device is a "zero order hold" (ZOH) that produces a "staircase" output as shown in Fig. D.1.

If the sampling rate is chosen to be greater than about six times the closed-loop bandwidth, the ZOH control signal is very nearly continuous. Thus the analog control logic described up to this point in the book can simply be "discretized" and used as digital logic. This method is often referred to as "designing in the s-plane." This is the method that has been used in synthesizing most of the digital control logic in current systems.

It is possible to produce satisfactory control systems sampling at rates that are comparable to the closed-loop bandwidth. To do this, some special synthesis methods are required involving the use of z-transforms. Consequently, these methods are often referred to as "designing in the z-plane."

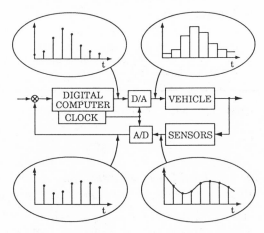

Figure D.1. Block diagram of a digital flight control system (ref. FPW).

D.2 Synthesis in the s-Plane

If the sample rate is high enough (say 6 to 20 times the closed-loop bandwidth), then compensator synthesis may be done in the s-plane: an analog compensator is synthesized; then this compensator is discretized using the method of Section D.4. If only output feedback is used, no discretization is necessary.

The design should be checked by simulation, using an analog computer to simulate the plant and a digital computer to simulate the controller. This requires A/D and D/A devices. The reason for using an analog computer is to check that the outputs behave properly in between sample times.

A preliminary check can be made using only digital simulation, which requires that the plant equations be discretized by the method of Section D.4. A necessary condition is, of course, that the closed-loop eigenvalues in the z-plane be stable—inside the unit circle.

Example

As an example, consider the altitude-velocity hold autopilot for the Navion synthesized in Section 12.5.1. The closed-loop bandwidth is about .7 Hz; hence if we use a sample rate of (say) 10 Hz, an s-plane design should be satisfactory. Using the output feedback gains of Table 12.5 with the discretized model, the closed-loop eigenvalues of the digitally controlled system were calculated and are shown in Table D.1; they are stable and close to the values one would expect from using the mapping

$$z(i) = \exp[Ts(i)],$$

where the $s(i)$ are the closed-loop eigenvalues in the s-plane from Table 12.5 (these values are also shown in Table D.1). Digital simulations of the responses to step commands in velocity and altitude gave results that were indistinguishable from those of the analog control system to the scales used in Figs. 12.16 and 12.17.

TABLE D.1 – *Closed-Loop Eigenvalues for Navion Longitudinal Digital Autopilot; Sample Time = .1 sec; Feedback of* (u, θ, h) *Only*

```
F=[-.045 .036 0 -.322 0;-.370 -2.02 1.76 0 0;...
   .191 -3.96 -2.98 0 0;0 0 1 0 0;0 -1 0 1.76 0];
G=[0 1;-.282 0;-11 0;zeros(2)];Ts=.1;
[phi,ga]=c2d(F,G,Ts);[phi ga]
   0.9954    0.0034   -0.0012   -0.0321    0.0000    0.0004    0.0998
  -0.0316    0.7896    0.1355    0.0005    0.0000   -0.1069   -0.0017
   0.0225   -0.3049    0.7157   -0.0003    0.0000   -0.9361    0.0011
   0.0011   -0.0167    0.0855    1.0000    0.0000   -0.0495    0.0000
   0.0017   -0.0906    0.0005    0.1760    1.0000    0.0012    0.0001
kr       =
   0.254     0.286    -0.123    -0.945    -0.230
   0.291    -0.010    -0.002     0.006     0.0324
ev=eig(phi-ga*kr)
   0.7792 +/- 0.3170i    0.9468 +/- 0.0576i    0.9733
exp(.1*ev)
   0.7578 +/- 0.3066i    0.9477 +/- 0.0567i    0.9736
```

D.3 Synthesis in the z-Plane

If one wishes to use sample rates that are comparable to the desired closed-loop bandwidth, then the s-plane design method is inadequate, since the discontinuities in the "staircase" control input are not small compared to the control authority. The design must be done in the z-plane, starting with the ZOH model of the vehicle for the desired sample rate. The lower limit on sample rate depends on desired closed-loop bandwidth and on passenger and structure tolerance of "rough running."

Synthesis in the z-plane is similar to that in the s-plane, except that the closed-loop poles must lie inside the unit circle instead of the left half-plane. Successive loop closures using root locus or frequency-response techniques parallel classical design in the s-plane. Simultaneous loop closures using linear-quadratic synthesis also parallel the corresponding techniques in the s-plane. Linear quadratic synthesis techniques are discussed in Sections D.5–D.9.

D.4 Conversion of Analog Systems to Digital Systems

A continuous (analog) system with ZOH inputs can be converted into a discrete-step (digital) system without any further approximations. This can be done in a straightforward manner using a state-variable representation of the system

$$\dot{x} = Fx + Gu. \tag{D.1}$$

We may write the solution of this system in the form

$$x(t) = \Phi(t)x(0) + \int_0^t \Phi(t - t')Gu(t')dt', \tag{D.2}$$

where

$$\Phi(t) = \exp(Ft) = \text{transition matrix},$$
$$x(0) = \text{initial state vector}.$$

For ZOH the control vector $u(t')$ is constant over the sample period T. Thus the integration can be carried out over a sample period to produce the "exact" discrete-step model

$$x(k + 1) = \Phi x(k) + \Gamma u(k), \tag{D.3}$$

where

$$x(k) = x(kT),$$
$$u(k) = u(kT),$$
$$\Phi = \Phi(T),$$
$$\Gamma = \int_0^T \Phi(t)Gdt.$$

Current professional software quickly and accurately finds Φ and Γ given F, G, and T. Several different methods are used to do this. Modal decomposition has the advantage that it is exact (no truncation of a series at N terms), but it will not handle repeated eigenvalues (this restriction can be removed using Jordan forms, cf. Walker (WA). Γ is computed as follows:

$$\Gamma = \int_0^T \Phi(t)dt \cdot G, \tag{D.4}$$

$$= T \cdot \left\{ \int_0^T \exp(\Lambda t)dt \right\} (M)^{-1}G, \tag{D.5}$$

where $\exp(\Lambda t)$ is the *block-diagonal* modal transition matrix, and M is the modal transformation matrix (cf. A.1, where it is, unfortunately, called "T"). For scalar blocks (real eigenvalues, $s = \sigma$),

$$\int_0^T \exp(\Lambda t)dt = \frac{\exp(\sigma T) - 1}{\sigma}. \tag{D.6}$$

For 2×2 blocks (complex eigenvalues, $s = \sigma \pm j\omega$),

$$\int_0^T \exp(\Lambda t)dt = \frac{1}{\sigma^2 + \omega^2} \begin{bmatrix} b_1 & b_2 \\ -b_2 & b_1 \end{bmatrix}, \tag{D.7}$$

where

$$\begin{bmatrix} b_1 \\ b_2 \end{bmatrix} = \sigma[\exp(\sigma T)\cos\omega T - 1] \pm \omega\exp(\sigma T)\sin\omega T. \tag{D.8}$$

Expanding in powers of the sample time T, we have the approximate relations

$$\Phi = I + FT + (FT)^2/2 + \cdots, \tag{D.9}$$

$$\Gamma = (I + FT/2 + (FT)^2/3 + \cdots)GT. \tag{D.10}$$

D.5 Conversion of Analog to Digital Performance Indices

Given the continuous performance index

$$J = \int_0^\infty [x^T(t)Ax(t) + 2x^T(t)Nu(t) + u^T(t)Bu(t)]dt, \tag{D.11}$$

we wish to determine the equivalent discrete-step performance index as an infinite sum over the sampled states and the ZOH control inputs

$$J = \sum_{k=0}^{\infty} [x^T(k)A_D x(k) + 2x^T(k)N_D u(k) + u^T(k)B_D u(k)], \qquad \text{(D.12)}$$

where the system model is (3).

Using (12) with ZOH will give almost exactly the same outputs at the sample times as the continuous outputs using (11) for a given set of initial conditions.

Using (2) in (11) and integrating over one sample period with u held constant (ZOH) gives the weighting matrices $A_D, B_D,$ and N_D as follows:

$$A_D = \int_0^T \Phi^T(t)A\Phi(t)dt, \qquad \text{(D.13)}$$

$$N_D = \int_0^T [\Phi^T(t)(A\Gamma(t) + N)]dt, \qquad \text{(D.14)}$$

$$B_D = \int_0^T [B + \Gamma^T(t)A\Gamma(t) + N^T\Gamma(t) + \Gamma^T(t)N]dt, \qquad \text{(D.15)}$$

where

$$\Gamma(t) = \int_0^t \Phi(t - t')Gdt',$$

Note $N_D \neq 0$ *even if* $N = 0$. A PC-MATLAB .m file for computing A_D, N_D and B_D is listed in Table D.2; it uses an algorithm suggested by C.F. Van Loan (Ref. VAL) .

TABLE D.2 – MATLAB.m File for Computing Equivalent Discrete-Step System and Quadratic Performance Index

```
function [FD,GD,AD,ND,BD]=cvrt(F,G,A,N,B,TS)
% Converts continuous system and quadratic performance
% index to equivalent discrete system and quadratic performance
% index.
% Jcont = integral(x'Ax + 2*x'Nu + u'Bu)dt, dx/dt = Fx + Gu.
% Jdisc = sum(x'ADx + 2x'NDu + u'BDu), x(k+1) = FDx(k) + GDu(k).
% TS = sample time.
[NS,NC]=size(G);
Z1=zeros(NS);Z2=zeros(NS,NC);
B2=sqrt(B);
S=[-F',eye(NS),Z1,Z2;Z1,-F',A,N/B2;Z1,Z1,F,G/B2;Z2',Z2',
    Z2',zeros(NC)];
```

```
SD=expm(S*TS);
K1=SD([1:NS],[3*NS+1:3*NS+NC]);
G2=SD([NS+1:2*NS],[2*NS+1:3*NS]);
H2=SD([NS+1:2*NS],[3*NS+1:3*NS+NC]);
FD=SD([2*NS+1:3*NS],[2*NS+1:3*NS]);
G3=SD([2*NS+1:3*NS],[3*NS+1:3*NS+NC]);
GD=G3*B2;
AD1=FD'*G2;AD=(AD1+AD1')/2;
ND=FD'*H2*B2;
BD1=G'*FD'*K1;
BD=TS*B+B2'*(BD1+BD1')*B2;
```

Expanding in powers of the sample time T, we have the approximate relations

$$A_D = AT + (AF + F^T A)T^2/2 \cdots, \tag{D.16}$$

$$N_D = NT + (F^T N + AG)T^2/2 + \cdots, \tag{D.17}$$

$$B_D = BT + (N^T G + G^T N)T^2/2 + \cdots. \tag{D.18}$$

The higher order terms in T are significant if the sample rate is small compared to the closed-loop bandwidth, or if the control authority is large, that is, B is small compared to A. In these cases, if the higher-order terms in T are neglected, the analog outputs of the resulting regulator tend to "ring" between sample times in an unacceptable manner.

D.6 Linear-Quadratic-Regulator (LQR) Synthesis

Consider the time-invariant discrete-step linear system described by (3). We wish to minimize the quadratic performance index (12). The solution to this minimization problem is discussed in many places (e. g. Ref. FR-1). Vaughn's method is based on finding the eigensystem of the equations

$$\begin{bmatrix} x(k+1) \\ \lambda(k+1) \end{bmatrix} = \begin{bmatrix} \mathcal{F}_D + \mathcal{G}_D \mathcal{A}_D & -\mathcal{G}_D \\ -\mathcal{F}_D^{-T} \mathcal{A}_D & -\mathcal{F}_D^{-T} \end{bmatrix} \begin{bmatrix} x(k) \\ \lambda(k) \end{bmatrix}, \tag{D.19}$$

where

$$\mathcal{F}_D = \Phi - \Gamma B_D^{-1} N_D^T,$$

$$\mathcal{G}_D = \Gamma B_D^{-1} \Gamma^T \mathcal{F}_D^{-T},$$

$$\mathcal{A}_D = A_D - N_D B_D^{-1} N_D^T.$$

Half of the eigenvalues of this system are inside and half are outside the unit circle in the *complex z*-plane. Furthermore, for each eigenvalue z_i, there is another eigenvalue $1/z_i$. Write the modal coordinate transformation as

$$
\begin{bmatrix} x(k) \\ \lambda(k) \end{bmatrix} = \begin{bmatrix} X_- & X_+ \\ \Lambda_- & \Lambda_+ \end{bmatrix} \begin{bmatrix} \xi_- \\ \xi_+ \end{bmatrix}, \tag{D.20}
$$

where the columns of X_- and Λ_- are the eigenvectors associated with the stable eigenvalues (inside the unit circle), and the columns of X_+ and Λ_+ are the eigenvectors associated with the unstable eigenvalues (outside the unit circle). The solution is then

$$
u(k) = -K_R \cdot x(k), \tag{D.21}
$$

$$
K_R = K_1 + B_D^{-1} N_D^T, \tag{D.22}
$$

$$
K_1 = (B_D + G_D^T S G_D)^{-1} G_D^T S F_D, \tag{D.23}
$$

$$
S = \Lambda_-(X_-)^{-1}, \tag{D.24}
$$

and the eigenvalues of the optimal closed-loop system (the eigenvalues of $(\Phi - \Gamma \cdot K_R)$ are the stable eigenvalues of (19)).

Current professional software is based on this algorithm. The commands "DREGULATOR" in MATRIXX and "dlqr" in MATLAB determine K_R, S, and the closed-loop eigenvalues, given Φ, Γ, A_D, N_D, and B_D.

MATLAB will only allow $N_D = 0$; to get around this, use the following: For the digital LQR problem (12)–(15), use the "dlqr" command as follows:

```
KR = dlqr(PHI-GAMMA*BD\ND', GAMMA, AD-ND*(BD\ND'), BD);
```
then

$$
\texttt{KR = KR+BD\textbackslash ND'}
$$

where

$$
u(k) = -KR \cdot x(k).
$$

Example

As an example, we consider an altitude and velocity hold autopilot for the Navion, with a sample rate of only 1 Hz where the desired bandwidth is about .7 Hz. Table D.3 shows the computation of the ZOH model of the Navion for a sample period of 1 second, and the conversion of the analog quadratic performance index to an equivalent digital quadratic performance index using CVRT (see Table D.2). The results are then used in DREGULATOR with velocity (u) and altitude (h) as outputs and elevator (δe) and throttle (δt) as controls, for the case diag(A) = $(1, 0, 0, 0, .0625)$, diag(B) = $(1, 9)$.

TABLE D.3 – MATLAB.m *File for Synthesis of Digital Altitude/Velocity SAS for Navion*

```
F=[-.045 .036 0 -.322 0;-.370 -2.02 1.76 0 0;...
    .191 -3.96 -2.98 0 0;0 0 1 0 0;0 -1 0 1.76 0];
G=[0 1;-.282 0;-11 0;zeros(2)];
H=[1 zeros(1,4);0 0 0 0 1];
A=diag([1 1/16]); N=zeros(5,2); B=diag([1 9]); Ts=1;
[Fd,Gd,Ad,Nd,Bd]=cvrt(F,G,H'*A*H,N,B,Ts)
Fd        =
    0.9424    0.0654   -0.0488   -0.3136    0.0000
   -0.0632   -0.0662    0.0320    0.0201    0.0000
    0.1445   -0.0595   -0.0825   -0.0312    0.0000
    0.0970   -0.3120    0.1751    0.9884    0.0000
    0.1257   -0.5656    0.1591    1.7457    1.0000
Gd        =
    0.2251    0.9738
   -1.6173   -0.0625
   -1.8378    0.0970
   -1.8226    0.0359
   -0.4921    0.0444
Ad        =
    0.9489    0.0242   -0.0201   -0.1491    0.0028
    0.0242    0.0089   -0.0024   -0.0281   -0.0199
   -0.0201   -0.0024    0.0010    0.0091    0.0033
   -0.1491   -0.0281    0.0091    0.0972    0.0548
    0.0028   -0.0199    0.0033    0.0548    0.0625
Nd        =
    0.0620    0.4743
    0.0063    0.0180
   -0.0032   -0.0146
   -0.0258   -0.1023
   -0.0067    0.0007
```

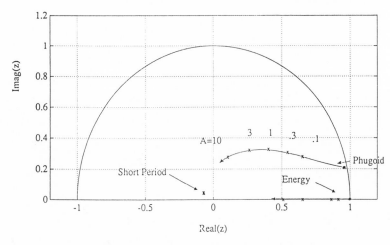

Figure D.2. Locus of closed-loop roots vs. LQ weight for altitude/velocity hold; Navion with sample time = 1.0 sec.

```
Bd          =
   1.0107      0.0774
   0.0774      9.9627
kr=dlrn(Fd,Gd,Ad,Bd,Nd)
   0.0721      0.1992     -0.0887     -0.6531     -0.1251
   0.2386     -0.0095     -0.0024      0.0056      0.0307
ev=eig(Fd-Gd*kr)
   0.7698      0.4041 +/- 0.3258i    -0.0697 +/- 0.0429i
```

Fig. D.2 shows the closed-loop root locus vs. $|A|$ as the elements of A are kept in the same ratio (see the Appendix to Chapter 9 of Ref. FPW for a discussion of this "reciprocal root locus"). Note that the short-period poles are changed only slightly from their open-loop values, whereas the phugoid and energy modes are substantially stabilized.

We simulated the closed-loop system, and the results are shown in Figs. D.3 and D.4. The simulations were done using the Matlab command "dlsim." The responses are satisfactory and not too different from the continuous feedback responses of Figs. 12.16 and 12.17, except for the discontinuities in the control inputs at the sample times.

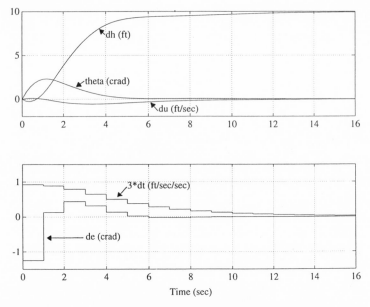

Figure D.3. Response of Navion with LQ digital regulator to commands for 10 ft increase in altitude, no change in airspeed; sample time = 1.0 sec.

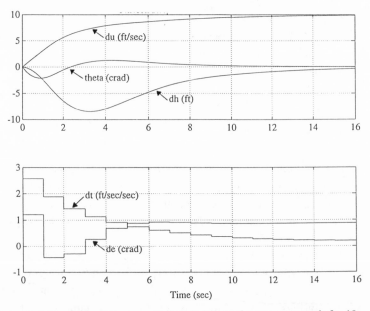

Figure D.4. Response of Navion with LQ digital regulator to commands for 10 ft/sec increase in airspeed, no change in altitude; sample time = 1.0 sec.

D.7 Linear-Quadratic-Gaussian (LQG) Estimator Synthesis

The system model is the ZOH model of the vehicle

$$x(k+1) = \Phi x(k) + w_d(k), \tag{D.25}$$

which is derived from the continuous model

$$\dot{x} = Fx + \Gamma w(t). \tag{D.26}$$

Thus, it follows that

$$\Phi = \exp(F \cdot T), \tag{D.27}$$

where T = the sample period. The measurements $z(k)$ are samples of the continuous sensor signals $z(t)$, where

$$z(k) = Hx(k) + v_d(k), \tag{D.28}$$

$$z(t) = Hx(t) + v(t), \tag{D.29}$$

and $(w_d(k)$, $v_d(k))$ are purely random sequences with covariance matrices (W, R_D).

The covariance matrix W is determined as follows (using duality and Appendix A of Ref. KAT):

$$W = \int_o^T \Phi(t)\Gamma Q \Gamma^T \Phi^T(t) dt, \tag{D.30}$$

where Q is the spectral density of the white noise process $w(t)$ and

$$\Phi(t) = \exp(Ft). \tag{D.31}$$

A MATLAB .m file for determining W given (F, Γ, Q, T) is listed in Table D.4 (algorithm suggested by C. F. Van Loan (Ref. VAL).

TABLE D.4 – MATLAB.m File for Determining the Equivalent Discrete-Step Disturbance Covariance Matrix

```
function [PHI,W]=CVRTQ(F,GA,Q,T)
% Determines covariance W of full-rank random sequence wd(k)
% that is equivalent to white noise w(t) having spectral
% density Q and distribution matrix GA, with sampling
% time T, i.e. xdot = Fx + GA*w is equivalent to x(k+1) =
% PHI*x(k) + wd(k) at t = k*T.
```

```
% W=Integral[PHI(t)*GA*Q*GA'*PHI'(t)dt] over 0 to T; 5/30/89.
[NS,ND]=size(GA);
S=[-F,GA*Q*GA';zeros(NS),F'];
C=expm(S*T);
G2=C([1:NS],[NS+1:2*NS]);
F3=C([NS+1:2*NS],[NS+1:2*NS]);
PHI=F3';
W1=F3'*G2;
W=(W1+W1')/2;
```

Expanding the expression for W in powers of the sample time T, we have the approximate relation

$$W = Q_1 T + (FQ_1 + Q_1 F^T)T^2/2 + \cdots, \tag{D.32}$$

where

$$Q_1 = \Gamma Q \Gamma^T. \tag{D.33}$$

The determination of R_D is less clear, since $v_d(k)$ is a sample of $v(t)$, not an integrated value. Here, instead of modeling $v(t)$ as white noise, we must think of it as colored noise with a variance V and a correlation time T_c that is small compared to the shortest time constant of the system. For the scalar case, $v(t)$ could be generated by the first-order system

$$T_c \dot{v} + v = n(t), \tag{D.34}$$

where $n(t)$ is a white-noise process with spectral density $2VT_c$. Discretizing this shaping filter, we have

$$v_d(k + 1) = \Phi_c v_d(k) + n_d(k), \tag{D.35}$$

where

$$\Phi_c = \exp(-T/T_c)$$

and the variance of the purely random sequence $n_d(k)$ is

$$
\begin{aligned}
R_D &= \int_0^T \Phi(t)[2VT_c/(T_cT_c)]\Phi(t)dt, \\
&= V[1 - \exp(-2T/T_c)],
\end{aligned} \tag{D.36}
$$

where T = sample period. Then, for $T/T_c \gg 1$, it follows that $v_d(k)$ is a purely random sequence with variance V.

Current professional software is based on this algorithm. The commands "DESTIMATOR" in MATRIXX and "dlqe" in MATLAB determine K, P, and the closed-loop eigenvalues, given Φ, Γ, A_D, N_D, and B_D. MATLAB (1987 version) allows only $N_D = 0$; however, the .m file in Table D.5 gets around this by using different arguments for the "dlqe" command.

The prediction estimator has the form

$$x_p(k+1) = \Phi x_p(k) + K_P \cdot [z(k) - H x_p(k)], \qquad (D.37)$$

where $x_p(k)$ is the prediction estimate of $x(k)$.

The current estimator has the form

$$x_c(k+1) = \Phi x_c(k) + K_C \cdot [z(k+1) - H\Phi x_c(k)]. \qquad (D.38)$$

where $x_c(k)$ is the current estimate of $x(k)$.

The terms "prediction" and "current" come from the fact that the prediction estimator gives $x_p(k+1)$ based on the measurements at the previous sample time, $z(k)$, whereas the current estimator gives $x_e(k+1)$ based on the current measurements $z(k+1)$. Note that $K_P = \Phi \cdot K_C$ (Ref. FPW).

D.8 LQG Compensator Synthesis

For the assumptions made in developing the optimal estimator, the expected value of (12) is minimized for the system (26) and (29), when only the measurements $z(k)$ are available, by feeding back the estimated states with the optimal gains, that is,

$$u(k) = -K_R x_p(k). \qquad (D.39)$$

This is the discrete-step "certainty-equivalence principle" (cf. JOS and GUN).

Current software quickly and accurately determines the transfer function matrix from $z(k)$ to $u(k)$ given the state feedback gains K_R and the prediction-estimator gains K_P (such as the LQGCOMP command in MATRIXX).

A MATLAB .m file for synthesizing the digital MIMO LQG prediction compensator is listed in Table D.5. The current compensator cannot usually be implemented because of computation delays.

TABLE D.5 – MATLAB.m File for Determining the Discrete MIMO LQG Prediction Compensator

```
function [FD,GD,SC,RMSY,RMSU,EV]=mimodcom(F,G,HY,A,N,B,GA,H,Q,V,T)
% Synthesizes discrete MIMO LQG prediction-compensator &
% determines RMS response of closed-loop system.
% T = sample time. Other inputs are defined by:
% dx/dt = Fx + Gu + GA*w;  z(k) = Hx(k) + v(k) = measurements,
% J = integral(x'Ax + 2x'Nu + u'Bu)dt; density(w)=Q; cov(v(k))=V.
% Outputs defined by: SC=[FC,KP;KR,0] where
% xp(k+1) = FC*xp(k) + KP*z(k), - u(k) = KR*xp(k),
% and FC = FD - GD*KR - KP*H, KP=FD*KE; x(k+1)=FD*x(k) + GD*u(k).
[NS,NU]=size(G);  [NZ,NS]=size(H);
[FD,GD,AD,ND,BD]=cvrt(F,G,A,N,B,T);
KR=dlqr(FD-GD*inv(BD)*ND',GD,AD-ND*inv(BD)*ND',BD);
KR=KR+BD\ND';
[FD,W]=cvrtq(F,GA,Q,T);
[KE,M,P]=dlqe(FD,eye(NS),H,W,V);
EVR=eig(FD-GD*KR);
EVE=eig(FD-FD*KE*H);
EV=[EVR,EVE];
SC=real([FD-GD*KR-FD*KE*H,FD*KE;KR,zeros(NU,NZ)]);
XP=dlyap(FD-GD*KR,FD*(M-P)*FD');
X=XP+M;
RMSU=real(sqrt(diag(KR*XP*KR')));
RMSY=real(sqrt(diag(HY*X*HY')));
```

Appendix E:
Simulation

E.1 Time Response of Continuous Systems

Control logic is always checked by simulation of the closed-loop system. This is done by simulating the response to initial conditions, step commands, impulsive disturbances, sinusoidal disturbances, and random disturbances. For checks on robustness, it is a good idea to simulate the response to the worst initial conditions, the worst disturbances, and the worst parameter changes.

Simulation of continuous systems can be done on analog computers, but currently it is usually done on digital computers through the use of one of the following techniques:

(a) Integration of the differential equations using an integration scheme such as Runge-Kutta.

(b) Discretization of the analog system using transition matrices and the zero-order-hold method.

(c) Finding residues in the partial fraction expansion of the appropriate transfer functions.

Method (a) is the only method available if nonlinear simulation is desired. If random disturbances are to be simulated, then either method (a) or (b) can be used. Method (c) can simulate only rational disturbances (disturbances represented by rational transfer functions, that is, exogenous systems with a finite number of states), but it is accurate and fast.

Software packages such as MATLAB and MATRIXX do these simulations and include graphics packages that plot them. For example, MATLAB has a command "lsim" that does linear simulation for continuous systems. The outputs are then easily plotted with the command "plot." These commands were used to produce many of the figures in the text.

We have plotted the response to *step output commands* in many of our figures in the text. A MATLAB.m file that does this is listed below; it first calculates the steady-state outputs and then uses the initial condition option of "lsim."

If dynamic compensation is used, it must be put into state variable form and added to the plant equations.

TABLE E.1 – MATLAB .m File for Step Output Commands

```
function [y,y1,u]=stepcmd(F,G,H,L,C,yc,H1,L1,t)
% Function to generate responses to step command y = yc at
% points defined by vector t; dx/dt = Fx + Gu, y = Hx + Lu,
% u = - C*x; dim(u) = dim(y); y1 = H1*x + L1*u.
[ns,nc]=size(G);
G1=zeros(ns,1);
Xas=[F,G;H,L]\[G1;yc];
Xs=Xas([1:ns]);
Us=Xas([ns+1:ns+nc]);
X0=-Xs;
y1s=H1*Xs+L1*Us;
H=[H;H1];L=[L;L1];
[n1,ns]=size(H1);
H2=[H-L*C;-C];L2=zeros(2*nc+n1,1);
T=length(t);
U1=zeros(T,1);
Y=lsim(F-G*C,G1,H2,L2,U1,t,X0);
y=Y(:,[1:nc])+ones(T,1)*yc';
y1=Y(:,[nc+1:nc+n1])+ones(T,1)*y1s';
u=Y(:,[nc+n1+1:2*nc+n1])+ones(T,1)*Us';
```

Response to Worst Initial Conditions

For a system whose only inputs are the initial conditions (ICs), the outputs are obviously proportional to the ICs. For

$$\dot{x} = Fx,$$
$$y = Hx,$$
$$x(0) = x_o,$$

it is of interest to find the response to the *worst IC vector of unit magnitude*, that is, the value of x_o that maximizes

$$J = \int_0^\infty y^T Q y \, dt,$$

with the constraint that

$$x_o^T x_o = 1.$$

For stable F and arbitrary x_o, it is straightforward to show that

$$J = x_o^T S x_o,$$

where S is the solution of the steady-state Lyapunov equation

$$SF + F^T S + H^T Q H = 0.$$

Obviously then, the unit x_o that maximizes J is the unit eigenvector of S that corresponds to the largest eigenvalue of S (S is a nonnegative definite matrix).

We have plotted the closed-loop response to the worst unit ICs in many places in the text; F is then replaced by $F - GC$ where

$$
\begin{aligned}
\dot{x} &= Fx + Gu, \\
y &= Hx, \\
u &= -Cx, \\
x(0) &= x_o,
\end{aligned}
$$

and $H^T Q H$ is replaced by $H^T Q H + C^T R C$ where

$$J = \int_0^\infty (y^T Q y + u^T R u) \, dt.$$

E.2 Time Response for Digital Systems

Simulation of analog systems that use zero-order-hold (ZOH) digital control logic can be done on a hybrid computer—an analog computer connected to a digital computer through D/A and A/D devices. However, for linear systems it is easier to do all of the simulation on a digital computer, using the ZOH model of the analog plant instead of an analog computer.

This latter procedure does not usually simulate the plant outputs between sample times and hence cannot detect undesirable "ringing" in these outputs. However, it is fairly simple to modify continuous simulation codes to plot several points in between sample times.

The MATLAB code "dlqr" simulates digital systems at the sample times.

E.3 Mean Square Response of Analog Systems with Random Inputs

The mean-square (MS) response of a stable linear continuous system to gaussian white-noise inputs is a useful measure of system behavior. If the system is described by

$$\dot{x} = Fx + \Gamma w, \tag{E.152}$$

$$y = Mx, \tag{E.153}$$

where

$$E[w(t)] = 0,$$
$$E[w(t)w^T(\tau)] = Q\delta(t - \tau),$$

then the steady MS state is given by solution of the Lyapunov equation

$$0 = FX + XF^T + \Gamma Q\Gamma^T, \tag{E.154}$$

where

$$X = E[xx^T],$$
$$Y = E[yy^T] = MXM^T.$$

The MATLAB command "lyap" calulates the solution to Lyapunov equation (3).

For state feedback, $u = -Cx$, where $\dot{x} = Fx + Gu + \Gamma w$ and noise-free measurements of the states are assumed,

$$\dot{x} = (F - GC)x + \Gamma w, \tag{E.155}$$

which is of the same form as (1).

For compensators with noisy measurements z,

$$z = Mx + v, \tag{E.156}$$

where

$$\dot{m} = Am + Bz, \tag{E.157}$$

$$u = -Cm - C_z z, \tag{E.158}$$

$$E[v(t)] = 0, \tag{E.159}$$

$$E[v(t)v^T(\tau)] = R\delta(t - \tau), \tag{E.160}$$

form an augmented system

$$
\begin{bmatrix} \dot{x} \\ \dot{m} \end{bmatrix} = \begin{bmatrix} F - GC_zM & -GC \\ BM & A \end{bmatrix} \begin{bmatrix} x \\ m \end{bmatrix} + \begin{bmatrix} \Gamma & -GC_z \\ 0 & B \end{bmatrix} \begin{bmatrix} w \\ v \end{bmatrix}, \quad \text{(E.161)}
$$

which is of the same form as (1).

E.4 Spectral Density of the Response of Analog Systems with Random Inputs

Additional information about system behavior is obtained by calculating the spectral densities of the outputs,

$$
S_i(\omega) = \sum_k |Y_{ik}(j\omega)|^2 Q_{kk}, \quad \text{(E.162)}
$$

where $Y_{ik}(s)$ = transfer function from w_k to y_i,, and $S_i(\omega)$ is the spectral density of y_i,. $S_i(\omega)$ gives the frequency content of the mean-square output (it is the Fourier transform with respect to τ of the auto-correlation function $E[y_i(t)y_i(t + \tau)]$).

The mean-square output may be computed from

$$
E[y_i(t)]^2 = \frac{1}{\pi} \int_0^\infty S_i(\omega)d\omega. \quad \text{(E.163)}
$$

$S_i(\omega)$ can be calculated using the MATRIXX command "PSD." For SISO systems the PSD is identical to the singular value decomposition (SVD) of the transfer function at each frequency ω; hence it can also be calculated using the MATRIXX command "FREQ" with the "SVD" option.

E.5 Mean-Square Response of Discrete-Step Systems

The mean-square (MS) response of a linear discrete-step system to gaussian purely random sequences is a useful measure of system behavior. The conversion of a white-noise process to an equivalent purely random sequence is discussed in Section D.4. If the system is described by

$$
x(k + 1) = \Phi \cdot x(k) + \Gamma \cdot w(k), \quad \text{(E.164)}
$$
$$
y(k) = M \cdot x(k), \quad \text{(E.165)}
$$

where

$$E[w(k)] = 0,$$

$$E[w(k)w^T(\ell)] = \begin{cases} Q_D & k = \ell \\ 0 & \neq \ell \end{cases},$$

then the steady MS state is given by

$$X = \Phi X \Phi^T + \Gamma Q_D \Gamma^T, \tag{E.166}$$

where

$$X = E[xx^T], \tag{E.167}$$

$$Y = E[yy^T] = MXM^T. \tag{E.168}$$

Systems with state feedback or dynamic compensators can be put into the form (13)–(14) in a manner similar to that described for analog systems.

Eqn. (15) can be put into the form (3) by the matrix transformation (from FPW)

$$A = (\Phi - I)(\Phi + I)^{-1}, \tag{E.169}$$

where I = unit matrix. Using (18), equation (13) becomes

$$0 = AX + XA^T + B, \tag{E.170}$$

where

$$B = \frac{1}{2}(I - A)\Gamma Q_D \Gamma^T (I - A)^T. \tag{E.171}$$

E.6 Digital Simulation with Random Inputs

Another good check on control logic is to calculate the response of the closed-loop system to random disturbances; this is called "Monte Carlo" simulation for obvious reasons. Such a simulation can be produced using a random number generator on a digital computer. These generators usually produce numbers that have a uniform distribution from 0 to 1. For simulations we want random vectors that have a gaussian distribution with a specified covariance matrix. This can be done as follows:

(1) Use the lower-triangular (Cholesky) square root of the covariance matrix as the random-input distribution matrix, and an input vector of uncorrelated gaussian scalars, each having a unit variance.

(2) Generate gaussian scalars by adding together a large number (12 or more) of uniformly distributed random numbers, then scale the number so that the distribution has unit variance.

This technique is useful to produce random initial conditions, random process disturbances, and random errors in the measurements.

Appendix F:
Modeling Flexible
Systems

F.1 Introduction

The determination of a dynamic model for a spacecraft or aircraft that includes structural deformations must usually be done with the aid of *finite-element computer codes*. However, some flexible vehicles can be subdivided into elements that can be modeled reasonably well as concentrated masses, rigid bodies, rods, cables, or uniform slender beam-columns. The rods and cables can be modeled as massless springs (see the hoop/column spacecraft problem at the end of this section), or, for more accuracy, as one-dimensional continuum elements in tension, compression, or torsion. The beam-columns can be modeled as massless leaf springs, or, for more accuracy, as one-dimensional Euler beam-columns.

Structural Damping

All structural materials have some inherent damping so the open-loop resonant poles are actually stable, not neutrally stable. As yet, there is no satisfactory quantitative theory to predict structural damping. Empirically most aerospace materials exhibit a fraction of critical damping at resonance poles of about .001, that is, the poles are at

$$s_n \cong (-.001 \pm j)\omega_n,$$

instead of $s_n = \pm j\omega_n$.

Actuator and Sensor Dynamics

Most actuators and sensors have "low-pass" dynamic characteristics, that is, above some characteristic frequency the ratio of output amplitude to input amplitude decreases. These characteristic frequencies are often well above the controller bandwidth and consequently have negligible effect on the design of compensation logic. However, these characteristic frequencies may well be comparable to the first or second structural resonance frequency and therefore *actuator/sensor dynamics should be examined in any dynamic model where structural flexibility is considered.*

The low-pass dynamic characteristics of actuators and sensors are helpful in avoiding destabilization of higher-frequency structural modes when using active control. This comes about because there is less excitation of the high-frequency structural modes; high-frequency system outputs are attenuated by sensor dynamics, and high-frequency control inputs are attenuated by the actuator dynamics.

Examples of actuator dynamics include null momentum wheels driven by D. C. motors (see Chapter 4) and control moment gyros (see Chapter 6).

Examples of sensor dynamics include gyros and accelerometers (see Chapter 2).

F.2 Continuum Model of a Uniform Beam

Consider the simplest (Bernoulli-Euler) model for a uniform vibrating beam with forces and torques acting only on the two ends (see Fig. 1):

$$EI\frac{\partial^4 u}{\partial x^4} + \sigma\frac{\partial^2 u}{\partial t^2} = 0, \tag{F.1}$$

where

$u(x,t)$ = displacement of the neutral axis of the beam (positive upward) at x at time t,

$\theta(x,t)$ = $\dfrac{\partial u}{\partial x}$ is the slope of the neutral axis,

$Q(x,t)$ = $-EI\dfrac{\partial^2 u}{\partial x^2}$ is the bending moment in the beam,

$F(x,t)$ = $EI\dfrac{\partial^3 u}{\partial x^3}$ is the transverse shear force in the beam,

E = Young's modulus of elasticity of the material,

I = second moment of cross-sectional area of beam about the neutral axis,

σ = mass per unit length of the beam,

x = distance along beam,

t = time.

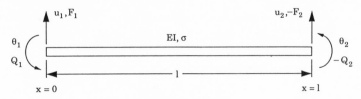

Figure F.1. Nomenclature for uniform beam model.

Let $u(x, s)$ be the Laplace transform of $u(x, t)$, then (1) may be written as

$$\frac{d^4 u(x, s)}{dx^4} - \lambda^4 u(x, s) = 0, \tag{F.2}$$

where $\lambda^4 \stackrel{\Delta}{=} -s^2$ and (u, x) are measured in units of ℓ, Q in units of EI/ℓ, F in units of EI/ℓ^2, and t in units of $(\sigma\ell^4/EI)^{\frac{1}{2}}$. Treating s as a parameter, (2) has the general solution

$$u(x, s) = A(s)sh(\lambda x) + B(s)s(\lambda x) + C(s)ch(\lambda x) + D(s)c(\lambda x), \tag{F.3}$$

where $sh(\ ) \equiv \sinh(\ )$, $ch(\ ) \equiv \cosh(\ )$, $s(\ ) \equiv \sin(\ )$, and $c(\ ) \equiv \cos(\ )$. If we use the definitions of $u, \theta, Q,$ and F at $x = 0$ in (3), we have

$$u_1(s) \equiv u(0, s) = C + D, \tag{F.4}$$

$$\theta_1(s) \equiv u'(0, s) = \lambda(A + B), \tag{F.5}$$

$$-Q_1(s) \equiv u''(0, s) = \lambda^2(C - D), \tag{F.6}$$

$$F_1(s) \equiv u'''(0, s) = \lambda^3(A - B). \tag{F.7}$$

These four equations may be used to express A, B, C, D in terms of $u_1, \theta_1, Q_1,$ and F_1, so that (3) becomes

$$u(x, s) = \frac{u_1}{2}(ch\lambda x + c\lambda x) + \frac{\theta_1}{2\lambda}(sh\lambda x + s\lambda x)$$
$$- \frac{Q_1}{2\lambda^2}(ch\lambda x - c\lambda x) + \frac{F_1}{2\lambda^3}(sh\lambda x - s\lambda x). \tag{F.8}$$

We may differentiate (8) three times with respect to x to obtain values of

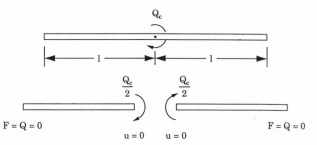

Figure F.2. Free-free uniform beam with a control torque in center.

$\theta(x, s), Q(x, s)$, and $F(x, s)$. In particular, for $x = 1$, we find that

$$
\begin{bmatrix} u_2 \\ \theta_2/\lambda \\ -Q_2/\lambda^2 \\ F_2/\lambda^3 \end{bmatrix} = \frac{1}{2} \begin{bmatrix} ch\lambda + c\lambda & sh\lambda + s\lambda & ch\lambda - c\lambda & sh\lambda - s\lambda \\ sh\lambda - s\lambda & ch\lambda + c\lambda & sh\lambda + s\lambda & ch\lambda - c\lambda \\ ch\lambda - c\lambda & sh\lambda - s\lambda & ch\lambda + c\lambda & sh\lambda + s\lambda \\ sh\lambda + s\lambda & ch\lambda - c\lambda & sh\lambda - s\lambda & ch\lambda + c\lambda \end{bmatrix} \begin{bmatrix} u_1 \\ \theta_1/\lambda \\ -Q_1/\lambda^2 \\ F_1/\lambda^3 \end{bmatrix}.
$$

(F.9)

Equation (9) is a *set of dynamic transfer functions* between the displacements, slopes, moments, and shears at the two ends of a uniform beam. We show below and in Section F.2 how (9) can be used to determine dynamic models of a spacecraft made up of beamlike elements, concentrated masses, and rigid body elements.

F.2.1 *Flexible Spacecraft with a Central Torquer*

Fig. F.2 shows a slender, flexible spacecraft modeled as a uniform beam with a control torque in the center; it may be thought of as two identical pinned-free beams joined together with a discontinuity in bending moment in the center equal to the control torque Q_c. The displacement and bending moment are antisymmetric about the center while the slope and shear force are symmetric, so the displacement is zero at the center and bending moment goes from $+Q_c/2$ to $-Q_c/2$ in passing from left to right across the center. Shear and bending moment are zero at both free ends.

Natural Motions

We first find the natural frequencies and mode shapes of the vibrating beam. The left half of the beam in Fig. F.2 is a uniform free-pinned beam of length ℓ. With $F_1 = Q_1 = 0, u_2 = 0, Q_2 = 0$, Eqn. (9) becomes four homogeneous equations for the four quantities $u_1, \theta_1, \theta_2, F_2$:

I and 2 still I unit apart ?

$$
\begin{bmatrix}
0 & 0 & ch\lambda + c\lambda & sh\lambda + s\lambda \\
1 & 0 & sh\lambda + s\lambda & ch\lambda + c\lambda \\
0 & 0 & ch\lambda - c\lambda & sh\lambda - s\lambda \\
0 & 1 & sh\lambda + s\lambda & ch\lambda - c\lambda
\end{bmatrix}
\begin{bmatrix}
\theta_2/\lambda \\
F_2/\lambda^3 \\
-u_1/2 \\
-\theta_1/2\lambda
\end{bmatrix}
\tag{F.10}
$$

For nontrivial solutions, the determinant of the coefficient matrix in (10) must be zero. In this case it simplifies to

$$
\sinh \lambda \cos \lambda - \cosh \lambda \sin \lambda = 0. \tag{F.11}
$$

This transcendental characteristic equation has an infinite number of discrete roots ω_n, $n = 0, 1, 2, \ldots$. The first few may be determined numerically using Newton's method; for $n \gg 1$, $\sinh \lambda \approx \cosh \lambda$, so that $\lambda_n \approx (n + 1/4)\pi$.

Each value of ω_n (or λ_n) may be substituted back into (10) to find the ratio of θ_1/λ to u_1 for that value of n. This may be substituted into (8) to find the mode shapes. In this case, it produces

$$
u(x, s) \sim (sh\lambda + s\lambda)(ch\lambda x + c) - (ch\lambda + c\lambda)(sh\lambda x + s\lambda x). \tag{F.12}
$$

Fig. F.3 shows the first four mode shapes; the first mode is the rigid-body mode with frequency = 0.

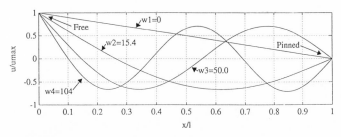

Figure F.3. First four antisymmetric modes of a free-free beam.

Transfer Functions

Considering the left half of the beam, we may use (9) with $F_1 = Q_1 = 0, u_2 = 0, Q_2 = \frac{1}{2}Q_c$:

$$\begin{bmatrix} 0 \\ \theta_2/\lambda \\ -Q_c/2\lambda^2 \\ F_2/\lambda^3 \end{bmatrix} = \frac{1}{2} \begin{bmatrix} ch\lambda + c\lambda & sh\lambda + s\lambda \\ sh\lambda - s\lambda & ch\lambda + c\lambda \\ ch\lambda - c\lambda & sh\lambda - s\lambda \\ sh\lambda + s\lambda & ch\lambda - c\lambda \end{bmatrix} \begin{bmatrix} u_1 \\ \theta_1/\lambda \end{bmatrix}, \qquad \text{(F.13)}$$

which are four equations for $\theta_2, F_2, u_1, \theta_1$ in terms of Q_c. Solving them, we find the transfer functions

$$\frac{\theta_2(s)}{Q_c(s)} = \frac{ch\lambda c\lambda + 1}{\Delta(s)}, \qquad \text{(F.14)}$$

$$\frac{\theta_1(s)}{Q_c(s)} = \frac{ch\lambda + c\lambda}{\Delta(s)}, \qquad \text{(F.15)}$$

$$\frac{u_1(s)}{Q_c(s)} = -\frac{sh\lambda + s\lambda}{\Delta(s)}, \qquad \text{(F.16)}$$

where

$$\Delta(s) = 2\lambda^2(sh\lambda c\lambda - ch\lambda s\lambda). \qquad \text{(F.17)}$$

The *co-located angle transfer function* (14) has an infinite number of discrete poles and zeros alternating on the imaginary axis of the s-plane $\left(s = j\omega \Rightarrow \lambda = \sqrt{\omega}\right)$. A proof of the fact that transfer functions from an actuator to a co-located output have alternating poles and zeros (for conservative systems) was given by Martin (Ref. MAR). This was known earlier in connection with electric circuits which contain no dissipative elements (the Foster reactance theorem, described in Ref. CAM).

Thus (14) may be written as the infinite product (in dimensional units)

$$\frac{\theta_c(s)}{Q_c(s)} = \frac{1}{m\ell^2/3}\frac{1}{s^2}\prod_{n=1}^{\infty}\frac{1+s^2/z_n^2}{1+s^2/\omega_n^2}. \tag{F.18}$$

As $s \to 0$, (18) tends to the rigid-body transfer function $(1/Js^2)$, since $m\ell^2/3 = J =$ moment of inertia of a rigid bar of length 2ℓ and mass $m \equiv \sigma(2\ell)$. The zeros and poles are

n	1	2	3	$n \geq 4$
$z_n\left(\sigma\ell^4/EI\right)^{1/2}$	3.516	22.03	61.70	$\left(n-\frac{1}{2}\right)^2\pi^2$
$\omega_n\left(\sigma\ell^4/EI\right)^{1/2}$	15.41	49.96	104.2	$\left(n+\frac{1}{4}\right)^2\pi^2$

The *separated angle transfer function* (15) has the same poles as (14), but it has zeros on the *real* axis of the s-plane. Hence (15) may be written as the infinite product (in dimensional units)

$$\frac{\theta_t(s)}{Q_c(s)} = \frac{1}{m\ell^2/3}\frac{1}{s^2}\prod_{n=1}^{\infty}\frac{1-s^2/z_n^2}{1+s^2/\omega_n^2}. \tag{F.19}$$

The zeros are

n	1	2	3	$n \geq 4$
$z_n\left(\sigma\ell^4/EI\right)^{1/2}$	4.935	44.41	123.37	$(2n+1)^2\pi^2/2$

The *separated displacement transfer function* (16) has the same poles as (14), but, like (15), it has zeros on the *real* axis of the s-plane. Hence (16) may be written as the infinite product (in dimensional units)

$$\frac{u_1(s)}{Q_c(s)} = -\frac{1}{m\ell/3}\frac{1}{s^2}\prod_{n=1}^{\infty}\frac{1-s^2/z_n^2}{1+s^2/\omega_n^2}. \tag{F.20}$$

The zeros are

n	1	2	3	$n \geq 4$
$z_n\left(\sigma\ell^4/EI\right)^{1/2}$	11.19	60.45	149.3	$2\left(n-\frac{1}{4}\right)^2\pi^2$

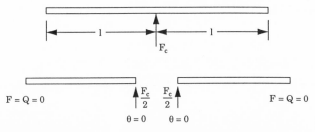

Figure F.4. Free-free uniform beam with control force in center.

Problem F.2.1 – *Translational Control of a Beam Using a Thruster in the Center*

A uniform beam in free space with a control force, F_c, in the center may be thought of as two identical beams joined together with a discontinuity in shear force in the center equal to the control force (see Fig. F.4). The displacement and bending moment are symmetric about the center while the slope and shear force are antisymmetric, so the slope is zero at the center, and the shear force goes from $-F_c/2$ to $+F_c/2$ in passing from left to right across the center. Shear and bending moment are zero at both ends.

(a) *Show that* the transfer function from F_c to displacement at the center, u_c, is given (in normalized units) by

$$\frac{u_c(s)}{F_c(s)} = \frac{-1}{2\lambda^3} \frac{1 + ch\lambda c\lambda}{ch\lambda s\lambda + sh\lambda c\lambda},$$

or, in dimensional units,

$$\frac{u_c(s)}{F_c(s)} = \frac{1}{\sigma(2\ell)s^2} \frac{1 + ch\lambda c\lambda}{(ch\lambda s\lambda + sh\lambda c\lambda)/\lambda},$$

$$\equiv \frac{1}{\sigma(2\ell)s^2} \prod_{n=1}^{\infty} \frac{1 + s^2/z_n^2}{1 + s^2/\omega_n^2}.$$

(b) Determine the poles and zeros of this transfer function, verifying that they alternate on the $j\omega$ axis.

Problem F.2.2 – *Attitude Control of a Beam Using a Reaction Wheel in the Center*

(a) *Show that* the Laplace transform of the equations of motion may be written as

$$\begin{bmatrix} 1 + JRs/N^2 & -1 \\ JP(s) & J_o\Delta(s) \end{bmatrix} \begin{bmatrix} q_w(s) \\ \dot{\theta}_c(s) \end{bmatrix} = \begin{bmatrix} -1 \\ 0 \end{bmatrix} e(s)/N,$$

where the nomenclature is the same as in Section 4.1 and

$$J_o \overset{\Delta}{=} m\ell^2/3,$$

$$P(s) \overset{\Delta}{=} ch\lambda c\lambda + 1,$$

$$\Delta(s) = \frac{3(ch\lambda s\lambda - sh\lambda c\lambda)}{\lambda^3},$$

$$\lambda^4 = -\frac{\sigma\ell^4}{EI}s^2.$$

(b) *Show that* the characteristic equation of the open-loop system ($e = 0$) may be written in the Evans's form

$$-\frac{N^2}{JR} = \frac{J_o s\Delta(s)}{JP(s) + J_o\Delta(s)}.$$

(c) *Show that* the open-loop system is stable, that is, $\dot{\theta}_c$ and all vibration modes are damped.

Problem F.2.3 – *Transfer Functions for a Uniform Bar in Torsion*

The equation of motion of a uniform bar in torsion (see Fig. 5) is

$$GJ\frac{\partial^2 \phi}{\partial x^2} - I\frac{\partial^2 \phi}{\partial t^2} = 0,$$

where

$\phi(x, t)$ = twist angle of the bar at x at time t,

$Q(x, t)$ = $-GJ\dfrac{\partial\phi}{\partial x}$ = twisting moment in the bar,

G = shear modulus of material,

J = polar second moment of cross-sectional area of bar about elastic axis,

I = polar moment of inertia per unit length of bar about elastic axis.

This is the one-dimensional wave equation where $c \overset{\Delta}{=} \sqrt{GJ/I}$ = wave speed.

(a) Measuring x in units of ℓ, Q in units of GJ/ℓ, and letting $\mu^2 \overset{\Delta}{=} -s^2\ell^2/c^2$, *show that*

$$\begin{bmatrix} \phi_2(s) \\ Q_2(s)/\mu \end{bmatrix} = \begin{bmatrix} \cos\mu & -\sin\mu \\ \sin\mu & \cos\mu \end{bmatrix} \begin{bmatrix} \phi_1(s) \\ Q_1(s)/\mu \end{bmatrix},$$

where $( )_1, ( )_2$ denote left, right ends of the bar.

Figure F.5. Nomenclature for uniform bar in torsion.

(b) A uniform bar in free space with a control torque in the center may be thought of as two identical bars joined together with a discontinuity in twisting moment in the center equal to the control torque, Q_c (same as Fig. F.2 except that bending moments are replaced by twisting moments). The twist angle, ϕ, is symmetric about the center so the twisting moment, Q, is antisymmetric; thus Q goes from $Q_c/2$ to $-Q_c/2$ in passing from left to right across the center. Q is equal to zero at both free ends. *Show that*

$$\frac{\phi_c(s)}{Q_c(s)} = \frac{-\cos\mu}{2\mu\sin\mu},$$

where ϕ_c is twist angle at the center.

(c) *Show that* the transfer function in (b) may be written (in dimensional quantities) as

$$\frac{\phi_c(s)}{Q_c(s)} = \frac{1}{2\ell I s^2}\prod_{n=1}^{\infty}\frac{1+s^2/z_n^2}{1+s^2/\omega_n^2},$$

where

$$z_n = \left(n+\frac{1}{2}\right)\pi c/\ell,$$
$$\omega_n = n\pi c/\ell.$$

Note for $|s\ell/c| \ll 1$, $\phi_c(s)/Q_c(s) \cong 1/I(2\ell)s^2$ where $I(2\ell)$ = total polar moment of inertia of rod (the rigid-body transfer function). Note that the poles and zeros alternate along the $j\omega$ axis.

Problem F.2.4 – *Transfer Functions for a Beam with Tip Masses*

Consider a beam in free space with a torque in the center Q_c, as in Fig. F.1, but with two equal concentrated masses at each free end. Determine the transfer function $\theta_c(s)/Q_c(s)$. *Hint*: the shear force at the right tip will be $F_2 = m\ddot{u}_2$ where m = tip mass.

F.2.2 *Flexible S/C with a Noncentral Torquer*

Fig. F.6 shows a slender, flexible spacecraft modeled as a uniform beam of length 2ℓ with a control torque at $x = \ell(1+\epsilon)$. It may be thought of as two

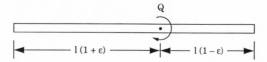

Figure F.6. Free-free uniform beam of length 2ℓ with a control torque $\epsilon\ell$ to the right of center.

beams joined together, one of length $\ell(1+\epsilon)$ and the other of length $\ell(1-\epsilon)$ with a discontinuity in bending moment at the torquer location equal to the control torque Q. Shear and bending moment are zero at both free ends, while shear, slope, and deflection are continuous at the torquer location.

Natural Motions

For $\epsilon = 0$ the torquer is at the center of the beam and can only excite the antisymmetic vibration modes, shown in in Fig. F.3. For $epsilon \neq 0$ the torquer can also excite the symmetric modes of vibration.

To calculate the symmetric modes, we may treat the left half of the beam as a uniform beam of length ℓ, with $F_1 = Q_1 = 0, \theta_2 = 0$, and $F_2 = 0$. Equation (9) becomes four homogeneous equations for the four quantities u_2, Q_2, u_1, θ_1 :

$$
\begin{bmatrix}
1 & 0 & ch\lambda + c\lambda & sh\lambda + s\lambda \\
0 & 0 & sh\lambda + s\lambda & ch\lambda + c\lambda \\
1 & 0 & ch\lambda - c\lambda & sh\lambda - s\lambda \\
0 & 1 & sh\lambda + s\lambda & ch\lambda - c\lambda
\end{bmatrix}
\begin{bmatrix}
u_2 \\
-Q_2/\lambda^3 \\
-u_1/2 \\
-\theta_1/2\lambda
\end{bmatrix}.
\tag{F.21}
$$

Where is λ in this equation?

For nontrivial solutions, the determinant of the coefficient matrix in (21) must be zero. In this case it simplifies to

$$
\sinh \lambda \cos \lambda + \cosh \lambda \sin \lambda = 0.
\tag{F.22}
$$

This transcendental characteristic equation has an infinite number of discrete roots ω_n, $n = 1, 2, \ldots$ The first few may be determined numerically using Newton's method; for $n \gg 1$, $\sinh \lambda \approx \cosh \lambda$, so that $\lambda_n \approx (n - 1/4)\pi$.

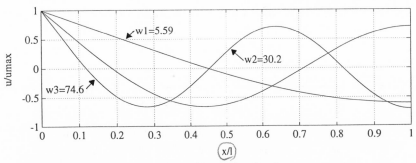

Figure F.7. First three symmetric modes of a free-free beam.

Each value of ω_n (or λ_n) may be substituted back into (21) to find the ratio of θ_1/λ to u_1 for that value of n. This may be substituted into (8) to find the mode shapes. In this case, it produces

$$u(x,s) \sim (ch\lambda - c\lambda)(ch\lambda x + c) - (sh\lambda + s\lambda)(sh\lambda x + s\lambda x). \qquad (F.23)$$

Fig. F.7 shows the first three mode shapes.

Transfer Functions

Considering the part of the beam to the left of the torquer in Fig. F.8, we may write the state of the beam just to the left of the torquer in terms of the state at the left tip:

$$x^{(2-)} = T(\lambda_1)\Lambda(\lambda_1)x^{(1)}, \qquad (F.24)$$

where

$$
\begin{aligned}
x &= [\, u \quad \theta \quad Q \quad F \,]^T \\
\Lambda(\lambda) &= \text{diag}[\, 1 \quad 1/\lambda \quad -1/\lambda^2 \quad 1/\lambda^3 \,] \\
\lambda_1 &= \lambda(1 + \epsilon),
\end{aligned}
$$

and the four-by-four matrix $T(\lambda)$ is defined in (9).

Similarly, we can write the state of the beam at the right tip in terms of the state just to the right of the torquer:

$$x^{(3)} = T(\lambda_2)\Lambda(\lambda_2)x^{(2+)}, \qquad (F.25)$$

where

$$\lambda_2 = \lambda(1 - \epsilon).$$

At the torquer, deflection, slope, and shear are continuous, while the torque is discontinuous, so that

$$x^{(2+)} = x^{(2-)} + BQ \tag{F.26}$$

where

$$B = [\,0 \quad 0 \quad 1 \quad 0\,]^T.$$

The inverse of $T(\lambda)$ is simply $T(-\lambda)$, so it is straightforward to solve (25) for $x^{(2+)}$. Substituting the result, along with (24), into (26) gives

$$[\Lambda(\lambda_2)]^{-1} T(-\lambda_2) x^{(3)} - T(\lambda_1)\Lambda(\lambda_1) x^{(1)} = BQ. \tag{F.27}$$

This gives the four transfer functions from Q to the four quantities $u_1, \theta_1, u_3, \theta_3$, since

$$\begin{aligned} x^{(1)} &= [u_1, \theta_1, 0, 0]^T, \\ x^{(3)} &= [u_3, \theta_3, 0, 0]^T. \end{aligned}$$

Given ϵ, the zeros of these transfer functions can be determined numerically. For $\epsilon = 0$, all the symmetric mode poles are cancelled by zeros.

F.3 Rigid Spacecraft with Flexible Appendages

Dynamic Model

The beam transfer function approach is useful in formulating dynamic models of rigid spacecraft with flexible appendages. Consider a spacecraft like the one shown schematically in Fig. F.8. The central body may be modeled as rigid, while the two flexible appendages may be modeled as simple beams cantilevered out from the central body. A free-body diagram of the spacecraft is also shown in Fig. F.8.

If a control torque Q_c is applied to the central body, the transverse displacements of the appendages will be antisymmetric. Taking the right appendage and using the nomenclature of Fig. F.1, we have

$$\begin{aligned} F_2 &= Q_2 = 0, \\ \theta_1 &= \theta_c, \\ u_1 &= a\theta_c. \end{aligned}$$

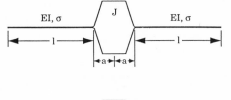

Figure F.8. Spacecraft with rigid central body and symmetric flexible appendages.

Substituting these relations into the last two rows of (9) gives

$$
\begin{bmatrix} 0 \\ 0 \end{bmatrix} = \frac{1}{2} \begin{bmatrix} ch\lambda - c\lambda & sh\lambda - s\lambda & ch\lambda + c\lambda & sh\lambda + s\lambda \\ sh\lambda + s\lambda & ch\lambda - c\lambda & sh\lambda - s\lambda & ch\lambda + c\lambda \end{bmatrix} \begin{bmatrix} a\theta_c \\ \theta_c/\lambda \\ -Q_1/\lambda^2 \\ F_1/\lambda^3 \end{bmatrix},
$$

(F.28)

which may be solved to yield transfer functions from $\theta_c(s)$ to $Q_1(s)$ and $F_1(s)$. In normalized units, these are

$$
\begin{bmatrix} Q_1(s)/\lambda \\ F_1(s)/\lambda^2 \end{bmatrix} = \frac{1}{ch\lambda c\lambda + 1} \begin{bmatrix} sh\lambda c\lambda - ch\lambda s\lambda + a\lambda sh\lambda s\lambda \\ -sh\lambda s\lambda - a\lambda(ch\lambda s\lambda + sh\lambda c\lambda) \end{bmatrix} \theta_c(s).
$$

(F.29)

The equation of motion of the central body is

$$
J\ddot{\theta}_c = Q_c - 2Q_1 - 2aF_1,
$$

(F.30)

where J = moment of inertia of central body about its center of mass.

Taking the Laplace transform of (30), substituting for $Q_1(s)$ and $F_1(s)$ from (29), and replacing s^2 by $(-EI\lambda^4/\sigma\ell^4)$, we have

$$
\frac{\theta_c(s)}{Q_c(s)} = \frac{-(ch\lambda c\lambda + 1)}{\Delta},
$$

(F.31)

where

$$\Delta \overset{\Delta}{=} J\lambda^4(ch\lambda c\lambda + 1) + 2\lambda(ch\lambda s\lambda - sh\lambda c\lambda)$$
$$+ 4a\lambda^2 sh\lambda s\lambda + 4a^2\lambda^3(ch\lambda s\lambda + sh\lambda c\lambda),$$

and J is in units of $\sigma\ell^3$, Q_c in units of EI/ℓ, and a in units of ℓ.

(31) reduces to the rigid-body transfer function for $J = 0, a = 0$, as it should. For $\ell = 0$, it reduces (in dimensional units) to

$$\frac{\theta_c(s)}{Q_c(s)} \to \frac{1}{Js^2}, \tag{F.32}$$

as it should.

The advantage of (31) over finite-element methods is that as many exact poles and zeros can be calculated as desired with little computational effort. Note that the zeros of (31), where $1 + ch\lambda c\lambda = 0$, are the natural frequencies of a single appendage cantilevered from a rigid base; $\theta_c = 0$ is precisely the condition at the end of a cantilevered beam attached to a rigid base.

Of course, only uniform appendages can be modeled easily in this manner, and real appendages are seldom uniform.

Attitude Control with Co-Located Sensor and Actuator

Attitude control of a spacecraft like the one in Fig. F.10, using an angle sensor and a torquer on the central body (co-located), is nearly the same as attitude control of a flexible beam with a torque in the center (Fig. F.2), since the poles and zeros will alternate on the $j\omega$ axis just as in Fig. F.2, but their locations will be different because $J \neq 0, a \neq 0$.

Problem F.3.1 – *Spacecraft with Flexible Solar Panels*

A spacecraft with flexible solar panels may be modeled approximately as a rigid body with two small masses on springs as in Fig. F.9.

(a) Assuming antisymmetric deflection, as shown in Fig. F.9, show that

$$\frac{\theta(s)}{Q(s)} = \frac{s^2 + 1}{s^2(s^2 + 1 + \epsilon)}, \quad \frac{y(s)}{Q(s)} = \frac{1}{s^2 + 1 + \epsilon},$$

where Q = torque on central body (in units of $J\omega^2$), J = moment of inertia of body without masses, $\omega^2 \overset{\Delta}{=} k/m$, time in units of $1/\omega$, y in units of b, and $\epsilon \overset{\Delta}{=} 2mb^2/J$.

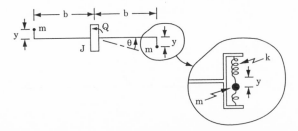

Figure F.9. Two-degree-of-freedom model of a spacecraft with rigid central body and symmetric flexible appendages.

(b) Show that simple rate feedback ($d\theta/dt$ to Q) will stabilize the vibration mode as well as damping out $d\theta/dt$.

(c) Using (b) as a first loop closure, design a second feedback loop θ to Q, which stabilizes the central body position and still preserves some damping in the vibration mode (an example of "low-authority, high-authority control"; cf. Section 9.5).

Problem F.3.2 – *Spacecraft with Appendages in Torsion*

Consider a spacecraft with a rigid central body and symmetric flexible appendages (as in Fig. F.7) except that the torque Q_c puts the appendages in *torsion* instead of bending. Referring to Problem F.1.3, determine the transfer function $\phi_c(s)/Q_c(s)$ where ϕ_c = attitude angle of the central body whose moment of inertia is J.

F.4 Truss Structures

Most of the space station designs being proposed are truss structures. These can be analyzed by finite-element computer programs, but they can also be analyzed using beam transfer functions in bending and in torsion. As an example, consider the *roll attitude control* of the *triangular truss structure* shown in Fig. F.10 using a control torque, Q_c, in the center of the element $\widehat{AB}$ (Ref. WIE).

The right half of Fig. F.10 shows a free-body diagram of the three beam-like elements of the truss, with the bottom element broken into two equal-length elements. The symbol $\odot$ indicates a vector pointing up out of the plane of the paper. We assume all three elements are identical, having bending stiffness EI, mass per unit length σ, and length 2ℓ. The joints are assumed to be ball-and-socket-like so that no bending or torsion moments can be transmitted from one

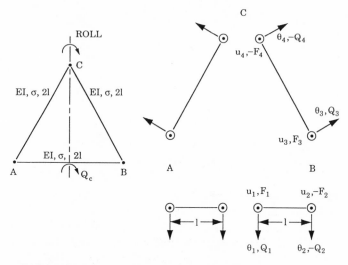

Figure F.10. Equilateral triangular truss with control torque.

element to another. Thus

$$Q_2 = Q_3 = Q_4 = 0.$$

In roll, the out-of-plane deflection of the members will be anti-symmetric across the centerline, so

$$u_1 = u_4 = 0.$$

At joint B, compatability of displacements requires

$$u_2 = u_3,$$

while the fact that F_2 and F_3 are internal forces in the structure requires

$$F_2 = F_3.$$

Finally, bending moment is antisymmetric in element $\widehat{AB}$, so the control torque will produce a discontinuity in bending moment in the center, or

$$Q_1 = -\frac{1}{2}Q_c.$$

Using the beam transfer functions from (9) for the right half of beam $\widehat{AB}$, we have the three relations

$$
\begin{bmatrix} u_2 \\ -Q_2/\lambda^2 \\ F_2/\lambda^3 \end{bmatrix} = T(\lambda) \begin{bmatrix} u_1 \\ \theta_1/\lambda \\ Q_c/2\lambda^2 \\ F_1/\lambda^3 \end{bmatrix}, \tag{F.33}
$$

where the θ_2 equation was not used, $u_1 = 0, Q_2 = 0, Q_1 = -Q_c/2$, and $T(\lambda)$ represents the first, third, and fourth rows of the 4×4 matrix in (9), omitting the second column.

Using the beam transfer functions from (9) again, for the beam $\widehat{BC}$ (which has length 2ℓ), we have the three relations

$$
\begin{bmatrix} u_4/2 \\ -Q_4/2\lambda^2 \\ F_4/2\lambda^3 \end{bmatrix} = T(2\lambda) \begin{bmatrix} u_3/2 \\ \theta_3/\lambda \\ F_3/2\lambda^3 \end{bmatrix}, \tag{F.34}
$$

where the θ_4 equation was not used and $u_4 = Q_4 = Q_3 = 0$.

Eliminating u_3 and F_3 with (30) and (31), Equations (33) and (34) comprise six linear equations for $u_2, F_2, \theta_1, F_1, \theta_3$, and F_4, in terms of the forcing function, Q_c. It is straightforward to solve for $\theta_1(s)$, giving

$$
-\frac{\theta_1(s)}{Q_c(s)} = \frac{1}{2}\frac{\frac{1}{2}(ch2\lambda s2\lambda - sh2\lambda c2\lambda)(ch\lambda s\lambda - sh\lambda c\lambda) - (sh2\lambda s2\lambda)(1 + ch\lambda c\lambda)}{(ch2\lambda s2\lambda - sh2\lambda c2\lambda)(sh\lambda s\lambda) + (sh2\lambda s2\lambda)(ch\lambda s\lambda - sh\lambda c\lambda)},
\tag{F1.35}
$$

which may be written in the infinite product form (in dimensional coordinates):

$$
\frac{\theta_1(s)}{Q_c(s)} = \frac{1}{Js^2} \prod_{n=1}^{\infty} \frac{1 + s^2/z_n^2}{1 + s^2/\omega_n^2}, \tag{F1.36}
$$

where $J = 2\sigma\ell^3 =$ roll moment of inertia of the whole structure, and the zeros and natural frequencies are

n	1	2	3	4	$\cdots$
$z_n(\sigma\ell^4/EI)^{1/2}$	1.586	3.747	10.846	18.135	$\cdots$
$\omega_n(\sigma\ell^4/EI)^{1/2}$	3.162	9.869	13.799	24.25	$\cdots$

References

ACK   IBM Ackermann, J. "Der Entwurf linearer Regelungssysteme im Zustands-raum." *Regelunstechnik und Prozessdatenverarbeitung* 7 (1972): 279–300.

AG   "Spacecraft Pointing and Control." AGARDograph, no. 260 (1982).

ASH   Ashkenazi, A., and Bryson, A. E., Jr. "Control Logic for Parameter Insensitivity and Disturbance Attenuation." *Journal of Guidance and Control* 5, no. 4 (July–August 1982): 383–88.

AST   Aström, K. J., and Wittenmark, B. *Computer-Controlled Systems.* Englewood Cliffs, N.J.: Prentice-Hall, 1984.

AT   Athans, M., and Falb, P. L. *Optimal Control.* New York: McGraw-Hill, 1966.

AU   Aubrun, J. N. "Theory of the Control of Structures by Low Authority Controllers." *Journal of Guidance and Control* 3, no. 4 (July–August, 1980).

BA   Battin, R. H., and Fraser, D. C. "Space Navigation and Guidance." AIAA Professional Study Series, 1970.

BE   Bernard, D. "Control System Design for Lightly Coupled Space Structures." Ph.D. diss. Stanford University, 1984.

BL   Blakelock, J. H. *Automatic Control of Aircraft and Missiles.* New York: John Wiley and Sons, 1965.

BRA   Bramwell, A.R.S. *Helicopter Dynamics.* New York: John Wiley and Sons, 1976.

BRG   Bracewell, R. N., and Garriott, O. K. "Rotation of Artificial Earth Satellites." *Nature* 82, no. 4638 (September 20, 1958): 760–62.

BRH   Bryson, A. E., Jr., and Ho, Y. C. *Applied Optimal Control.* Washington, D.C.: Hemisphere, 1975.

BR-1   Bryson, A. E., Jr. "Nonlinear Feedback Solution for Minimum-Time Rendezvous with Constant Thrust Acceleration." *Proc. 16th Inter. Astro. Congress.* Athens, 1965.

BR-2   Bryson, A. E., Jr. "Control of Random Systems." *Proc. 13th Intl. Congress Theo. and Appl. Mech.* (Moscow, 1972). Berlin: Springer-Verlag, 1973, 1–19.

BR-3   Bryson, A. E., Jr. "Some Connections between Modern and Classical Control Concepts." *Journal of Dynamic Systems, Measurement, and Control* (June 1979).

BU   Burt, E.G.C. "On the Attitude Control of Earth Satellites." *8th Angle-American Aero. Conf.* London, September 1961.

CA-1   Cannon, R. H., "Gyroscopic Coupling in Space Vehicle Attitude Control Systems." *Trans. ASME, Jour. Basic Engr.* (1961).

CA-2   Cannon, R. H. "Basic Response Relations for Satellite Attitude Control Using Gyros." *Automatica* (1963), 535–44.

CCV   "B-52 CCV Control System Synthesis." AFFDL-TR-74-92, vol. II, Air Force Flight Dynamics Laboratory, Wright Patterson AFB, Ohio (1975).

CH   Chalk, C. R.; Neal, T. P.; Harris, T. M.; Pritchard, F. E.; and Woodcock, R. J. "Background Information and User Guide for Military Specification of Flying Qualities of Aircraft." Air Force Flight Dynamics Lab. Tech Report 69-72, August 1969.

DA   Davis, L. D., Jr.; Follin, L. D., Jr.; and Blitzer, L. *The Exterior Ballistics of Rockets*. Princeton, N.J.: Van Nostrand, 1958.

DAV   Davenport, W. W. *Gyro! The Life and Times of Lawrence Sperry*. New York: Scribners, 1978.

DE   DeBra, D. B., and Cannon, R. H. "Momentum Vector Considerations in Wheel-Jet Satellite Control." Guidance, Control and Navigation Conference, American Rocket Society, Stanford University, August 1961.

DO   Doyle, J. C., and Stein, G. "Multivariable Feedback Design: Concepts for a Classical/Modern Synthesis." *IEEE Trans.*, vol. AC-26, no. 1 (February 1981).

ECB   El Ghaoui, L.; Carrier, A.; and Bryson, A. E. "LQ Minimax Methods for Controller Analysis and Design." Plenary Lecture, AIAA Guidance, Navigation, and Control Conference, Portland, Oregon, Aug. 1990; submitted to *Journal Guidance, Control, and Dynamics*, 1991.

ED-1   Edwards, J. W. "Unsteady Aerodynamic Modeling and Active Aeroelastic Control." Dept. of Aero. /Astro. Report No. 504, Stanford University, February 1977.

ED-2   Edwards, J. W., Breakwell, J. V., and Bryson, A. E., Jr. "Active Flutter Control Using Generalized Unsteady Aerodynamic Theory." *Journal Guidance and Control*, vol. 1, no. 1 (1978), 32–40.

EV   Evans, W. R. "Graphical Analysis of Control Systems." *Trans. AIEE*, vol. 67 (1948), 547–51.

FA   Fath, A. F. "Computational Aspects of the Linear Optimal Regulator Problem." *IEEE Trans. Auto. Control*, vol. AC-14 (October 1969),547–50.

FPW   Franklin, G. F.; Powell, J. D.; and Workman, M. L. *Digital Control of Dynamic Systems*. Second edition. Reading, Mass.: Addison-Wesley, 1990.

FRA   Francis, J.G.F. "The QR Transformation, Parts I and II." *Computer Journal* 4 (1961): 265–71; vol. 5 (1961): 332–45.

GE   Gevarter, W. B. "Basic Relations for Control of Flexible Vehicles." *AIAA Journal* 8, no. 4 (April 1970).

GEM   Gessow, A., and Myers, G. C. *Aerodynamics of the Helicopter*. 1952. Reprint. New York: Ungar Publishing Company, 1967.

GO   Golub, G. H., and Van Loan, C. F. *Matrix Computations*. Baltimore, MD: Johns Hopkins University Press, 1983.

GR   Greensite, A. L. *Analysis and Design of Space Vehicle Flight Control Systems*. 2 volumes, New York: Spartan Books, 1970.

GUN   Gunckel, T. F., and Franklin, G. F. "A General Solution for Linear Sampled-Data Control Systems." *Trans. ASME* 85D (1963); 197.

GUP   Gupta, N.; Lyons, M. G.; Aubrun, J. N.; and Margulies, G. "Modeling, Control, and System Identification Methods for Flexible Structures." *Spacecraft Pointing and Control*, AGARDograph, no. 260 (1982).

HA-1   Hall, W. E., and Bryson, A. E., Jr. "Optimal Control and Filter Synthesis by Eigenvector Decomposition." Dept. of Aero./Astro. Report No. 436, Stanford University, November 1971.

HA-2   Hall, W. E., and Bryson, A. E., Jr. "Modal Methods in Optimal Control Synthesis." In *Control and Dynamic Systems*, vol. XVI, edited by C. Leondes. New York: Academic Press, 1980, 53–80.

HA-3   Hall, W. E., Jr., and Bryson, A. E., Jr. "Inclusion of Rotor Dynamics in Controller Design for Helicopters." *AIAA Journal of Aircraft* 10, no. 4 (April 1973): 200–206.

HE-1   Heffley, R. K., and Jewell, W. F. "Aircraft Handling Qualities." Systems Technology Inc. Tech. Report 1004-1, Hawthorne, Calif., 1972.

HE-2   Heffley, R. K.; Jewell, W. F.; Lehman, J. M.; and Van Winkle, R. A. "A Compilation and Analysis of Helicopter Handling Qualities Data." NASA Contractor Report 3144, August 1979.

HO   Holley, W. E. "Wind Modeling and Lateral Control for Automatic Landing." *Journal Spacecraft and Rockets* 14, no. 2 (February 1977): 62–72.

HU   Hughes, P. C. *Spacecraft Attitude Dynamics*. New York: Wiley, 1986.

IO   Iorillo, A. J. "Analyses Related to the Hughes Gyrostat Systems." Hughes Aircraft Company Report 70438B, December 1967.

JA   James, H. M.; Nichols, N. B.; and Phillips, R. S. *Theory of Servomechanisms*. New York: McGraw-Hill, 1947.

JO   Jones, R. T. "Operational Treatment of the Nonuniform Lift Theory to Airplane Dynamics." NACA Tech. Note 667, 1938.

JOH   Johnson, C. R. "TACSAT I Nutation Dynamics." AIAA 3rd Communications Satellite Systems Conference, Paper No. 70-455, April 1970.

JOS   Joseph, P. D., and Tou, J. T. "On Linear Control Theory." *Trans. AIEE*, part III, vol. 80, no. 18 (1961).

KAI   Kailath, T. *Linear Systems*. Englewood Cliffs, N.J.: Prentice-Hall, 1980.

KAL-1   Kalman, R. E. "A New Approach to Linear Filtering and Prediction Problems." *Trans. ASME, Jour. Basic Engr.* 82D (March 1960).

KAL-2   Kalman, R. E., and Bucy, R. "New Results in Linear Filtering and Prediction." *Trans. ASME, Jour. Basic Engr.* 83D (1961).

KAL-3   Kalman, R. E., and Bertram, J. E. "Control System Analysis and Design via the Second Method of Lyapunov." *ASME Journal of Basic Engineering* 82, no. 2, 371–93 (1960).

KAN   Kant, I. R., ed. "Theory and Applications of Optimal Control in Aerospace Systems." AGARDograph, no. 251 (July 1981).

KANE   Kane, T. R.; Likins, P. W.; and Levinson, D. A. *Spacecraft Dynamics*. New York: McGraw-Hill, 1983.

KAP   Kaplan, M. H. *Modern Spacecraft Dynamics and Control*. New York: John Wiley and Sons, 1976.

KAT   Katz, P. *Digital Control Using Microprocessors*. Englewood Cliffs, N.J.: Prentice-Hall, 1981.

KAY   Kayton, M., and Fried, W. R., eds. *Avionics Navigation Systems*. New York: John Wiley and Sons, 1969.

KO   Kortum, W. and Bryson, A. E., Jr. "Estimation of the Local Attitude of an Orbiting Spacecraft." *Automatica* 7 (1971): 163–80.

KU   Kuo, B. C. *Digital Control Systems*. 2d ed. New York: Holt, Rinehart, and Winston, New York, 1980.

KW   Kwakernaak, H., and Sivan, R. *Linear Optimal Control Systems*. New York: Wiley-Interscience, 1972.

LY   Ly, Uy-Loi "A Design Algorithm for Robust Low-Order Controllers." Dept. of Aero./Astro. Report No. 536, Stanford University, November 1982.

MA-1   MacFarlane, A.G.J. *Dynamical System Models*. London: Harrap & Co., 1970.

MA-2   MacFarlane, A.G.J. "An Eigenvector Solution of the Optimal Linear Regulator." *Jour. Electron. Control* 14 (June 1963): 643–54.

MAR   Martin, Gary D. "Attitude Control of Flexible Spacecraft." *Journal Guidance and Control* 3 no. 1, (January 1980): 37–41.

MAT   "Numeric Computation and Visualization Software." The Math-Works, Inc., 24 Prime Park Way, Natick, MA 01760-1500.

MB   Mills, Raymond A., and Bryson, Arthur E. "Parameter-Robust Control Design Using a Minimax Method." *Journal Guidance, Control, and Dynamics* 15, no. 5: 1068–1075, 1991.

MC   McRuer, D., Ashkenas, I., and Graham, D. *Aircraft Dynamics and Automatic Control*. Princeton, N.J.: Princeton University Press, 1973.

MI   Miller, J. E., ed. "Space Navigation, Guidance, and Control." AGAR-Dograph, no. 105 (1966).

PE-1   Perkins, C. D., and Hage, R. E. *Airplane Performance Stability, and Control*. New York: John Wiley and Sons, 1949.

PE-2   Perkins, C. D. "Development of Airplane Stability and Control." *Journal of Aircraft* 7, no. 4 (1970): 290–301.

PO   Potter, J. E. "Matrix Quadratic Solutions." *SIAM Jour. Appl. Math.* 14 (May 1966): 496–501.

PU   Puckett, A. E., and Ramo, S., eds. *Guided Missile Engineering*. New York: McGraw-Hill, 1959.

RO   Roger, K. L.; Hodges, G. E.; and Felt, L. "Active Flutter Suppression—A Flight Demonstration." *Journal of Aircraft* 7 (June 1975): 551–56.

ROC   Rock, S. M. "Transient Motion of an Airfoil; An Experimental Investigation in a Small, Subsonic Wind Tunnel." Dept. of Aero./Astro. Report No. 513, Stanford University, May 1978.

ROS   Roskam, J. *Airplane Flight Dynamics and Automatic Flight Controls*. 2 volumes. Lawrence, Kans.: Roskam Aviation Engineering Corp., 1979.

RY   Rynaski, E. A., and Whitbeck, R. F. "Theory and Application of Linear Optimal Control." Air Force Flt. Dyn. Lab. Tech. Report 65-28, October 1965.

SA   Salvatore, J. O. Hughes Aircraft Company Lectures on "Attitude Control—System Design and Dynamics." Calif. Inst. of Tech., 1975.

SC   Scott, E. D., and Rodden, J. J. "Performance of Gravity-Gradient VCMG Systems." AIAA Guidance, Control, and Flight Mech. Conf., 1969.

SO   Sorenson, A. A., and Williams, I. J. "Spacecraft Attitude Control." *Quest Magazine* (Summer 1982).

STE   Stengel, R. F. "Some Effects of Parameter Variations on the Lateral-Directional Stability of Aircraft." *Journal of Guidance and Control* 3 (March–April 1980).

STO   Stoltz, P. M. "Unsteady Aeroelastic Modeling and Trailing-Edge Flap Control of an Experimental Wing in a Two-Dimensional Wind Tunnel." Dept. of Aero./Astro. Report No. 527, Stanford University, May 1981.

STU   Stubbs, G. S.; Penchuk, A.; and Schlundt, R. W. "Digital Autopilots for Thrust Vector Control of the Apollo CSM/LM Vehicles." AIAA Guidance, Control, and Flight Mech. Conf., 1969.

SY   *Symposium on Passive Gravity-Gradient Stabilization*, NASA SP-107, Ames Research Center, May 1965.

TE   Teper, G. L. "Aircraft Stability and Control Data." Systems Technology Inc. Tech. Report 176-1, Hawthorne, Calif., 1969.

THE   Theodorsen, T., and Garrick, E. E. "Non-Stationary Flow about a Wing-Aileron-Tab Combination Including Aerodynamic Balance." NACA Report 736, 1942.

THO   Thomson, W. T. *Introduction to Space Dynamics.* New York: John Wiley and Sons, 1961.

TI   Timoshenko, S., and Young, D. *Advanced Dynamics.* New York: McGraw-Hill Book Co., Inc., 1948.

TR   Trankle, T. L., and Bryson, A. E., Jr., "Control Logic to Track Outputs of a Command Generator." *Journal Guidance and Control* 1, no. 2, (1978): 130–35.

VA   Vaughan, D. R. "A Non-Recursive Solution of the Discrete Matrix Riccati Equation." *IEEE Trans. on Automatic Control* (October 1970).

VAL   Van Loan, C. F. *Trans. IEEE*, vol. AC-23 (June 1978).

VAN   Vander Velde, W. E. "Space Vehicle Flight Control." Part 7 of *Space Navigation, Guidance, and Control*, AGARDograph, no. 105 (1966).

VE   Velman, J. R. "Orbit and Attitude Control." Lecture 16 in volume 1 of *Geosynchronous Spacecraft Case Histories.* 3 volumes, Hughes Aircraft Company, 1981.

WA   Walker, R. A. "Computing the Jordan Form for Control of Dynamic Systems." Dept. Aero/Astro Report 528, Stanford University, 1981.

WE   Wertz, J. R., ed. *Spacecraft Attitude Determination and Control*. Dordrecht, Netherlands: Reidel, 1978.

WH   Wheeler, P. *Advances in Astronautics*, AIAA vol. 13 (1960).

WHI   Whittaker, H. P., and Potter, J. E. "Optimization of the Use of Automatic Flight Control Systems for Manned Aircraft." MIT Instrumentation Lab. Rpt. R-558, Aug. 1966.

WI   Wilkinson, J. H.; Martin, R. S.; and Peters, G. "The QR Algorithm for Real Hessenberg Matrices." *Numer. Math.* 14 (1970): 219–33.

WIE   Wie, B. "On the Modeling and Control of Flexible Space Structures." Dept. of Aero. /Astro. Report No. 525, Stanford University, June 1981.

WIL   Williams, D. D. "Torques and Attitude Sensing for Spin-Stabilized Synchronous Satellites." In *Torques and Attitude Sensing in Earth Satellites*, edited by S. F. Singer. New York: Academic Press, 1964.

WIN   Wing, W. G. "Attitude and Heading References." Chapter 10 of *Avionics Navigation Systems*, edited by Kayton and Fried. New York: John Wiley and Sons, 1969.

WR   Wrigley, W.; Hollister, W. M.; and Denhard, W. G. *Gyroscopic Theory, Design, and Instrumentation*. Cambridge, Mass.: MIT Press, 1969.

WY   Wykes, J. H. "Structural Dynamic Stability Augmentation and Gust Alleviation of Flexible Aircraft." AIAA Paper 68-1067, AIAA Annual Meeting, October 1968.

YO   Yocum, J. F., and Slafer, L. I. "Control System Design in the Presence of Severe Structural Dynamics Interactions." *Journal Guidance and Control* 1, no. 2 (1978): 109–16.

Index